Eckhard Worch
**Hydrochemistry**
De Gruyter Graduate

# Also of Interest

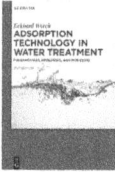

Eckhard Worch

# Hydrochemistry

———

Basic Concepts and Exercises

2nd Edition

**DE GRUYTER**

**Author**
Prof. Dr. Eckhard Worch
Technische Universität Dresden
Institute of Water Chemistry
01062 Dresden
Germany
Eckhard.Worch@tu-dresden.de

ISBN 978-3-11-075876-4
e-ISBN (PDF) 978-3-11-075878-8
e-ISBN (EPUB) 978-3-11-075884-9

**Library of Congress Control Number: 2022940343**

**Bibliographic information published by the Deutsche Nationalbibliothek**
The Deutsche Nationalbibliothek lists this publication in the Deutsche Nationalbibliografie;
detailed bibliographic data are available on the Internet at http://dnb.dnb.de.

© 2023 Walter de Gruyter GmbH, Berlin/Boston
Cover image: Eckhard Worch
Typesetting: Integra Software Services Pvt. Ltd.
Printing and binding: CPI books GmbH, Leck

www.degruyter.com

# Preface to the Second Edition

The first edition of *Hydrochemistry* was published in 2015. Gratifyingly, the textbook has been well received by the readers, in particular students and lecturers. Therefore, encouraged by the publisher, I decided to provide an improved and extended edition of the book.

The general structure of the book has been retained. However, in order to support the active acquisition of knowledge, the number of examples and problems to be solved by the reader has been significantly increased compared to the first edition. The text is now supplemented by 63 examples within the text. The number of problems to be solved by the reader has been increased to 70.

The problem of the increase in the atmospheric $CO_2$ concentration and the resulting impact on aquatic systems receive now more attention at various points in the book. In the respective examples and problems, the atmospheric $CO_2$ concentration has been updated.

Some chapters have been extended by additional topics, for instance, new aspects of water quality (Chapter 1), temperature dependence of equilibrium constants (Chapter 5), coupling of gas–water partitioning and chemical reaction (Chapter 6), relationship between photosynthesis and the carbonic acid system in surface waters (Chapter 7), establishment of redox half-reaction equations (Chapter 10), and sorption-influenced subsurface solute transport (Chapter 12). Furthermore, some minor revisions of the text have been carried out.

I hope this edition will be as well received as the previous one. Thanks to the staff of the publishing house, De Gruyter. I appreciate the fruitful cooperation.

Eckhard Worch<br>June 2022

# Preface to the First Edition

Yet another book on hydrochemistry? This question could arise if one has in mind such famous and well-known standard textbooks as *Aquatic Chemistry* by W. Stumm and J. J. Morgan and *Water Chemistry* by M. M. Benjamin. However, students who have just started studying hydrochemistry often feel overwhelmed with the amount of information presented in these comprehensive monographs of about 1 000 and 700 pages, respectively. What they are looking for is a shorter textbook that provides an easy step-by-step introduction to hydrochemistry, that requires little prior knowledge in chemistry, and that facilitates easier access to the more comprehensive textbooks. This book is intended to close this gap. It is based on over 20 years of experience in teaching hydrochemistry at a beginner level for students of various science and engineering study courses.

https://doi.org/10.1515/9783110758788-202

This textbook introduces the elementary basics of hydrochemistry in a compacted form with a special focus on reaction equilibria in aquatic systems and their mathematical description. It is designed as an introductory textbook for students of all environment-related courses who are taking their first course in hydrochemistry.

After a short introduction in Chapter 1, basic information on the chemical substance "water" and its extraordinary properties are provided in Chapter 2. Chapter 3 deals with the concentration measures that are needed to quantify the content of water constituents and with the activities as the effective concentrations in reactions. Then, some general properties of aqueous solutions that are independent of the chemical nature of the solutes and that depend only on their concentrations (the so-called colligative properties) are discussed in Chapter 4. Chapter 5 leads to the main part of the book on the different chemical equilibria in aqueous systems, and provides some basics on the chemical equilibrium and its mathematical description by the law of mass action. From Chapters 6 to 12, all important types of chemical equilibria relevant for the hydrosphere are discussed: gas–water partitioning (Chapter 6), acid/base equilibria (Chapter 7), precipitation/dissolution equilibria (Chapter 8), calco-carbonic equilibrium (Chapter 9), redox equilibria (Chapter 10), complex formation equilibria (Chapter 11), and sorption equilibria (Chapter 12).

The text is supplemented by a large number of examples (>50) to facilitate the understanding of the theoretical considerations. Furthermore, each chapter (except the introduction) includes a section with numerous problems (>60 in total) to be solved by the reader. Complete and detailed solutions to all problems are documented in Chapter 13 in order to give the reader the opportunity to verify the own solutions. An appendix comprises a list of important constants, a short introduction to logarithm rules necessary for the equilibrium calculations, and a useful list of the most important equations presented in the textbook.

I would be pleased if this book would find a broad acceptance by students and by all who are interested in hydrochemistry.

Last but not least, I would like to thank all those who contributed to this book by some means or other. Special thanks to my family and especially my wife, Karola, for her patience during the times of intensive writing. Thanks to my students, whose questions and discussions have been an important source of inspiration. Thanks to the staff of the publishing house, De Gruyter. I appreciate the useful support and the fruitful cooperation during the work on this book.

Eckhard Worch
November 2014

# Contents

# 1 Introduction

Water is the basis of all life. For humans, water is essential for survival and, therefore, irreplaceable. Water is also a habitat for aquatic organisms. Furthermore, industry and agriculture are unthinkable without water. Accordingly, the preservation of aquatic ecosystems and the protection of water resources belong to the most important goals of sustainable development.

The quality of water is determined by its constituents, which is the totality of the substances dissolved or suspended in water. A substantiated assessment of water quality, therefore, requires in-depth knowledge about the occurrence and behavior of these constituents. That explains the importance of hydrochemistry (also referred to as water chemistry or aquatic chemistry) as a scientific discipline that deals with the constituents of water and their reactions within the natural water cycle and within the cycle of water use.

Although water, with a total mass of about $1.38 \times 10^{18}$ t, is the most common molecular substance on the Earth, its distribution between individual reservoirs is strongly unbalanced (Table 1.1). Approximately 97% of the total water is salt water of the oceans that cannot be consumed by humans directly and is also not suitable for many other purposes due to its relatively high salt content. Although the production of drinking and service water from ocean water is, in principle, possible (e.g., by reverse osmosis), it is unfavorable due to the high energy demand for the treatment processes. Therefore, for drinking and service water production, freshwater is typically used as raw water resource. The most important freshwater resources are the glaciers and polar ice caps as well as groundwater and surface water. From these reservoirs, only parts of surface water and groundwater can be utilized with acceptable technical effort for drinking water or service water purposes. It is estimated that less than 1% of the huge global water resources is actually available for human use.

**Table 1.1:** Global water resources (data from Trenberth et al. 2007).

| Water resource | Volume ($10^3$ km$^3$) | Portion (%) |
|---|---|---|
| Oceans | 1 335 040 | 96.95 |
| Polar ice caps and glaciers | 26 350 | 1.91 |
| Groundwater | 15 300 | 1.11 |
| Lakes and rivers | 178 | 0.013 |
| Soil moisture | 122 | 0.009 |
| Permafrost | 22 | 0.0016 |
| Atmosphere | 12.7 | 0.0009 |
| Total | 1 377 024.7 | 100.0 |

https://doi.org/10.1515/9783110758788-001

However, the usable freshwater inventories are constantly renewed by the hydrological cycle. Figure 1.1 shows the global water cycle in a very simplified form. Approximately 413 000 km$^3$ of water is transferred by evaporation into the atmosphere each year, and the same volume is returned by precipitation to the mainland and to the oceans. The cycle thus resembles a distillation plant whose energy demand, covered by the Sun, can be calculated from the enthalpy of vaporization of water (Chapter 2, Section 2.2.3) to be about 10$^{21}$ kJ/a. For the renewal of freshwater resources, the proportion of evaporated water from the oceans that is transported through the atmosphere to the land is of particular importance. This proportion is approximately 40 000 km$^3$/a.

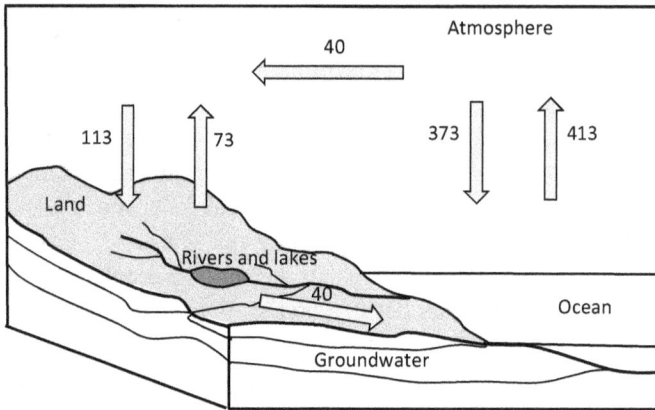

**Figure 1.1:** The hydrological cycle (volumetric flow rates in 10$^3$ km$^3$/a). Data from Trenberth et al. (2007).

A part of the rain water reaches directly the surface water bodies (streams, rivers, lakes, and reservoirs) or flows on the land surface into the surface water. Another part of the precipitate, the seepage water, infiltrates into the soil. There it is absorbed by plants or further transported into the ground where it finally reaches the groundwater level. Surface water can also infiltrate into the subsurface where it becomes groundwater. After more or less long residence times and flow paths, groundwater returns to the surface in the form of springs or directly by exfiltration into surface water. The hydrological cycle is completed by evaporation processes and the water runoff to the sea. The evaporation can take place both from the ground and from the water surfaces as well as from the plants. The runoff to the sea occurs via creeks, rivers, and streams.

The different concentration levels of water constituents that we can find in different water bodies can be explained by considering the water cycle as a distillation process as mentioned previously. During evaporation, mainly pure water is transferred to the atmosphere, whereas the solutes (except the very volatile substances) remain in the aqueous phase. Therefore, we can find relatively high concentrations in the oceans (the residue of the distillation). However, the water that is evaporated does not remain pure water over longer times because during the transport through the atmosphere, substances (gases, vapors, or aerosols) are dissolved. The main input of solutes, however, occurs after the precipitation to the mainland. Dissolution of solids from soils, sediments, aquifers, as well as wastewater effluents leads to an increase in the concentrations of inorganic and organic water constituents. Nevertheless, the concentrations, in particular those of inorganic solutes, in freshwater are typically much lower than in ocean water. The highest ion concentrations in freshwater are in the mg/L range, whereas the highest ion concentrations in seawater are in the g/L range.

It follows from the previous discussion that naturally occurring liquid water is never pure $H_2O$, but contains dissolved, colloidally dispersed and coarsely dispersed constituents. Water bodies differ strongly regarding the nature and especially the concentrations of the various constituents.

The oceans, which cover about 70% of the Earth's surface, contain an average of about 3.5% (35 000 ppm) dissolved salts. In ocean water, we can find all the naturally occurring elements, but most of them only in low concentrations. Only a few ions contribute to the high salt content. These are primarily sodium and chloride ions; their total concentration alone amounts to approximately 30 g/kg (3%). The six major constituents of seawater that make up more than 99% of the total salinity are listed in Table 1.2. The composition of seawater is relatively constant with only very small differences between and within the big open oceans. In contrast, marginal seas with limited water exchange with the open oceans often show stronger deviations from the total salinity and composition of the big oceans due to the impact of freshwater inflow (leading to dissolution) and evaporation (leading to concentration). Depending on the relative strength of these effects, the salt concentration in the marginal seas can be higher or lower than that of the open oceans (e.g., Baltic Sea, 0.3–1.9% salinity, and Mediterranean Sea, 3.6–3.9% salinity).

**Table 1.2:** Major seawater components.

| Component | Concentration in mg/kg (ppm) | Concentration in mmol/kg | Concentration in mmol/L (calculated with a density of 1.025 g/cm³) |
|---|---|---|---|
| $Cl^-$ | 19 350 | 545.79 | 559.43 |
| $Na^+$ | 10 760 | 468.03 | 479.73 |
| $SO_4^{2-}$ | 2 710 | 28.21 | 28.92 |
| $Mg^{2+}$ | 1 290 | 53.08 | 54.41 |
| $Ca^{2+}$ | 412 | 10.28 | 10.54 |
| $K^+$ | 400 | 10.23 | 10.49 |
| Sum | 34 922 | 1 115.62 | 1 143.52 |

The oceans play an important role for the climate because they are able to store large amounts of heat (Chapter 2, Section 2.2.3). They compensate for temperature fluctuations and act as a heat buffer. The oceans also have a weakening impact on the climate change. They can uptake and store large amounts of the greenhouse gas carbon dioxide, $CO_2$ (for the solubility of $CO_2$, see Chapter 6). About 30% of the carbon dioxide released to the atmosphere is absorbed by the oceans. Without this $CO_2$ absorption, the $CO_2$ concentration increase in the atmosphere and the resulting global warming would be much faster. However, the oceans are not able to absorb the total anthropogenic $CO_2$ and therefore they cannot completely stop the global warming. Moreover, a further warming of the oceans would decrease the possible $CO_2$ uptake due to the temperature dependence of the gas solubility (Chapter 6). Another effect that needs to be mentioned is that the $CO_2$ absorption influences the hydrochemical state of the oceans, in particular, the carbonic acid system and the calco-carbonic equilibrium. The resulting effects (referred to as ocean acidification) are a shift of pH to lower values and a decrease of the carbonate/hydrogencarbonate ratio (for details, see Chapters 7 and 9). This makes it more difficult for shell-building marine animals (e.g., corals and oysters) to form their shells from calcium carbonate.

Due to the huge amount of water in the oceans and the slowly proceeding processes, the impact of climate change on the oceans is not as pronounced and easily observable as the changes in the atmosphere. The pH decrease in comparison to the preindustrial period is estimated to be 0.1 units (NOAA 2022a), and the increase of the mean surface water temperature of the oceans since 1880 is estimated to be about 0.06 °C per decade and 0.11 °C per decade since 1980 (NOAA 2022b).

In contrast to seawater, the inorganic composition of freshwaters is characterized by smaller and strongly variable concentrations. The composition of groundwater is mainly influenced by the geological background. It is dominated by the cations $Ca^{2+}$, $Mg^{2+}$, $Na^+$, and $K^+$ and the anions $HCO_3^-$, $SO_4^{2-}$, and $Cl^-$ (Table 1.3). These major ions typically occur in the lower mmol/L or higher µmol/L range. Dissolved gases such as carbon dioxide and oxygen play a significant role in the interaction of water with solid phases. When seepage water on its way to the groundwater level flows through the soil layers, it is enriched with carbon dioxide in particular, which is a product of biological degradation processes in the soil. This carbon dioxide enhances and accelerates chemical processes of dissolution of minerals by seepage and groundwater. Dissolved oxygen determines the redox intensity (or the redox potential, see Chapter 10) and thus the dissolution or precipitation of certain ions (such as iron or manganese). The solid layers above the groundwater protect it largely from pollutants. However, in recent years, negative anthropogenic influences on the groundwater quality such as through input of nitrate, pesticides, and chlorinated hydrocarbons have increased.

The composition of spring water is initially the same as that of groundwater from which it originated, but as surface water (creeks and rivers), it is then subject to extensive changes due to natural and anthropogenic inputs as well as through

**Table 1.3:** Major and minor ions in freshwater.

| Major ions | |
| --- | --- |
| **Cations** | **Anions** |
| $Na^+$ | $HCO_3^-$ |
| $K^+$ | $SO_4^{2-}$ |
| $Ca^{2+}$ | $Cl^-$ |
| $Mg^{2+}$ | |
| **Minor ions** | |
| **Cations** | **Anions** |
| $NH_4^+$ | $NO_3^-$ |
| $Fe^{2+}$ (under reducing conditions) | $H_2PO_4^-/HPO_4^{2-}$ |
| $Mn^{2+}$ (under reducing conditions) | $Br^-$ |
| $Al^{3+}$ (in the medium pH range in the form of hydroxo complexes) | |

gas exchange with the atmosphere. Especially in urban areas, the rivers receive substantial amounts of inorganic and organic wastewater constituents. Also runoff from agricultural areas (fertilizers and pesticides) and input from other diffuse (nonpoint) sources can contribute to the pollution of rivers and lakes. Rivers have a biological self-cleaning force, mainly based on the oxidative degradation of organic water constituents by microorganisms. These oxidation processes require an adequate oxygen supply. Although the oxygen uptake from the atmosphere by rivers is favored through the turbulence of the flowing water, an excessive organic load can increase the risk of widespread oxygen depletion coupled with a decrease in the water quality.

River and lake sediments act as pollutant sinks, for example, for heavy metals or weakly soluble organic compounds, and thus are often heavily loaded with harmful substances. The introduction of complexing agents or changes in temperature or redox potential can remobilize bound substances from the sediments. Frequently, the concentrations of anthropogenic pollutants are too high for a direct use of river water as raw water for drinking water supply. In such cases, bank filtrate, extracted from wells that are located in a certain distance from the river bank, is often used as raw water source. During the subsurface transport of the river water from the river to the extraction wells, the pollutants are eliminated to some extent by filtration, biodegradation, and sorption processes.

The material balances of standing water bodies are strongly influenced by biological processes and, in moderate (temperate) latitudes, also influenced by the change of stagnation and circulation periods, resulting from the density anomaly of water (density maximum at 4 °C, see Chapter 2, Section 2.2.1). During the stagnation periods in summer and winter, the lakes or reservoirs form separate layers due to the vertical temperature and density differences (lake stratification). The temperature at the bottom of the water body is about 4 °C (density maximum), whereas the temperatures in the upper part of the water body are lower (winter) or higher (summer) than 4 °C (density lower than the maximum). The formation of the separate layers impedes the vertical transfer of dissolved species, whereas sedimentation of particles is still possible. In spring and autumn, the temperature and density differences vanish and circulation becomes possible. Accordingly, vertical concentration profiles are found during the stagnation periods in summer and winter, whereas the circulation in spring and autumn leads to a uniform distribution of the dissolved substances.

In the summer stagnation period, the photosynthetic production of biomass and oxygen from inorganic substances (water, carbon dioxide, nitrate, phosphate) dominates in the upper water layer (epilimnion). The dissolved oxygen, however, cannot be transferred into the deeper water layer (hypolimnion) due to the stratification, whereas dead biomass particles are transported to the bottom of the lake by sedimentation. During the following microbial mineralization of the dead biomass on the bottom of the lake, the organic substances are converted back to inorganic material, which is associated with a depletion of dissolved oxygen in the lower layers of the lake. Since oxygen is not replenished during the stagnation phase, the redox state in the hypolimnion often changes to reducing conditions with the result that reduced compounds, such as $NH_4^+$, $Mn^{2+}$, $Fe^{2+}$, $H_2S/HS^-$, or $CH_4$, increasingly occur. During the reduction and dissolution of manganese(IV) oxide and iron (III) oxide hydrate under reducing conditions, substances bound to these solids (e.g., other metals or phosphate) are also remobilized from the sediment. In particular, high inputs of anthropogenic nutrients (especially phosphorus) intensify the photosynthetic production in the epilimnion and the subsequent mineralization and oxygen consumption in the hypolimnion with the negative impacts described previously. This intensified process is known as eutrophication of lakes.

In the last decades, an increasing number of anthropogenic organic substances have been found in surface water and, to a lower extent, also in groundwater. They originate from point sources (e.g., municipal or industrial wastewater treatment plants) or nonpoint (diffuse) sources (e.g., agriculture). The concentrations of the anthropogenic organic substances range from ng/L to µg/L. These compounds are therefore also referred to as organic micropollutants or organic trace pollutants. The micropollutants belong to very different groups (with respect to their chemical structure or application), for instance, halogenated solvents, pesticides, polycyclic aromatic hydrocarbons (PAHs), petroleum-derived hydrocarbons, BTEX aromatics (benzene, toluene, ethylbenzene, and xylene), phenols, per- and polyfluoroalkyl substances (PFAS), synthetic complexing agents, pharmaceuticals and their metabolites, personal care products, sweeteners, and corrosion inhibitors. The fate of the micropollutants depends on their stability. Some of them are persistent; others are transformed or mineralized by oxidation processes (chemical, biochemical, or photochemical oxidation). Hydrophobic compounds can be immobilized by sorption processes onto solid material (e.g., soils, sediments, or aquifer material), whereas hydrophilic (polar) substances are mobile and can be widely distributed in the hydrosphere. Among the organic water pollutants, persistent, mobile, and toxic substances (PMT substances) are of particular importance for the water quality. It is therefore an urgent task of water protection to reduce their entry into water bodies.

In the recent past, a new problem has come into focus: the occurrence of microplastics. Microplastics are any plastic particles with sizes smaller than 5 mm, where a distinction can be made between primary and secondary microplastics. Primary microplastics are introduced into the environment with particle sizes that are already smaller than 5 mm (e.g., microfibers from clothing, microbeads from personal care products, or other small pellets from various products). Secondary microplastics arise from breakdown of larger plastic material (e.g., plastic bottles or plastic bags) by natural weathering processes.

Even the few examples presented in the previous paragraphs illustrate that the composition of natural waters is influenced by a variety of physical, chemical, and biological processes (Table 1.4), which often act in a very complex manner inside the water phase but also at the boundaries to other compartments such as the atmosphere or solid phases like soils or sediments. Therefore, the composition of water not only varies depending on the type of the water body (oceans, groundwater, lakes, rivers, reservoirs) but, particularly in freshwater, also within a considered water body type. Every river, every lake, or every groundwater body has an individual composition, determined by the natural environment and further modified by anthropogenic inputs. The assessment of water quality, also with respect to possible water use, requires at first knowledge about the nature and concentration of the water constituents. To provide this knowledge is the task of chemical analytics, particularly of water analytics. However, an interpretation of the data provided by chemical analyses is not possible without understanding the hydrochemical and hydrobiological processes and without knowledge about the properties and effects of the water constituents.

Table 1.4: Physicochemical and biochemical processes in water bodies.

| Physicochemical processes | Biochemical processes |
| --- | --- |
| Absorption and desorption of gases | Photosynthesis |
| Acid/base reactions | Aerobic mineralization of organic |
| Dissolution and precipitation of solids | substances |
| Redox processes | Other biochemical redox processes |
| Complex formation | (e.g., nitrification, denitrification, or |
| Sorption onto solid material | desulfurication) |

Besides some general facts about water and aqueous solutions, this textbook presents an introduction to the main hydrochemical processes that are relevant for understanding the occurrence and the fate of water constituents with a focus on reaction equilibria.

# 2 Structure and Properties of Water

## 2.1 Structure of Water

The isolated triatomic $H_2O$ molecule has an angled shape. The O–H bond lengths are 96 pm (1 picometer (pm) = $1 \times 10^{-12}$ m), and the angle between the bonds is 104.5° (Figure 2.1). The bonds between hydrogen and oxygen are single covalent bonds, each consisting of a bonding electron pair (one electron from oxygen and one electron from hydrogen). The bonding partners oxygen and hydrogen have different tendencies to attract the bonding electrons toward themselves. This tendency is quantified by the measure "electronegativity" which has, in the Pauling scale, the value 3.44 for oxygen and 2.2 for hydrogen, respectively. Accordingly, oxygen has a higher tendency to attract the bonding electrons, which leads to a higher electron density at the oxygen atom in comparison to that at the hydrogen atom. As a consequence, the atoms possess partial charges (oxygen: $\delta = -0.34$; hydrogen: $\delta = +0.17$). Here, partial charge means that there is only a shift in the electron density but not a complete gain or loss of an electron as in the case of univalent ions (e.g., $Cl^-$ and $Na^+$) which have full charges (−1, +1).

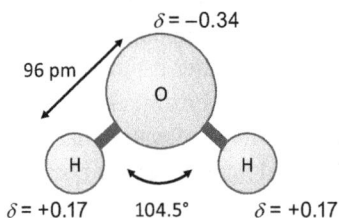

$\delta = -0.34$

96 pm

O

H        H

$\delta = +0.17$    104.5°    $\delta = +0.17$

**Figure 2.1:** Schematic representation of the water molecule.

As a consequence of the occurrence of opposite charge centers within a certain distance, dipole moments (dipole moment = charge × distance) exist along the bond axes and, because the molecule is not linear, a total dipole moment also exists for the molecule. Vector addition of the dipole moments along the axes leads to a total dipole moment of 1.84 D, where Debye (D) is the unit that is typically used for dipole moments; 1 D is equivalent to $3.33 \times 10^{-30}$ C·m, where the charge is given as coulomb (C) and the distance is given as meter (m). The relatively high value of the dipole moment indicates that water is a strongly polar molecule.

The molecular geometry of $H_2O$ can be explained by a $sp^3$ hybridization of the orbitals of the oxygen atom within the water molecule (Figure 2.2). The oxygen atom possesses eight electrons. Six of them are valence electrons, which reside in the outermost electron shell and can participate in the formation of chemical bonds. Whereas the ground state of a single oxygen atom is characterized by the existence of one 2s and three 2p orbitals, which contains the six valence electrons (left-hand side of Figure 2.2),

https://doi.org/10.1515/9783110758788-002

the electron configuration of the oxygen atom in the water molecule is different and can be explained by four $sp^3$ hybrid orbitals ("mixed" orbitals) of equal energy that contain the six valence electrons (right-hand side of Figure 2.2). The 1s orbital of the oxygen is not involved in bonding and remains unchanged. Two of the four $sp^3$ hybrid orbitals are occupied by two electrons each and two are occupied by only one electron each. Each of the half-filled $sp^3$ orbitals of the oxygen can accept an electron from a half-filled 1s orbital of hydrogen which results in two covalent bonds. The orbitals filled with two electrons from oxygen do not take part in bonding. The respective electrons are referred to as lone (nonbonding) electron pairs.

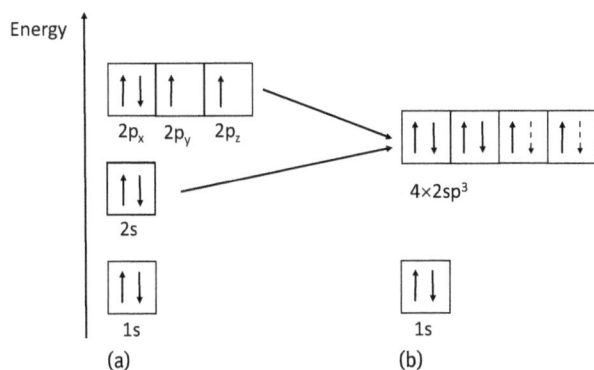

Figure 2.2: Electronic configuration of the oxygen atom in the ground state (a) and in the water molecule (b); hybridization of the 2s and 2p orbitals. The arrows indicate the electrons with their spins and the dashed arrows are the electrons from the two hydrogen atoms in the water molecule. The numbers (1, 2) indicate the principal quantum numbers and the letters (s, p) stand for the azimuthal quantum numbers of the orbitals.

The four $sp^3$ hybrid orbitals are arranged in such a manner that the distance between them becomes maximal. Therefore, the hybrid orbitals are directed to the corners of a tetrahedron as shown in Figure 2.3. Due to the higher space requirement of the lone electron pairs in comparison to the bonding pairs of electrons, the tetrahedral structure is deformed, which explains the slight compression of the tetrahedral angle (109.5°) to the value of the bond angle (104.5°).

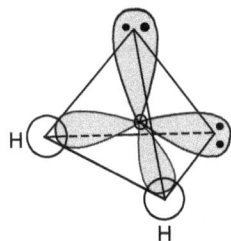

Figure 2.3: Tetrahedral orbital geometry of the water molecule.

In condensed phases (ice and liquid water), the distances between the water molecules are relatively short and electrostatic interactions between the partial charges of the oxygen and hydrogen atoms of neighboring molecules are possible (Figure 2.4). This type of electrostatic interaction is referred to as a hydrogen bond. The formation of hydrogen bonds determines the structure and the properties of the condensed phases ice and liquid water.

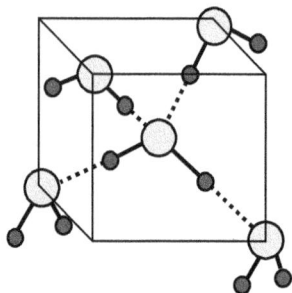

**Figure 2.4:** Tetrahedral arrangement of water molecules due to the formation of hydrogen bonds (dotted lines).

The ideal ice crystal is a network of oxygen atoms, each surrounded tetrahedrally by four hydrogen atoms, two of them covalently bonded within the water molecule and the two others (from neighbor molecules) bonded via hydrogen bonds. The bonding energy of the covalent bonds is stronger and the bond length is shorter than those of the hydrogen bonds. Due to the intermolecular interactions, the bond length of the covalent bonds within the ice crystal is slightly longer than that of the isolated water molecule (101 pm instead of 96 pm) and the bond angle is very close to the tetrahedron angle of 109.5°. The hydrogen bond length in the ice structure is about 180 pm. The resulting hexagonal ice structure is shown in Figure 2.5. This ice modification (referred to as ice Ih) is the stable structure under normal pressure conditions. As shown in Figure 2.5, ice Ih has an open structure with large voids in the interior. At very high pressures and/or very low temperatures, other ice modifications occur.

Currently, the structure of liquid water is not fully understood. It is beyond controversy that hydrogen bonds play an important role not only for the ice structure but also for the structure and the properties of liquid water. In the past, two different types of structure models have been developed: mixture models and continuum models. Mixture models treat water as a mixture of two different types of species. An illustrative, but not uncontroversial mixture model assumes the existence of a dynamic equilibrium between associates (clusters) with ice-like structure and free $H_2O$ molecules (cluster model). Although such a simple model can explain some of the unique properties of water, nowadays more sophisticated continuum models that are in better accordance with X-ray and neutron scattering experiments and with computer simulations are preferred. Following these models, water can be considered a dynamic macromolecular less-ordered hydrogen-bonded network in

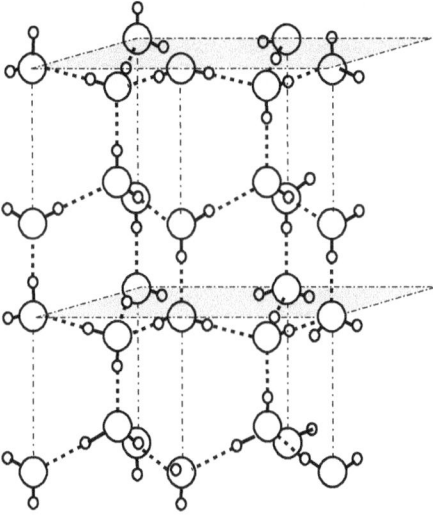

**Figure 2.5:** Crystal structure of ice (hexagonal ice Ih).

which the average coordination number of a water molecule is 4. In contrast to the fixed ice structure, a stronger bending of bonds is possible.

The change of the water structure during temperature increase and phase transformation can be interpreted as a transition from a long-range order in the ice crystal over a pronounced short-range order in the liquid water toward the existence of mainly isolated molecules in the gas state. As we will see in the following section, the occurrence of hydrogen bonds explains a number of exceptional properties of water.

## 2.2 Properties of Water

### 2.2.1 Density

The density, $\rho$, is defined as the ratio of mass, $m$, and volume, $V$:

$$\rho = \frac{m}{V} \tag{2.1}$$

The density of ice and water depends on the temperature as shown in Figure 2.6. Ice has an open structure (see Figure 2.5), which explains its relatively low density. In contrast to most other solids, ice shows a volume contraction during melting. This can be explained with partial collapsing of the bulky ice structure where some of the bonds are broken or bended. The density of the liquid water at 0 °C (0.9998 g/cm³) is about 9% higher than that of ice at the same temperature (0.9166 g/cm³). Accordingly, freezing of water leads to a volume expansion. In closed spaces where the volume cannot expand, the pressure strongly increases, often to a value where the walls of the closed space are destroyed. This effect explains, for instance, the alteration of porous rocks or other porous materials in winter, when the water in the pores freezes.

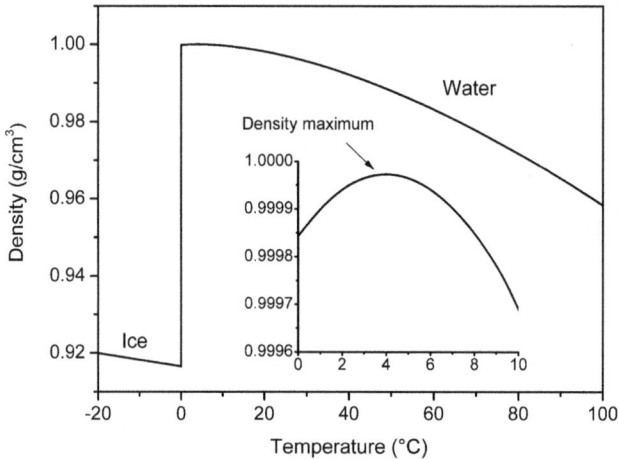

**Figure 2.6:** Density of ice and water as a function of temperature.

Liquid water shows a density maximum at about 4 °C (exactly 3.98 °C). The occurrence of a density maximum is a specific property of water and not observed for other liquids. It is therefore referred to as density anomaly of water. Typically, the volume of a liquid increases with increasing temperature, due to the volume expansion caused by increasing molecular motion. The density maximum observed in the case of water is the result of two opposing effects. At low temperatures, the collapsing of the residual ice structures with the related volume contraction (density increase) dominates the overall effect, whereas at temperatures higher than 4 °C the common volume expansion of liquids dominates.

One of the consequences of the density anomaly of water is the alteration of circulation and stratification that can be observed in standing water bodies in temperate zones of the Earth (Figure 2.7). Typically, stratification occurs in summer and winter, whereas circulation occurs in spring and autumn. The water with the highest density (temperature of about 4 °C) is always found at the bottom of the lake, whereas water with temperatures lower (in winter) or higher (in summer) than the temperature of the density maximum is located near the surface due to its lower density.

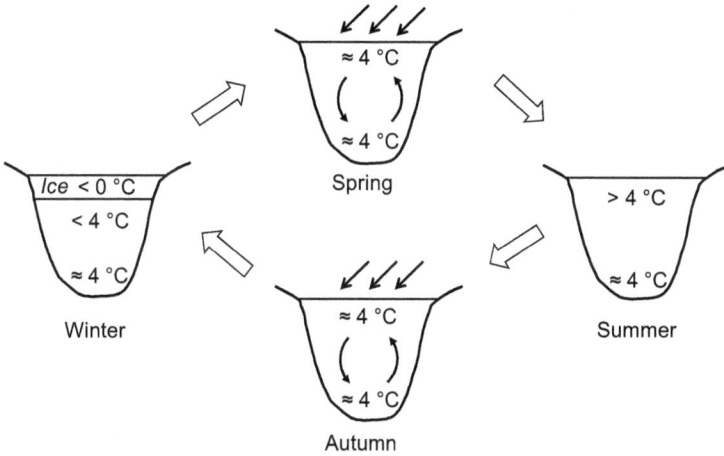

**Figure 2.7:** Circulation (spring and autumn) and stagnation (summer and winter) in dimictic lakes and reservoirs.

In winter, the water freezes at first at the water surface because the coldest water in the lake with the lowest density is located here. The formed ice prevents an energy input into the water body by wind or storm with the consequence that the stratification remains stable despite the small difference in density between 0 and 4 °C. In summer, the temperature differences and therefore also the density differences between the warmer water layers at the top and the colder (4 °C) layers at the bottom of the lake are larger due to the higher outside temperatures. Therefore, usually no disturbance of the thermal stratification can be observed even in strong winds.

Between these periods of stagnation, phases of full circulation occur in spring and autumn due to the heating or cooling of the upper water layers and the associated reduction of the temperature and density gradients in the water body. With the temperature equalization, the density differences vanish and circulation becomes possible. The energy needed to initiate a circulation of the water body comes from winds and storms that frequently occur in spring and autumn. Such lakes with two mixing periods per year are referred to as dimictic lakes. During the circulation, a transport of dissolved substances from the top to the bottom of the lake occurs, and also from the bottom to the top. In this manner, for instance, oxygen that is introduced into the water body from the air can be transported to the deeper layers of the lake where it is needed for degradation of organic material. In contrast, transport of dissolved substances is not possible during the stagnation periods. This can lead, under certain circumstances, to serious problems with respect to the water quality. A well-known example is the eutrophication of lakes. Introduction of high amounts of nutrients (in particular, phosphate) into the lake can cause strong photosynthetical biomass production rates in the summer. After die-off, the dead biomass is transported by sedimentation (that is not hampered by stratification) to the bottom of the lake and consumes oxygen for the oxidative degradation. Since the dissolved oxygen content in the lower layers cannot be refreshed during the stagnation period, the redox state changes to reducing conditions under which, for instance, iron(III) hydroxide is reduced and released from the sediment as $Fe^{2+}$ together with other ions previously sorbed onto $Fe(OH)_3$ (e.g., phosphate or heavy metals). Another example is manganese dioxide ($MnO_2$) that is reduced to $Mn^{2+}$. More details about such redox reactions can be found in Chapter 10.

Of course, the sequence of circulation and stagnation phases in the described manner can only occur under the condition that the outer temperatures show a corresponding seasonal trend (two times crossing the temperature of the density maximum). Although this condition is typically fulfilled in temperate climates, not all lakes show the described sequence. Deviations can be found in the case of shallow lakes with frequent mixing by temperature convection or wind (polymictic lakes) and in the case of deep lakes where the deeper water layers are excluded from circulation (meromictic lakes). Reasons for an incomplete circulation can be an insufficient energy input by wind and storm (e.g., lakes with small surface area in comparison to the depth, steep-sided lakes, and lakes in deep forests) or a high salinity and therefore also a high density in the deeper layers of the lake.

Deviations from the described alteration of stagnation and circulation are also found in regions of the Earth where the temperatures are always above or below the temperature of the density maximum (tropical or polar zones).

## 2.2.2 Phase Diagram – Melting Point and Boiling Point

The conditions under which the different forms of water (ice, liquid water, and water vapor) are stable can be best illustrated by means of a phase diagram. A phase in the chemical sense is a domain that is homogeneous with respect to physical and chemical properties and that is confined from other phases by phase boundaries where the properties are changing. The term "phase" is often used as synonym for state of matter (solid, liquid, and gas). This is acceptable for the gas state and the liquid state of single components where only one phase exists in the respective states. In the solid state, however, more than only one phase can occur even in single-component systems (e.g., different ice modifications in the case of water).

The phase diagram of water (Figure 2.8) shows the stability areas of the different phases (solid ice, liquid water and gaseous water vapor) with dependence on temperature and pressure. The curves represent the conditions under which phase transitions occur: fusion (melting, solid-to-liquid transition), boiling (vaporization, liquid-to-vapor transition), and sublimation (solid-to-vapor transition). The inverse processes are referred to as freezing (crystallization, liquid-to-solid transition), condensation (vapor-to-liquid transition), and deposition (vapor-to-solid transition). Along the curves, two phases are in equilibrium, and there is only one degree of freedom (univariant curves). If, for instance, the pressure is given, the temperature of the phase transition is fixed. It follows from the diagram that the boiling point increases with increasing pressure, whereas the melting point decreases marginally with increasing pressure. The intercept points of the line at $p = 1$ bar (normal atmospheric pressure) with the equilibrium curves give the melting point of ice (0 °C) and the boiling point of water (100 °C) under ambient conditions.

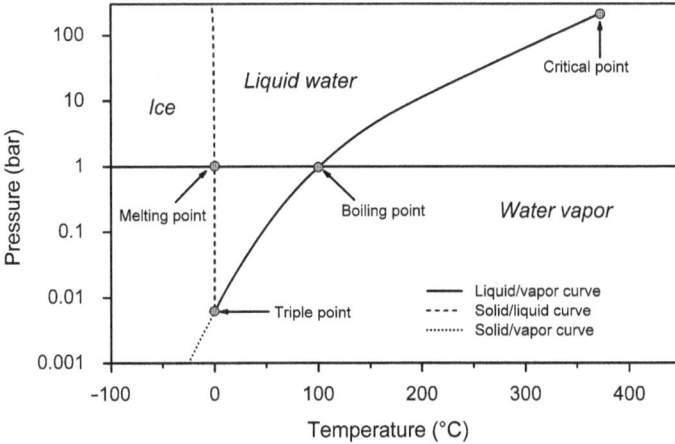

**Figure 2.8:** Phase diagram of water.

There is only one point in the diagram where all three phases coexist. This point is referred to as the triple point or invariant point. Here, pressure and temperature are uniquely specified. The respective data for water are $\vartheta_{tp} = 0.01$ °C (= 273.16 K) and $p_{tp} = 0.0061$ bar.

The liquid–vapor equilibrium curve ends at the critical point ($\vartheta_{crit} = 374.15$ °C, $p_{crit} = 221.3$ bar), where the differences in the physical properties between liquid water and vapor vanish and the liquid and the gas phases become indistinguishable. Water at pressures and temperatures higher than the critical point is referred to as supercritical water (general term: supercritical fluid). Supercritical water is used in practice as an efficient extracting agent.

The areas in the diagram describe the temperature and pressure conditions under which only ice, liquid, or vapor exist. Within these areas, $T$ and $p$ are both free to change (divariant areas).

It has to be noted that under extreme conditions (temperatures lower than −100 °C and/or pressures >100 MPa), further ice phases, different from the normal ice (Ih) shown in Section 2.1, exist. The stability areas of these other solid phases are not shown in Figure 2.8.

The occurrence of hydrogen bonding in the solid and liquid states is the reason for the relatively high phase transition temperatures of water. The melting point of 0 °C and the boiling point of 100 °C (at atmospheric pressure) differ significantly from the general trend of the dihydrogen compounds of the elements in the same group of the periodic table of elements (group 16 or – antiquated – main group 6): oxygen (O), sulfur (S), selenium (Se), tellurium (Te), and polonium (Po). Normally, the melting point and the boiling point are expected to decrease with decreasing molecule size (Figure 2.9). Thus, the melting point and the boiling point of water should be lower than those of dihydrogen sulfide, $H_2S$ (−85.7 °C, −60.2 °C). If this would be true, it would mean that water could not occur on the Earth in liquid form. Only the exceptionally high phase transition temperatures of water enable the existence of liquid water and also life in that form we know. The high phase transition temperatures are indicators that, in comparison to the other compounds, stronger interactions in the condensed phases must be overcome. These stronger interactions result from the hydrogen bonds that act additionally to the van der Waals interactions between the molecules. In contrast to water, the other compounds show no or only negligible hydrogen bonding, since the molecules are not as polar as water. The electronegativities of S, Se, Te, and Po are 2.58, 2.55, 2.1, and 2, respectively. When compared with the electronegativity of oxygen (3.44), these values are much closer to the electronegativity of hydrogen (2.2) and therefore the partial charges are very low. Accordingly, in the case of the other compounds, only the weaker van der Waals forces have to be overcome during melting and boiling. Also, the hydrogen compounds of the neighboring nonmetallic elements in the same period of the periodic table (C, N, F) have much lower melting and boiling points ($CH_4$: −182 °C and −162 °C; $NH_3$: −78 °C and −33 °C; HF: −83 °C and 19.5 °C) although at least $NH_3$ and HF are able to form hydrogen bonds (electronegativities: C: 2.55, N: 3.04, F: 3.98). It can be assumed that the tetrahedral arrangement of the hydrogen bonds in solid and liquid water causes a special stability.

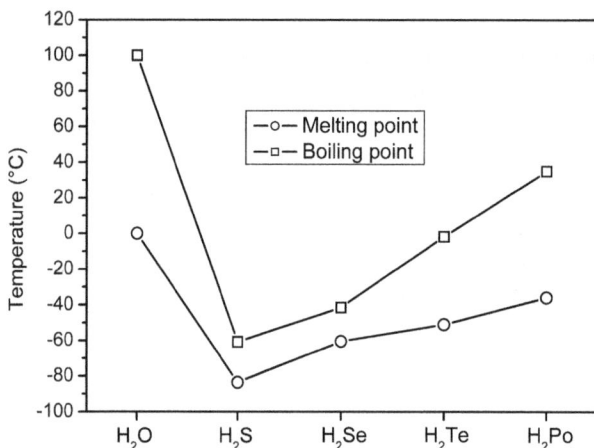

**Figure 2.9:** Melting and boiling points of the dihydrogen compounds of the elements in the 16th group of the periodic table of elements.

## 2.2.3 Energetic Quantities

In chemical thermodynamics, the heat transfer during a process carried out at constant pressure is referred to as enthalpy, and given as the difference between the initial and the final states. Accordingly, the amounts of energy that are necessary for melting a solid or boiling a liquid are expressed as enthalpy of fusion, $\Delta H_{fus}$, and enthalpy of vaporization, $\Delta H_{vap}$, respectively. For water, these quantities have relatively high values and amount to $\Delta H_{fus} = 6.01$ kJ/mol (= 334 kJ/kg) and $\Delta H_{vap} = 40.66$ kJ/mol (= 2 259 kJ/kg). These amounts of energy are stored during melting of ice and boiling of liquid water, respectively. During the reverse processes, condensation and freezing, the same amounts of energy are released. Accordingly, the enthalpy of condensation, $\Delta H_{cond}$, and the enthalpy of freezing, $\Delta H_{freez}$, have the same values as $\Delta H_{vap}$ and $\Delta H_{fus}$ but, by definition, a negative sign (exothermic processes). The reason for the relatively high values is the existence of strong hydrogen bonds in the solid and liquid phases that have to be overcome during the phase transitions solid/liquid and liquid/gas. It has to be noted that the given heat of vaporization is related to the phase transition at the boiling point 100 °C. Vaporization at lower temperatures is also possible, but requires higher energies (e.g., $\Delta H_{vap} = 44.02$ kJ/mol at 25 °C).

Ice, water, and water vapor are also able to store energy without phase transition. This amount of energy is given by the heat capacity at constant pressure, $c_p$. For a considered phase, $c_p$ is given by the change of the enthalpy with temperature:

$$c_p = \left(\frac{\partial H}{\partial T}\right)_p \tag{2.2}$$

Accordingly, $c_p$ describes the increase (or decrease) of the enthalpy of the substance when its temperature is increased (or decreased). In integrated form, we obtain:

$$\Delta H = \int_{T_1}^{T_2} c_p \, dT \tag{2.3}$$

and under neglecting the slight temperature dependence of $c_p$:

$$\Delta H = \bar{c}_p (T_2 - T_1) \tag{2.4}$$

For liquid water (between 0 and 100 °C), the mean molar heat capacity, $\bar{c}_p$, is 75.5 J/(K·mol) which is equivalent to the mean specific (mass-related) heat capacity of 4.2 kJ/(K·kg). This means that 4.2 kJ energy is stored by 1 kg of water if its temperature is increased by 1 K. The temperature-dependent values of $c_p$ show only a slight variation between 76.0 and 75.3 J/(K·mol) with the highest values at 0 and 100 °C and the minimum at about 35 °C.

The mean heat capacity of ice in the temperature range between 0 and −50 °C is 34.6 J/(K·mol) and that of water vapor between 100 and 200 °C is 35.8 J/(K·mol). The value of the specific heat capacity of liquid water (4.2 kJ/(K·kg)) is much higher than that of many other liquids (e.g., methanol: 2.54 kJ/(K·kg); ethanol: 2.45 kJ/(K·kg); benzene: 1.73 kJ/(K·kg)).

The total change of the enthalpy of water with temperature under consideration of the phase transition enthalpies is shown in Figure 2.10. To express the change of the enthalpy, a reference state has to be defined. In the diagram, 0 °C is used as the reference state. For higher temperatures, $\Delta H$ becomes positive, and for lower temperatures, it becomes negative.

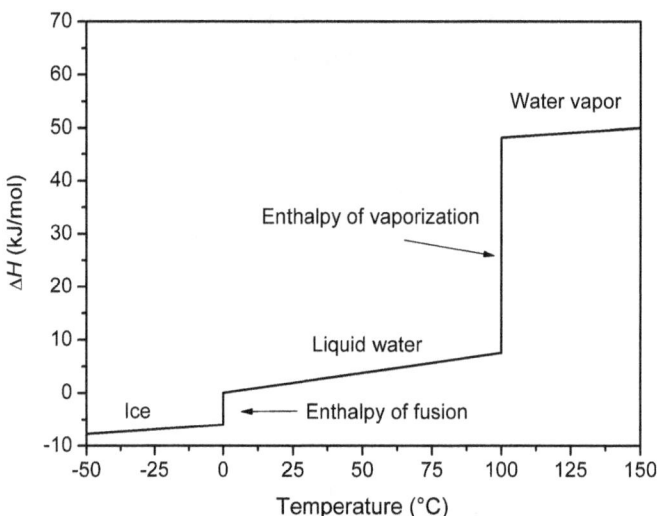

Figure 2.10: Enthalpy–temperature diagram of water.

In summary, it can be stated that the high values of the energetic quantities such as enthalpy of fusion, enthalpy of vaporization, and heat capacity characterize water as an excellent medium for heat storage. This is the reason why the temperature in water bodies does not vary as strongly as the temperature in the air. Since water can store high amounts of heat, it is also an efficient and widely used cooling medium.

**Example 2.1**

What is the enthalpy that is stored in 1 mol $H_2O$ during a temperature increase from −10 to +30 °C? The mean molar heat capacity of ice and liquid water is 34.6 J/(K · mol) and 75.5 J/(K · mol), respectively. The enthalpy of fusion is 6.01 kJ/mol.

**Solution:**

Three quantities contribute to the total increase of the enthalpy: the enthalpy stored in ice during heating from −10 °C to the melting point 0 °C, the enthalpy of fusion, and the enthalpy stored in liquid water during heating from 0 to 30 °C. Taking into account equation (2.4), we can write:

$$\Delta H_{total} = \bar{c}_{p, ice}(T_2 - T_1) + \Delta H_{fus} + \bar{c}_{p, water}(T_3 - T_2)$$

$T_1$ is the starting temperature (−10 °C = 263.15 K), $T_2$ is the melting point of ice (0 °C = 273.15 K), and $T_3$ is the end temperature (30 °C = 303.15 K). With the given data, we get:

$$\Delta H_{total} = (34.6 \text{ J}/(K \cdot mol)) \ (10 \text{ K}) + (6.01 \times 10^3 \text{ J}/mol) + (75.5 \text{ J}/(K \cdot mol)) \ (30 \text{ K})$$

$$\Delta H_{total} = 346 \text{ J}/mol + 6\,010 \text{ J}/mol + 2\,265 \text{ J}/mol = 8\,621 \text{ J}/mol = 8.621 \text{ kJ}/mol$$

The total enthalpy that is stored in 1 mol $H_2O$ during heating from −10 to +30 °C is 8.621 kJ.

## 2.2.4 Viscosity

Viscosity is a measure of the molecule mobility, or in other words, a measure of the internal resistance to flow. The higher the viscosity of a liquid is, the lower is the mobility of its molecules. The viscosity can be expressed as dynamic or as kinematic viscosity. Both are related by the density:

$$\eta = \nu \rho \tag{2.5}$$

where $\eta$ is the dynamic viscosity, $\nu$ is the kinematic viscosity, and $\rho$ is the density. The SI unit of the dynamic viscosity is Pa · s (= kg/(m · s) = N · s/m²). Frequently, the derived unit mPa · s (= $10^{-3}$ Pa · s) is used, which is identical with the antiquated unit centipoise (cP). The unit of the kinematic viscosity is m²/s. Other common units are cm²/s (= stokes, St) or mm²/s (= centistokes, cSt).

The mobility of the molecules and therefore also the viscosity depends on the strength of the intermolecular interactions. Since the hydrogen bonds in liquid water reduce the mobility, it can be expected that the viscosity of water is relatively high. However, not only hydrogen bonds but also van der Waals forces, which typically increase with increasing molecule size, influence the mobility. Furthermore, steric effects may play a role. Therefore, when compared with other liquids, the viscosity of water is not exceptionally high (Table 2.1). In particular, polyhydric alcohols, which combine larger molecule sizes with the possibility to form a high number of hydrogen bonds (due to the existence of more than one OH group per molecule), have considerably higher viscosities.

**Table 2.1:** Dynamic viscosities of selected liquids.

| Liquid | Dynamic viscosity at 20 °C (mPa · s) |
| --- | --- |
| Water | 1.003 |
| Cyclohexane | 0.283 |
| Benzene | 0.647 |
| Tetrachloromethane | 0.986 |
| Methanol | 0.584 |
| Ethanol | 1.190 |
| Ethane-1,2-diol (ethylene glycol) | 19.9 |
| Propane-1,2,3-triol (glycerol) | 1 412 |

The viscosity of water decreases exponentially with increasing temperature (Figure 2.11). The temperature dependence of the viscosity can be explained by an increase of molecule mobility with increasing temperature due to a decrease in the strength of interactions between the molecules.

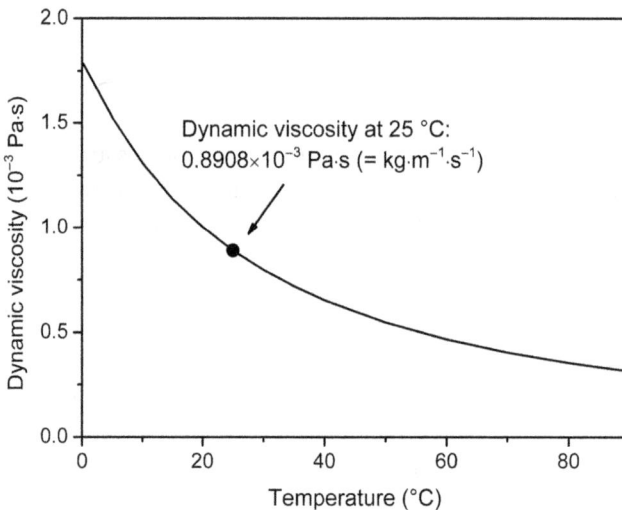

**Figure 2.11:** Temperature dependence of the dynamic viscosity of water.

## 2.2.5 Surface Tension

The surface tension of a liquid describes the energy that is necessary to increase its surface. It follows from the chemical thermodynamics that the change of the Gibbs free energy, $dG$, of a substance (including its surface) during a process can be expressed by the state equation:

$$dG = -S\,dT + V\,dp + \sigma\,dA \qquad (2.6)$$

where $S$ is the entropy, $T$ is the absolute temperature, $V$ is the volume, $p$ is the pressure, $\sigma$ is the surface tension, and $A$ is the surface area. For an isothermal–isobaric process ($dT = 0$, $dp = 0$), the equation simplifies to:

$$dG = \sigma\,dA \qquad (2.7)$$

from which we can derive the definition of the surface tension as the specific free surface energy or, in other words, as the energy necessary for the generation of a unit of area:

$$\sigma = \frac{dG}{dA} \qquad (2.8)$$

Accordingly, the unit of the surface tension is N/m (= N·m/m²= kg/s²) or mN/m (= $10^{-3}$ N/m).

If we further consider the thermodynamic principle that a process proceeds spontaneously in that direction that decreases the Gibbs energy, then we can conclude that due to the condition:

$$dG < 0 \quad \text{if} \quad dA < 0 \qquad (2.9)$$

the contraction of the surface is a thermodynamically favored process. The formation of spherical drops which is a typical behavior of liquids is one of the consequences of the surface tension, because a sphere is the geometrical object with the smallest surface related to the volume.

The effect of surface tension is depicted in Figure 2.12 in a simplified manner. On the surface, the attractive forces between the molecules are imbalanced. Whereas the attractive forces of molecules in the interior of the liquid act into all directions, the attractive forces of molecules at the surface are only directed to the inner of the liquid resulting in a pull into the interior and a tendency to make the surface as small as possible.

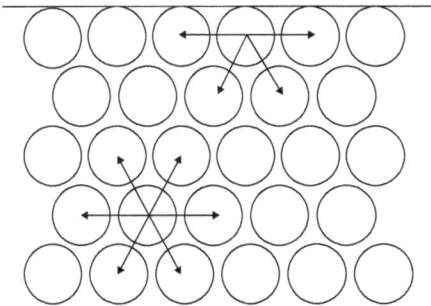

Figure 2.12: Schematic representation of the effect of surface tension.

Since in water the intermolecular attractive forces are particularly high due to the formation of hydrogen bonds, the surface tension of water is very high in comparison to other liquids (Table 2.2). Except for mercury, a liquid metal with quite different structure, the other liquids shown in the table have surface tensions that are in the range of only one-third of the surface tension of water.

**Table 2.2:** Surface tensions of selected liquids.

| Liquid | Surface tension at 20 °C (mN/m) |
|---|---|
| Water | 72.8 |
| Tetrachloromethane | 26.8 |
| Methanol | 22.5 |
| Ethanol | 22.6 |
| Benzene | 28.9 |
| Cyclohexane | 25.0 |
| Mercury | 476.0 |

The surface tension of water decreases with increasing temperature as shown in Figure 2.13.

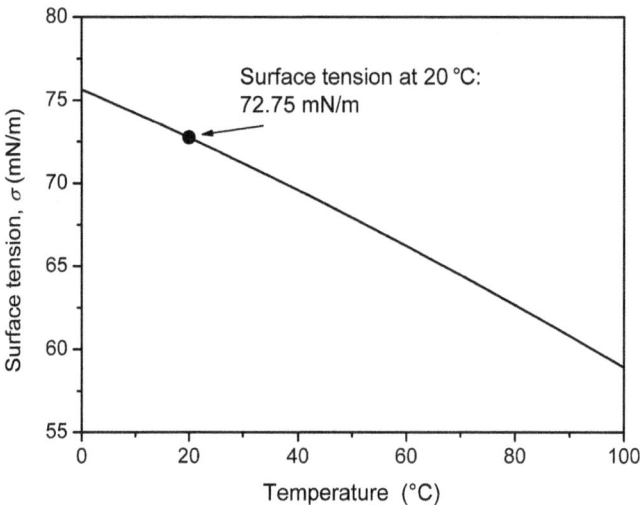

**Figure 2.13:** Temperature dependence of the surface tension of water.

Since the surface tension of water is very high, all effects connected to surface tension are also pronounced. Below, two of such effects will be discussed as examples.

The capillary action (capillarity) is one of the effects that are directly related to the surface tension. The capillary action is the ability of a liquid to rise in a small tube (capillary) without the support of external forces and in opposition to gravity.

The capillary rise, $h$, is directly proportional to the surface tension, $\sigma$, and given by the equation:

$$h = \frac{2\,\sigma\,\cos\theta}{\rho\,g\,r} \tag{2.10}$$

where $\theta$ is the contact angle between liquid and capillary wall, $\rho$ is the density, $g$ is the gravity of Earth, and $r$ is the capillary radius. For hydrophilic capillary surfaces, the contact angle between the water meniscus and the capillary wall is lower than 90° and approaches 0° as a limiting value. As can be derived from equation (2.10), besides the surface tension, the capillary radius is the other main influence factor that determines the capillary rise. The smaller the capillary, the higher is the resulting capillary rise. Capillary action is relevant for the vertical water transport in porous media, in soil, as well as in plants and trees. For instance, capillary rise leads to the formation of a capillary fringe above the water table (top of the groundwater) in the subsurface.

---

**Example 2.2**
Calculate the maximum capillary rise of water at 20 °C in capillaries with inner radii of 0.1 and 0.05 mm. The surface tension of water at 20 °C is 72.8 mN/m (1 N/m $= 1$ kg/s$^2$), the density of water is 1 g/cm$^3$, and the gravity of Earth is $g = 9.81$ m/s$^2$.

**Solution:**
The maximum capillary rise is expected to occur if the contact angle is 0° ($\cos\theta = 1.0$). For $r = 0.1$ mm $= 1 \times 10^{-4}$ m and $\rho = 1$ g/cm$^3 = 1 \times 10^3$ kg/m$^3$, we find:

$$h = \frac{2\,\sigma\,\cos\theta}{\rho\,g\,r} = \frac{2\,(72.8 \times 10^{-3}\ \text{kg/s}^2)\,(1.0)}{(1 \times 10^3\ \text{kg/m}^3)\,(9.81\ \text{m/s}^2)\,(1 \times 10^{-4}\ \text{m})} = 0.148\ \text{m}$$

and for $r = 0.05$ mm $= 5 \times 10^{-5}$ m, we obtain:

$$h = \frac{2\,\sigma\,\cos\theta}{\rho\,g\,r} = \frac{2\,(72.8 \times 10^{-3}\ \text{kg/s}^2)\,(1.0)}{(1 \times 10^3\ \text{kg/m}^3)\,(9.81\,\text{m/s}^2)\,(5 \times 10^{-5}\ \text{m})} = 0.297\ \text{m}$$

The capillary rises for $r = 0.1$ mm and 0.05 mm are $h = 14.8$ cm and $h = 29.7$ cm, respectively.

---

Another effect related to surface tension is the increase of the vapor pressure of a liquid with increasing curvature of its surface. Accordingly, small droplets have a higher vapor pressure than the bulk liquid with a noncurved surface.

The increase of the vapor pressure can be calculated by the Kelvin equation:

$$\frac{p(\text{droplets})}{p_0} = \exp\left(\frac{2\sigma\,V_m}{r\,R\,T}\right) \tag{2.11}$$

where $p(\text{droplets})$ is the vapor pressure of the droplets, $p_0$ is the vapor pressure of the liquid with noncurved surface, $\sigma$ is the surface tension, $V_m$ is the molar volume, $r$ is the droplet radius, $R$ is the gas constant, and $T$ is the absolute temperature. We can derive from this equation that the effect of the increase of the vapor pressure is greater, the higher the surface tension and the smaller the droplet radius. For water with its relatively high surface tension, we can expect that this effect is particularly pronounced in comparison to other liquids.

---

**Example 2.3**

Compute the relative increase of the vapor pressure for water droplets with $r = 1 \times 10^{-4}$ mm and $r = 1 \times 10^{-5}$ mm at 20 °C ($\sigma = 72.8$ mN/m, $R = 8.3145$ J/(mol·K) $= 8.3145$ N·m/(mol·K)). The molecular weight of water is 18 g/mol and its density is about 1 g/cm$^3$.

**Solution:**

The molar volume of water can be calculated from the molecular weight and the density:

$$V_m = \frac{M}{\rho} = \frac{18 \text{ g/mol}}{1 \text{ g/cm}^3} = 18 \text{ cm}^3/\text{mol} = 1.8 \times 10^{-5} \text{ m}^3/\text{mol}$$

If we introduce the given data into the Kelvin equation, we receive for $r = 1 \times 10^{-4}$ mm ($= 1 \times 10^{-7}$ m):

$$\frac{p(\text{droplets})}{p_0} = \exp\left(\frac{2\sigma V_m}{rRT}\right) = \exp\left(\frac{2 \, (7.28 \times 10^{-2} \text{ N/m}) \, (1.8 \times 10^{-5} \text{ m}^3/\text{mol}}{(1 \times 10^{-7} \text{ m}) \, (8.3145 \text{ N} \cdot \text{m/(mol} \cdot \text{K)}) \, (293.15 \text{ K})}\right)$$

$$\frac{p(\text{droplets})}{p_0} = 1.01$$

and for $r = 1 \times 10^{-5}$ mm ($= 1 \times 10^{-8}$ m):

$$\frac{p(\text{droplets})}{p_0} = \exp\left(\frac{2\sigma V_m}{rRT}\right) = \exp\left(\frac{2 \, (7.28 \times 10^{-2} \text{ N/m}) \, (1.8 \times 10^{-5} \text{ m}^3/\text{mol})}{(1 \times 10^{-8} \text{ m}) \, (8.3145 \text{ N} \cdot \text{m/(mol} \cdot \text{K)}) \, (293.15 \text{ K})}\right)$$

$$\frac{p(\text{droplets})}{p_0} = 1.11$$

This means that the vapor pressure is increased by 1% in the case of droplets with $r = 1 \times 10^{-4}$ mm and by 11% in the case of droplets with $r = 1 \times 10^{-5}$ mm in comparison to the compact water without curved surface.

---

The effect of the increased vapor pressure of small droplets influences the condensation of water vapor and is responsible for the fact that under certain conditions, air can be supersaturated with water vapor. This metastable state has relevance, for instance, for condensation processes in the troposphere and for the formation of clouds and fog. As an example, we want to consider warm air above the ground that is saturated with water vapor. In this context, the term "saturation" means that the partial pressure (for partial pressure, see also Section 3.5 in Chapter 3) of water vapor in the air equals the vapor pressure $p_0$ of water at the given temperature. This warm, vapor-saturated air rises to higher and cooler layers of the troposphere and cools down. Due to its temperature dependence, the vapor pressure decreases and is now lower than the partial pressure of the water in the air. Consequently, condensation should occur to reestablish the equality of vapor pressure and partial pressure. The condensation starts with the formation of very small droplets consisting of only some molecules. Since these small microdroplets have an increased vapor pressure, they vaporize again instantaneously. In other words, the system is supersaturated with respect to the vapor pressure of the bulk liquid but not with respect to the vapor pressure of the microdroplets. Therefore, condensation occurs only if the degree of supersaturation is very high or condensation nuclei, for instance, dust particles, trigger the condensation. The latter explains the frequent formation of fog in areas with polluted air.

## 2.2.6 Relative Permittivity

The electrical property "permittivity" is a measure of the electric polarizability of a medium in an electric field. It is directly related to the interaction force between point charges (Coulomb force). The Coulomb force, $F_C$, between the charges $q_1$ and $q_2$, which are located in a distance of $r$, is given by Coulomb's law:

$$F_C = \frac{1}{4\pi\varepsilon}\frac{q_1\,q_2}{r^2} \tag{2.12}$$

where $\varepsilon$ is the permittivity of the medium, which can be expressed by the vacuum permittivity, $\varepsilon_0$, and the relative permittivity of the medium, $\varepsilon_r$:

$$\varepsilon = \varepsilon_0\,\varepsilon_r \tag{2.13}$$

The vacuum permittivity is a fundamental constant. Its value is $8.854 \times 10^{-12}$ A·s/(V·m). The dimensionless quantity $\varepsilon_r$, also known by its antiquated name "dielectric constant," is a substance property that depends on the temperature. Table 2.3 shows the relative permittivity of water for different temperatures.

**Table 2.3:** Relative permittivity (dielectric constant) of water at different temperatures.

| Temperature (°C) | Relative permittivity |
|---|---|
| 0 | 88.00 |
| 5 | 86.40 |
| 10 | 84.11 |
| 15 | 82.22 |
| 20 | 80.36 |
| 25 | 78.54 |
| 30 | 76.75 |

Coulomb's law describes the simple case where electrostatic interactions exist between two point charges. However, the situation in aqueous ionic solutions is more complex and can be described by the concept of ionic atmospheres. Formation of ionic atmospheres means that each ion is surrounded by counter ions (ions with opposite charge). A well-known theoretical approach to describe this complex situation was proposed by Debye and Hückel (Debye–Hückel theory of ionic solutions). Despite the more complex character of the Debye–Hückel theory, the permittivity plays the same important role for the interaction force as in Coulomb's law.

The practical relevance of the Debye–Hückel theory consists of the fact that it provides the theoretical basis for the estimation of activity coefficients that are necessary to describe reaction equilibria under nonideal conditions (conditions where interactions influence the behavior of the ions). Accordingly, the relative permittivity also occurs in the equations that can be used for estimating activity coefficients (Chapter 3, Section 3.8).

## 2.3 Water as a Solvent

Water is a good solvent for many solids, liquids, and gases. This is the reason why we cannot find pure water in nature. Seawater and freshwater, such as ground or surface water, always contain certain amounts of dissolved species of natural or anthropogenic origin. The solubilities of inorganic and organic compounds in water differ depending on their chemical nature. The examples shown in Table 2.4 demonstrate the broad range of aqueous solubilities.

**Table 2.4:** Aqueous solubilities of selected compounds at 25 °C.

| Solute | Aqueous solubility (mg/L) |
|---|---|
| NaCl | 358 000 |
| Phenol | 86 600 |
| $CaSO_4 \cdot 2H_2O$ | 2 000 |
| Benzene | 1 780 |
| Trichloroethene | 1 090 |
| Tetrachloroethene | 141 |
| Atrazine | 33 |
| Phenanthrene | 1.29 |
| Benzo[a]pyrene | 0.0038 |

The general condition for dissolution is that the change of the Gibbs free energy during the dissolution process is negative. The change of Gibbs free energy of solution, $\Delta G_{sol}$, can be expressed by:

$$\Delta G_{sol} = \Delta H_{sol} - T \Delta S_{sol} < 0 \tag{2.14}$$

where $\Delta H_{sol}$ is the enthalpy of solution, $\Delta S_{sol}$ is the entropy of solution, and $T$ is the absolute temperature.

From equation (2.14), we can derive that the solubility is influenced by enthalpic and entropic effects. Typically, the entropy, which is a measure of the disorder (or randomness) in a given system, increases during a dissolution process. Accordingly, $\Delta S_{sol}$ is positive and the second term in equation (2.14) is negative. The enthalpy of solution, $\Delta H_{sol}$, summarizes the energetic effects of three different processes occurring during dissolution. Breaking the bonds within the solute is the first process that is relevant for dissolution. This means that in salts and other crystalline substances (e.g., oxides and hydroxides), the lattice energy and in nonelectrolytes the intermolecular interactions (van der Waals forces) have to be overcome. This process requires energy and is therefore referred to as endothermic. The second process is to break the attraction forces within the solvent (van der Waals forces and hydrogen bonds in water) to form cavities for the solute to enter. This process also requires energy and is therefore also endothermic. The third process comprises the formation of interaction forces between solvent and solute, for instance, ion–dipole interactions in the case of ionic species dissolved in water. This exothermic process is generally referred to as solvation. In the specific case of water as a solvent, the term "hydration" is typically used instead of solvation. The sum of the enthalpies of the three processes can be positive or negative. As we can further derive from equation (2.14), dissolution can occur despite a positive enthalpy of solution if the entropy term is greater than the enthalpy term.

For ionic solids, the enthalpy of solution is given by:

$$\Delta H_{sol} = \Delta U + \Delta H_{hyd, cat} + \Delta H_{hyd, an} \tag{2.15}$$

where $\Delta U$ is the molar lattice energy (here defined as the energy needed to separate a mole of a crystalline solid into ions), $\Delta H_{hyd,cat}$ is the molar enthalpy of hydration of the cation, and $\Delta H_{hyd,an}$ is the molar enthalpy of hydration of the anion. It has to be noted that the enthalpy necessary to break the interaction forces between the solvent molecules is here already included in the hydration enthalpies. Figure 2.14 shows schematically the dissolution of an ionic solid and the hydration of the ions.

The interactions between the ions in a crystal are typically very strong and often the lattice energy cannot be counterbalanced by the hydration enthalpies of the ions. Therefore, the enthalpy of solution of ionic solids is frequently positive. This means that, in this case, dissolution is an endothermic process and, accordingly, the solubility increases with increasing temperature. However, there are also numerous ionic solids where the dissolution is exothermic.

It has to be noted that the lattice energy is sometimes defined in an opposite manner as the release of energy during formation of 1 mol of an ionic solid. In this case, the absolute value of the energy remains the same but it receives a negative sign. Accordingly, the first term in equation (2.15) has to also be written with a negative sign so that it remains positive.

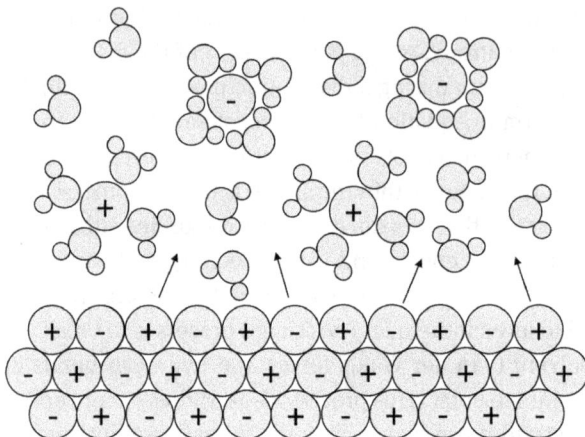

**Figure 2.14:** Dissolution of an ionic solid and hydration of the formed ions.

**Example 2.4**

What are the enthalpies of solution of sodium chloride (NaCl) and lithium chloride (LiCl)? The enthalpies of hydration are −398 kJ/mol for Na$^+$, −508 kJ/mol for Li$^+$, and −376 kJ/mol for Cl$^-$. The lattice energy is 786 kJ/mol for NaCl and 853 kJ/mol for LiCl.

**Solution:**

NaCl:

$$\Delta H_{sol} = \Delta U + \Delta H_{hyd, cat} + \Delta H_{hyd, an}$$

$$\Delta H_{sol} = 786 \text{ kJ/mol} - 398 \text{ kJ/mol} - 376 \text{ kJ/mol} = 12 \text{ kJ/mol}$$

LiCl:

$$\Delta H_{sol} = \Delta U + \Delta H_{hyd, cat} + \Delta H_{hyd, an}$$

$$\Delta H_{sol} = 853 \text{ kJ/mol} - 508 \text{ kJ/mol} - 376 \text{ kJ/mol} = -31 \text{ kJ/mol}$$

The dissolution of NaCl is endothermic and that of LiCl is exothermic. Accordingly, the solubility of NaCl increases and that of LiCl decreases with increasing temperature.

As can be seen from Example 2.4, NaCl has a positive enthalpy of dissolution. Nevertheless, its solubility is very high (see Table 2.4). This demonstrates the important role of the entropy term (equation (2.14)), which also increases with increasing temperature.

The interactions between the ions in the solid state (lattice energy) and the interactions in the solution (enthalpies of hydration) depend on the size and the charge of the cations and anions. The variety of possible combinations leads to quite different solubilities. Therefore, we can find a broad range of solubilities for ionic solids, from very high solubilities of alkaline earth halogenides (e.g., NaCl) to very low solubilities of heavy metal sulfides or hydroxides (see also Chapter 8).

As shown in Table 2.4, organic substances with covalent bonds also show more or less pronounced solubilities. Here, higher solubilities are found for molecules with polar groups that can interact with the solvent water to overcome the relatively weak intermolecular interactions of the pure solute. In general, the solubility of nonelectrolytes increases with increasing polarity. However, nonpolar substances, which are not able to interact significantly with the solvent, also show a weak solubility. This can be explained by the entropy increase due to the dispersion of the dissolved molecules within the liquid.

Sometimes the dissolution is connected with a chemical reaction, in particular with a proton transfer. Some strongly polar compounds with covalent bonds are able to form ions during dissolution. In this case, the dipole interactions with the polar solvent are strong enough to provide the energy necessary for ionization. Such compounds are referred to as potential electrolytes. Typical examples are HCl, $NH_3$, and organic acids (carboxylic acids and phenols) and bases (amines). In the case of acids, the protons released from the potential electrolytes are accepted by water molecules as demonstrated for HCl as an example:

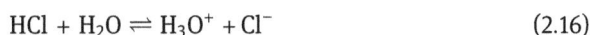

$$HCl + H_2O \rightleftharpoons H_3O^+ + Cl^- \tag{2.16}$$

In the case of bases, protons from water are accepted by the base and the protonated base is formed, here shown for aniline, a basic aromatic amine:

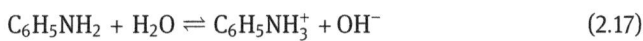

$$C_6H_5NH_2 + H_2O \rightleftharpoons C_6H_5NH_3^+ + OH^- \tag{2.17}$$

Due to the proton transfer during dissolution, the solubility becomes pH-dependent.

## 2.4 Problems

**2.1.** Within the hydrological cycle, 413 000 km$^3$ water is evaporated from the oceans per year. What is the energy that is needed for vaporization of this volume at 17 °C (which is the estimated mean surface water temperature of the oceans)? The heat of evaporation of water at 17 °C is $\Delta H_{vap} = 44.3$ kJ/mol. The density of water is 1 g/cm$^3$ and the molecular weight is 18 g/mol.

**2.2.** What is the enthalpy that is released when 100 L of water freezes? The enthalpy of fusion of water is 6.01 kJ/mol, the density of water is approximately 1 g/cm$^3$, and the molecular weight of water is 18 g/mol.

**2.3.** Calculate the enthalpy that is necessary to heat 1 L of water ($M = 18$ g/mol) from 10 to 60 °C. The density at 10 °C is approximately 1 g/cm$^3$, and the mean molar heat capacity is 75.5 J/(K·mol).

**2.4.** The dynamic viscosity of water at 10 °C is $1.308 \times 10^{-3}$ Pa·s and the density is approximately 1 g/cm$^3$. What is the kinetic viscosity at the same temperature?

**2.5.** Compare the capillary rise of water in a capillary with an inner radius of 0.06 mm at 10 and 30 °C. The density at both temperatures is approximately 1 g/cm$^3$. The surface tension of water is 74.23 mN/m at 10 °C and 71.2 mN/m at 30 °C. The contact angle is assumed to be 0° in both cases. The gravity of Earth is 9.81 m/s$^2$.

**2.6.** What is the droplet radius at which the water vapor pressure is increased by 50% in comparison to the bulk liquid? The temperature is assumed to be 10 °C, and the surface tension at this temperature is 74.23 mN/m. The molar volume of water amounts to $V_m = 18$ cm$^3$/mol and the value of the gas constant is $R = 8.3145$ J/(mol·K).

# 3 Concentrations and Activities

## 3.1 Introduction

As we have seen in Chapter 1, there is no pure water in nature or in the urban water cycle. All natural waters and waters in the urban water cycle (e.g., drinking water or wastewater) contain water constituents in different quantities. Therefore, we need measures to quantify the content of these water constituents. We need such measures in particular for all material balances and equilibrium calculations. Such measures that are used to quantify the composition of water are known as concentrations. Unfortunately, there are a number of different concentration measures in use. Therefore, it is necessary to know the exact definitions of the concentration measures and the possibilities of converting them. Besides the general concentration measures used in different fields of science, there are also some specific measures that are, in particular, applied in water chemistry.

The conventional concentrations, which can be measured by analytical methods, have to be distinguished from the effective concentrations that act in reactions. These effective concentrations are referred to as activities and have to be applied for very exact equilibrium calculations.

In this chapter, the most important concentration measures are presented together with some specific applications. Moreover, general aspects of determination and application of activities are discussed.

## 3.2 Concentrations

The most common measures used to describe the amount of water constituents in an aqueous system are the mass concentration, the molar concentration, and the equivalent concentration.

The mass concentration, $\rho^*$, is defined by:

$$\rho^*(X) = \frac{m(X)}{V} \tag{3.1}$$

where $m(X)$ is the mass of the substance X in the solution and $V$ is the volume of the solution. Common units are g/L, mg/L ($10^{-3}$ g/L), µg/L ($10^{-6}$ g/L), and ng/L ($10^{-9}$ g/L). An alternative symbol for the mass concentration is $\beta$.

The molar concentration, $c$, describes the ratio of the substance amount, $n$, and the volume of the solution. It is also referred to as molarity:

$$c(X) = \frac{n(X)}{V} \tag{3.2}$$

https://doi.org/10.1515/9783110758788-003

Since the substance amount (number of moles) can be derived from the ratio of the mass of the substance and its molecular weight, $M$:

$$n\left(X\right) = \frac{m(X)}{M(X)} \tag{3.3}$$

the relationship between molar concentration and mass concentration is given by:

$$c(X) = \frac{m(X)}{V\,M(X)} = \frac{\rho^*(X)}{M(X)} \tag{3.4}$$

Common units of $c$ are mol/L and mmol/L ($10^{-3}$ mol/L). Often, the abbreviation M is used for mol/L. For instance, 0.5 M NaCl means a NaCl solution with the concentration of $c = 0.5$ mol/L.

---

**Example 3.1**

The mass concentration of calcium ions in a water sample is 140 mg/L. What is the molar concentration? $M(Ca) = 40.1$ g/mol.

**Solution:**

Since the mass concentration is given in mg/L, it is reasonable to express the molecular weight in mg/mmol (40.1 g/mol = 40.1 mg/mmol). According to equation (3.4), we obtain:

$$c(Ca^{2+}) = \frac{\rho^*(Ca^{2+})}{M(Ca^{2+})} = \frac{140\ mg/L}{40.1\ mg/mmol} = 3.49\ mmol/L$$

---

The equivalent concentration is a specific concentration measure that is useful for dissolved ions in all cases where charge balances play a role. The equivalent concentration can be considered a molar concentration normalized to the charge of the ion. An equivalent (more precisely ion equivalent) can be thought of as that fraction of a real ion that carries the charge +1 or −1. As a consequence, each positive equivalent compensates (or neutralizes) exactly one negative equivalent. For instance, $Al^{3+}$ can be formally divided into three fractions, each of them carrying the charge +1 (Figure 3.1). Accordingly, each $Al^{3+}$ ion corresponds to three equivalents and the molar concentration of the equivalents is three times higher than the molar concentration of $Al^{3+}$. Accordingly, the definition of the equivalent concentration is given by:

$$c\left(\frac{1}{z}X\right) = \frac{n\left(\frac{1}{z}X\right)}{V} \tag{3.5}$$

where $z$ is the (absolute) charge number of the considered ion, and $\frac{1}{z}X$ represents the univalent fractions of the ions or, in other words, the equivalents.

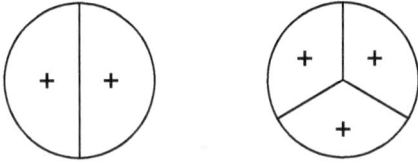

Ca$^{2+}$: two equivalents    Al$^{3+}$: three equivalents

**Figure 3.1:** Graphical representation of the definition of ion equivalents.

Since the number of equivalents is measured in moles and the ion charge number, $z$, is dimensionless, the unit of the equivalent concentration is the same as the unit of the molar concentration (mol/L, mmol/L). Therefore, it is necessary to use unique symbols for the equivalent concentration to avoid confusion, for instance, $c(1/2\ Ca^{2+})$ or $c(1/3\ Al^{3+})$. Sometimes, the term "normality" (abbreviation N) is also used to express the equivalent concentration, for instance, 0.01 N instead of $c(1/z\ X) = 0.01$ mol/L.

For the conversion of molar and equivalent concentrations, the following relationship has to be used:

$$c\left(\frac{1}{z}X\right) = z\,c(X) \tag{3.6}$$

---

**Example 3.2**
In a water sample, the following mass concentrations were found: $\rho^*(Ca^{2+}) = 35.5$ mg/L, $\rho^*(Na^+) = 8$ mg/L, and $\rho^*(SO_4^{2-}) = 27$ mg/L. What are the equivalent concentrations of these ions? $M(Ca) = 40.1$ g/mol, $M(Na) = 23$ g/mol, $M(S) = 32.1$ g/mol, $M(O) = 16$ g/mol.

**Solution:**
At first, the molar concentrations have to be calculated by means of equation (3.4). The required molecular weight for sulfate can be found by summing up the molecular weights of the atoms contained in the ion:

$$c(Ca^{2+}) = \frac{\rho^*(Ca^{2+})}{M(Ca^{2+})} = \frac{35.5\ \text{mg/L}}{40.1\ \text{mg/mmol}} = 0.89\ \text{mmol/L}$$

$$c(Na^+) = \frac{\rho^*(Na^+)}{M(Na^+)} = \frac{8\ \text{mg/L}}{23\ \text{mg/mmol}} = 0.35\ \text{mmol/L}$$

$$c(SO_4^{2-}) = \frac{\rho^*(SO_4^{2-})}{M(SO_4^{2-})} = \frac{27\ \text{mg/L}}{(32.1+4\times 16)\ \text{mg/mmol}} = \frac{27\ \text{mg/L}}{96.1\ \text{mg/mmol}} = 0.28\ \text{mmol/L}$$

To find the equivalent concentrations, the molar concentrations have then to be multiplied with the value of the charge (absolute value without sign):

$$c\left(\frac{1}{2}Ca^{2+}\right) = z(Ca^{2+})\ c(Ca^{2+}) = 2\ (0.89\ \text{mmol/L}) = 1.78\ \text{mmol/L}$$

$$c\left(\frac{1}{1}Na^+\right) = z(Na^+)\,c(Na^+) = 1\,(0.35\text{ mmol/L}) = 0.35\text{ mmol/L}$$

$$c\left(\frac{1}{2}SO_4^{2-}\right) = z(SO_4^{2-})\,c(SO_4^{2-}) = 2\,(0.28\text{ mmol/L}) = 0.56\text{ mmol/L}$$

As an alternative to the molar concentration, the molality is sometimes also used. The molality, abbreviated as $b$ or $m$ or $c_m$, relates the dissolved substance amount to the mass of the solvent:

$$b(X) = \frac{n(X)}{m_{solvent}} \tag{3.7}$$

The advantage over volume-related concentration measures is that the mass in contrast to the volume does not depend on temperature and, consequently, also the values of the mass-related concentrations do not depend on temperature. In water chemistry, however, where only small temperature ranges are considered, the temperature dependence of volume-related concentrations can be neglected and the molality is only used for specific applications (e.g., calculation of the freezing point depression or boiling point elevation, see Chapter 4).

In general, the conversion of molarity (volume-related) into molality (mass-related) and vice versa is not as simple as the mutual conversion of volume-related concentrations. One of the preconditions necessary for conversion is the knowledge of the density of the solution. On the other hand, if only dilute aqueous solutions are considered, a simple approximate conversion is possible, because in this case the density of the aqueous solution can be set equal to the density of the solvent water (about 1 kg/L). The admissibility of this assumption for natural waters can be demonstrated by considering ocean water with its relatively high salt concentration. Despite the relatively high salt content, the density of ocean water is only about 1.03 kg/L, which is not much different from 1 kg/L. Consequently, the density of freshwater with lower salt concentration is even closer to 1 kg/L. Accordingly, the following approximations can be made:

$$\rho_{solution} \approx \rho_{solvent} = 1\text{ kg/L} \rightarrow 1\text{ L solution} \approx 1\text{ kg solution} \tag{3.8}$$

$$m_{solutes} \ll m_{solvent} \rightarrow m_{solvent} \approx m_{solution} \rightarrow 1\text{ kg solvent} \approx 1\text{ kg solution} \approx 1\text{ L solution} \tag{3.9}$$

From these approximations, we finally find:

$$\frac{b}{\text{mol/kg}} \approx \frac{c}{\text{mol/L}} \tag{3.10}$$

This means that, for a dilute aqueous solution, the numerical value of the molality equals the numerical value of the molarity.

Fractions, for instance mass fractions, mole fractions, or volume fractions, are another type of measure that can be used to describe the composition of solutions. Here, the dimensions of solute and solution are the same (e.g., mass/mass or volume/volume). Using the same units for solute and solution (e.g., g/g), the fraction ranges from 0 to 1 and can be given alternatively as percentage (fraction multiplied with 100%), ranging from 0% to 100%. Whereas the volume fraction is seldom used for aqueous solutions, the use of mass fraction:

$$w(X) = \frac{m(X)}{m_{solution}} = \frac{m(X)}{\sum m} \tag{3.11}$$

or mole fraction:

$$x(X) = \frac{n(X)}{n_{solution}} = \frac{n(X)}{\sum n} \tag{3.12}$$

can be reasonable in specific cases. $\sum m$ is the sum of the masses of solvent and solutes (= mass of solution) and $\sum n$ is the sum of the substance amounts (moles) of solvent and solutes (= substance amount of solution). According to the typical concentration levels in aqueous systems, the values of the mass or mole fractions are very low.

In analogy to the definition of mass percent (parts per hundred, e.g., g/100 g), some specific units have been derived that allow better description of low or very low concentrations: parts per million (ppm), parts per billion (ppb), and parts per trillion (ppt):

$$1\,ppm = 1\,g/10^6\,g = 1\,mg/kg \tag{3.13}$$

$$1\,ppb = 1\,g/10^9\,g = 1\,\mu g/kg \tag{3.14}$$

$$1\,ppt = 1\,g/10^{12}\,g = 1\,ng/kg \tag{3.15}$$

It has to be noted that ppm, ppb, and ppt can also be defined on the basis of the volume fraction (volume/volume, typically indicated by the subscript v, e.g., $ppm_v$) but that is more common for gas mixtures than for aqueous solutions.

Although ppm, ppb, and ppt describe ratios of the same dimension (mass/mass), they are often used in practice as alternative units for mass concentrations (mass/volume). This is not exact, but can be accepted for dilute aqueous solutions, where 1 kg solution approximately corresponds to 1 L solution:

$$1\,ppm = 1\,mg/kg \approx 1\,mg/L \tag{3.16}$$

$$1\,ppb = 1\,\mu g/kg \approx 1\,\mu g/L \tag{3.17}$$

$$1\,ppt = 1\,ng/kg \approx 1\,ng/L \tag{3.18}$$

## 3.3 Conversion of Concentration Units

### 3.3.1 Introduction and Basic Equations

To convert concentration units, the respective definition equations have to be combined with other basic equations such as the definition of density, $\rho$, the relationship between substance mass, $m$, and substance amount (number of moles), $n$, as well as the balance equations for the mass and the substance amount:

$$\rho = \frac{m}{V} \tag{3.19}$$

$$n = \frac{m}{M} \tag{3.20}$$

$$m_{\text{solution}} = m_{\text{solvent}} + \sum_{j=1}^{N} m_j \tag{3.21}$$

$$n_{\text{solution}} = n_{\text{solvent}} + \sum_{j=1}^{N} n_j \tag{3.22}$$

where $V$ is the volume, $M$ is the molecular weight, $m_j$ are the masses of the solutes, and $n_j$ are the substance amounts of the solutes.

In the following subsections, conversion equations are summarized for most of the concentration measures presented in Section 3.2. For dilute aqueous solutions (e.g., concentrations in the mg/L or mmol/L range or lower, as can be found in freshwater), some of the complex equations can be simplified by considering the following assumptions: the density of the solution is approximated by the density of the solvent and the mass or amount of solution is set equal to the mass or amount of the solvent due to the negligible contribution of the solutes to the total mass or to the total substance amount of the solution. The quantity 55.56 mol/L that frequently occurs in the conversion equations is the formal molar concentration of pure water (density divided by molecular weight):

$$c(\text{pure water}) = \frac{\rho(\text{water})}{M(\text{water})} = \frac{1000 \text{ g/L}}{18 \text{ g/mol}} = 55.56 \text{ mol/L} \tag{3.23}$$

### 3.3.2 Conversion of Mass Concentration

Mass concentration into molar concentration:

$$c_i = \frac{\rho_i^*}{M_i} \tag{3.24}$$

Mass concentration into molality:

$$b_i = \frac{\rho_i^*}{M_i \left( \rho_{\text{solution}} - \sum\limits_{j=1}^{N} \rho_j^* \right)} \tag{3.25}$$

$$b_i \approx \frac{\rho_i^*}{M_i \, (1\,000\,\text{g/L})} \quad \text{(dilute aqueous solutions)} \tag{3.26}$$

Mass concentration into mole fraction:

$$x_i = \frac{\rho_i^* \, M_{\text{solvent}}}{M_i \left( \rho_{\text{solution}} - \sum\limits_{j=1}^{N} \rho_j^* + M_{\text{solvent}} \sum\limits_{j=1}^{N} \frac{\rho_j^*}{M_j} \right)} \tag{3.27}$$

$$x_i \approx \frac{\rho_i^*}{M_i \, (55.56\,\text{mol/L})} \quad \text{(dilute aqueous solution)} \tag{3.28}$$

Mass concentration into mass fraction:

$$w_i = \frac{\rho_i^*}{\rho_{\text{solution}}} \tag{3.29}$$

$$w_i \approx \frac{\rho_i^*}{1\,000\,\text{g/L}} \quad \text{(dilute aqueous solution)} \tag{3.30}$$

### 3.3.3 Conversion of Molar Concentration

Molar concentration into mass concentration:

$$\rho_i^* = c_i \, M_i \tag{3.31}$$

Molar concentration into molality:

$$b_i = \frac{c_i}{\rho_{\text{solution}} - \sum\limits_{j=1}^{N} c_j \, M_j} \tag{3.32}$$

$$b_i \approx \frac{c_i}{1\,000\,\text{g/L}} \quad \text{(dilute aqueous solution)} \tag{3.33}$$

Molar concentration into mole fraction:

$$x_i = \frac{c_i \, M_{\text{solvent}}}{\rho_{\text{solution}} - \sum\limits_{j=1}^{N} c_j \, M_j + M_{\text{solvent}} \sum\limits_{j=1}^{N} c_j} \tag{3.34}$$

$$x_i \approx \frac{c_i}{55.56 \text{ mol/L}} \quad \text{(dilute aqueous solution)} \tag{3.35}$$

Molar concentration into mass fraction:

$$w_i = \frac{c_i M_i}{\rho_{\text{solution}}} \tag{3.36}$$

$$w_i = \frac{c_i M_i}{1\,000 \text{ g/L}} \quad \text{(dilute aqueous solution)} \tag{3.37}$$

### 3.3.4 Conversion of Molality

Molality into mass concentration:

$$\rho_i^* = \frac{b_i M_i \rho_{\text{solution}}}{1 + \sum\limits_{j=1}^{N} b_j M_j} \tag{3.38}$$

$$\rho_i^* \approx b_i M_i \, (1\,000 \text{ g/L}) \quad \text{(dilute aqueous solution)} \tag{3.39}$$

Molality into molar concentration:

$$c_i = \frac{b_i \rho_{\text{solution}}}{1 + \sum\limits_{j=1}^{N} b_j M_j} \tag{3.40}$$

$$c_i \approx b_i \, (1\,000 \text{ g/L}) \quad \text{(dilute aqueous solution)} \tag{3.41}$$

Molality into mole fraction:

$$x_i = \frac{b_i M_{\text{solvent}}}{1 + M_{\text{solvent}} \sum\limits_{j=1}^{N} b_j} \tag{3.42}$$

$$x_i \approx b_i \, (18 \text{ g/mol}) \quad \text{(dilute aqueous solution)} \tag{3.43}$$

Molality into mass fraction:

$$w_i = \frac{b_i M_i}{1 + \sum\limits_{j=1}^{N} b_j M_j} \tag{3.44}$$

$$w_i \approx b_i M_i \quad \text{(dilute aqueous solution)} \tag{3.45}$$

### 3.3.5 Conversion of Mole Fraction

Mole fraction into mass concentration:

$$\rho_i^* = \frac{x_i\, M_i\, \rho_{\text{solution}}}{x_{\text{solvent}}\, M_{\text{solvent}} + \sum\limits_{j=1}^{N} x_j\, M_j} \tag{3.46}$$

$$\rho_i^* \approx x_i\, M_i\, (55.56\,\text{mol/L}) \quad \text{(dilute aqueous solution)} \tag{3.47}$$

Mole fraction into molar concentration:

$$c_i = \frac{x_i\, \rho_{\text{solution}}}{x_{\text{solvent}}\, M_{\text{solvent}} + \sum\limits_{j=1}^{N} x_j\, M_j} \tag{3.48}$$

$$c_i \approx x_i\,(55.56\,\text{mol/L}) \quad \text{(dilute aqueous solution)} \tag{3.49}$$

Mole fraction into molality:

$$b_i = \frac{x_i}{M_{\text{solvent}}\left(1 - \sum\limits_{j=1}^{N} x_j\right)} \tag{3.50}$$

$$b_i \approx \frac{x_i}{0.018\,\text{kg/mol}} \quad \text{(dilute aqueous solution)} \tag{3.51}$$

Mole fraction into mass fraction:

$$w_i = \frac{x_i\, M_i}{x_{\text{solvent}}\, M_{\text{solvent}} + \sum\limits_{j=1}^{N} x_j\, M_j} \tag{3.52}$$

$$w_i \approx \frac{x_i\, M_i}{18\,\text{g/mol}} \quad \text{(dilute aqueous solution)} \tag{3.53}$$

### 3.3.6 Conversion of Mass Fraction

Mass fraction into mass concentration:

$$\rho_i^* = w_i\, \rho_{\text{solution}} \tag{3.54}$$

$$\rho_i^* \approx w_i\, (1\,000\,\text{g/L}) \quad \text{(dilute aqueous solution)} \tag{3.55}$$

Mass fraction into molar concentration:

$$c_i = \frac{w_i \rho_{solution}}{M_i} \tag{3.56}$$

$$c_i \approx \frac{w_i \, (1\,000 \text{ g/L})}{M_i} \quad \text{(dilute aqueous solution)} \tag{3.57}$$

Mass fraction into molality:

$$b_i = \frac{w_i}{M_i \left(1 - \sum\limits_{j=1}^{N} w_j\right)} \tag{3.58}$$

$$b_i \approx \frac{w_i}{M_i} \quad \text{(dilute aqueous solution)} \tag{3.59}$$

Mass fraction into mole fraction:

$$x_i = \frac{w_i}{M_i \left(\dfrac{w_{solvent}}{M_{solvent}} + \sum\limits_{j=1}^{N} \dfrac{w_j}{M_j}\right)} \tag{3.60}$$

$$x_i \approx \frac{w_i \, (18 \text{ g/mol})}{M_i} \quad \text{(dilute aqueous solution)} \tag{3.61}$$

## 3.4 Element-Related Concentrations

For elements that occur in the form of different species, it is sometimes convenient (e.g., for material balances and stoichiometric calculations) to report the concentration of the element (e.g., carbon, nitrogen, sulfur, phosphorus, chlorine) instead of the species concentration. The conversion can be easily done by means of the molecular weights of the compound and the respective element as well as the stoichiometric species composition.

---

**Example 3.3**

The mass concentration of nitrate in surface water is found to be 40 mg/L. The molecular weights of $NO_3^-$ and N are 62 g/mol and 14 g/mol, respectively. What is the mass concentration of nitrogen and what are the molar concentrations of N and $NO_3^-$?

**Solution:**

Since the nitrate ion contains one atom of nitrogen, the mass fraction of nitrogen in nitrate is:

$$w(N) = \frac{M(N)}{M(NO_3^-)} = \frac{14 \text{ g/mol}}{62 \text{ g/mol}} = 0.226$$

To find the mass concentration of nitrogen for the given concentration of nitrate, we have to multiply the nitrate mass concentration with the mass fraction of nitrogen in nitrate:

$$\rho^*(N) = w(N)\,\rho^*(NO_3^-) = (0.226)\,(40\ mg/L) = 9.04\ mg/L$$

The molar concentration of nitrate is:

$$c(NO_3^-) = \frac{\rho^*(NO_3^-)}{M(NO_3^-)} = \frac{40\ mg/L}{62\ mg/mmol} = 0.645\ mmol/L$$

The molar concentration of nitrogen has the same value (0.645 mmol/L) as the molar concentration of nitrate, because 1 mol N is equivalent to 1 mol $NO_3^-$. The same result can be also found by dividing $\rho^*(N)$ by $M(N)$:

$$c(N) = \frac{\rho^*(N)}{M(N)} = \frac{9.04\ mg/L}{14\ mg/mmol} = 0.646\ mmol/L$$

The small deviation is due to round-off errors.

Frequently, the element concentrations are further specified. Thus, the nitrogen concentration can be reported as nitrate-N (nitrogen present as nitrate, $NO_3^-$), nitrite-N (nitrogen present as nitrite, $NO_2^-$), ammoniacal-N (nitrogen present as ammonium, $NH_4^+$, and ammonia, $NH_3$) or $N_{org}$ (organically bound N). In the case of carbon, the element content is frequently given as dissolved organic carbon (DOC), dissolved inorganic carbon (DIC), total (dissolved + particulate) organic carbon (TOC), or total (dissolved + particulate) inorganic carbon (TIC). In the case of sulfur, it is common to distinguish between sulfate-S, sulfide-S, and $S_{org}$.

**Example 3.4**
What is the DOC concentration (in mg/L) in a solution of 0.5 mmol/L acetic acid ($CH_3COOH$)? The molecular weights of C and $CH_3COOH$ are 12 g/mol and 60 g/mol, respectively.

**Solution:**
Acetic acid contains 2 atoms of carbon per molecule. The mass fraction of carbon is therefore:

$$w(C) = \frac{2\,M(C)}{M(CH_3COOH)} = \frac{2\,(12\ g/mol)}{60\ g/mol} = 0.4$$

The mass concentration of the acetic acid is:

$$\rho^*(CH_3COOH) = c(CH_3COOH)\,M(CH_3COOH) = (0.5\ mmol/L)(60\ mg/mmol)$$
$$\rho^*(CH_3COOH) = 30\ mg/L$$

To find the mass concentration of organic carbon, the mass concentration of the acetic acid has to be multiplied with the mass fraction of carbon:

$$\rho^*(C) = \rho^*(DOC) = \rho^*(CH_3COOH)\,w(C) = (30\ mg/L)\,(0.4) = 12\ mg/L$$

Alternatively, we can solve the problem in this way:

It follows from the stoichiometric composition of $CH_3COOH$ that 1 mol acetic acid contains 2 mol C. This means that in our example 0.5 mmol/L acetic acid contains 1 mmol/L C and with the molecular weight of C (12 mg/mmol) we get:

$$\rho^*(C) = \rho^*(DOC) = c(C)\,M(C) = (1\text{ mmol/L})\,(12\text{ mg/mmol}) = 12\text{ mg/L}$$

## 3.5 Gas-Phase Concentrations

Water is often in contact with a gas phase. Surface water, for instance, is in contact with the atmospheric air. To describe transfer processes between the gas and the aqueous phase (Chapter 6), we need adequate concentration measures for the gas phase. In principle, the same concentration measures as defined in Section 3.2 for solutions (e.g., mass concentration, molar concentration) can also be used to describe the composition of a gas phase. As an alternative, the partial pressure, $p_i$, of a gas component can be used as a concentration measure. The use of the partial pressure is often necessary for equilibrium calculations (see Chapters 5 and 6) due to the corresponding definition of the equilibrium constants.

The partial pressure (typically given in bar) is the pressure of a considered gas mixture component that this component would have if it would be alone in the given volume. As can be derived from the name, partial pressures are fractions of the total pressure and their sum equals the total pressure (Dalton's law of partial pressures):

$$p_{total} = p_1 + p_2 + p_3 + \cdots \tag{3.62}$$

The ratio of partial and total pressure equals the gas-phase mole fraction, $y_i$. Accordingly, the partial pressure can also be expressed by the product of the gas-phase mole fraction and the total pressure:

$$p_i = y_i\, p_{total} \tag{3.63}$$

It can be derived from equation (3.63) that the partial pressure is directly proportional to the mole fraction if the total pressure is constant. If we restrict our considerations to ideal gases, which is an acceptable approximation for the most gas phases relevant in water chemistry (e.g., atmospheric gases), the mole fraction can be set equal to the volume fraction. This equality results from the fact that the same substance amount of each ideal gas occupies the same volume (e.g., 1 mol of each ideal gas occupies 22.4 L under atmospheric pressure and at 0 °C).

**Example 3.5**
What is the partial pressure of oxygen in the atmosphere (20.9 vol% $O_2$) if the total pressure is
1 bar?

**Solution:**
The volume fraction can be found by dividing volume percent by 100%. Given that the mole fraction
can be set equal to the volume fraction, we receive:

$$y(O_2) = \frac{20.9\%}{100\%} = 0.209$$

With the mole fraction and the total pressure, we can calculate the partial pressure:

$$p(O_2) = y(O_2)\, p_{total} = 0.209\,(1\ bar) = 0.209\ bar$$

The partial pressure is directly related to the molar concentration of the considered
component in the gas mixture. The respective relationship can be derived from the
ideal gas law (state equation of an ideal gas). This ideal gas law reads:

$$p_i V = n_i R T \tag{3.64}$$

where $p_i$ is the partial pressure, $V$ is the total volume, $n_i$ is the substance amount (number of moles), $T$ is the absolute temperature (Kelvin temperature), and $R$ is the ideal (or
universal) gas constant ($R = 8.3145\ J/(mol \cdot K) = 8.3145 \times 10^{-2}\ L \cdot bar/(mol \cdot K)$). Given that
the molar concentration is the ratio of substance amount and volume, equation (3.64)
can also be written as:

$$p_i = \frac{n_i}{V} R T = c_i R T \tag{3.65}$$

This means that the partial pressure of a gas component at a given temperature is
directly proportional to its molar concentration in the gas phase.

## 3.6 Electroneutrality Condition and Ion Balance

Independent of its content of charged ions, water is always electrically neutral.
Therefore, the total number of positive charges must exactly counterbalance the
total number of negative charges. If the number of charges that occur in water is
expressed by the equivalent concentrations of the dissolved ions (Section 3.2), the
following electroneutrality condition can be established:

*Sum of the equivalent concentrations of the cations = Sum of the equivalent con-centrations of the anions*

$$\sum_{\text{cations}} c_i\, z_i = \sum_{\text{anions}} c_i\, z_i \tag{3.66}$$

Figure 3.2 shows schematically an ion balance that fulfills the electroneutrality condition. It has to be noted that the electroneutrality condition requires only that the sum of the equivalent concentrations of the cations must equal the sum of the equivalent concentrations of the anions, whereas the individual concentration ratios are not fixed.

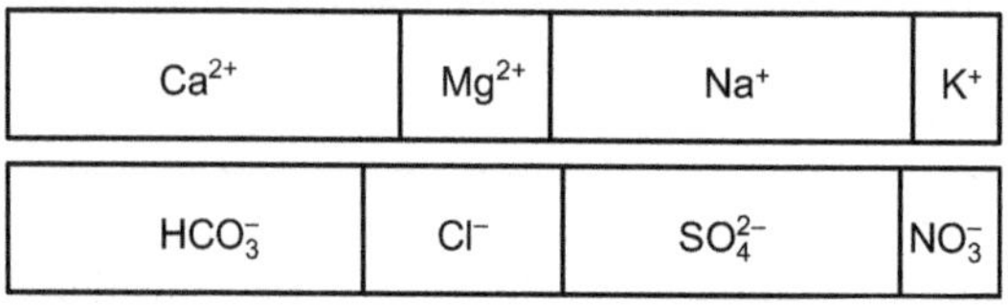

| Ca$^{2+}$ | Mg$^{2+}$ | Na$^+$ | K$^+$ |
|---|---|---|---|

| HCO$_3^-$ | Cl$^-$ | SO$_4^{2-}$ | NO$_3^-$ |
|---|---|---|---|

**Figure 3.2:** Graphical representation of the ion balance (here restricted to the major ions). The bars represent the equivalent concentrations. According to the electroneutrality condition, the sum of the cation equivalent concentrations must equal the sum of the anion equivalent concentrations, whereas the specific ratios of the single cations and anions are variable and different for different waters.

Equation (3.66) can be used to check the plausibility of water analyses that include all major ions ("complete" water analyses). For each complete water analysis, equation (3.66) must be fulfilled within the limits of experimental errors. To quantify the deviations from the electroneutrality condition, a balance error can be defined as follows:

$$\text{Balance error} = \frac{\left| \sum_{\text{cations}} c_i z_i - \sum_{\text{anions}} c_i z_i \right|}{0.5 \left( \sum_{\text{cations}} c_i z_i + \sum_{\text{anions}} c_i z_i \right)} \times 100\% \tag{3.67}$$

The error should not be higher than 5%. A larger deviation is an indicator of an erroneous or incomplete analysis.

## 3.7 Hardness as a Specific Concentration Measure

Hardness is a water quality parameter that comprises the concentrations of earth alkaline ions. Among the earth alkaline ions, calcium and magnesium ions are of particular practical relevance due to their relatively high concentration levels in natural waters. The other earth alkaline ions occur only in much lower concentrations and their contribution to the total earth alkaline ion concentration can be neglected. Therefore, in a good approximation, hardness can be set equal to the sum of the concentrations of calcium and magnesium ions, where the concentration of magnesium ions in natural waters is typically only a fraction of the concentration of calcium ions. Hardness is responsible for a number of problems that may occur during the use of drinking or process water, for instance, high consumption of detergents, precipitation of calcium carbonate in tubes and on heating installations (scaling), problems in cooking specific meals (e.g., beans and peas), and negative influence on the taste of tea and coffee. In particular scaling in tubes (clogging) and on heating elements (reduction of heat transfer, heat accumulation) can lead to technical problems and often to the necessity of expensive repairs.

Under practical aspects, it is reasonable to distinguish between two types of hardness, the temporary hardness (carbonate hardness) and the permanent hardness (non-carbonate hardness).

Carbonate hardness is defined as the concentration of calcium and magnesium ions that is equivalent to the concentrations of hydrogencarbonate (bicarbonate) and carbonate ions. Non-carbonate hardness is the calcium and magnesium concentration that is equivalent to anions other than hydrogencarbonate/carbonate (e.g., $Cl^-$, $SO_4^{2-}$, and $NO_3^-$).

The theoretical background of these definitions is the different stability during heating of the water. Carbonate hardness precipitates and forms solid carbonates (scale) during heating and is therefore referred to as temporary hardness, whereas non-carbonate hardness does not precipitate during moderate heating (due to the better solubility of the chlorides, sulfates, and nitrates of calcium and magnesium) and is therefore referred to as permanent hardness. This will be explained below in more detail by using calcium as the main constituent of hardness as an example. Moreover, we will restrict our further considerations to hydrogencarbonate, because the hydrogencarbonate concentration in natural and tap waters with pH values in the medium range is much higher than the carbonate concentration (see also Chapter 7).

It has to be noted that under certain conditions (high pH, low hydrogencarbonate/carbonate concentration) $Mg^{2+}$ can precipitate as hydroxide rather than as carbonate. An exact assessment which precipitate can be expected is possible by comparing the ion activity products (based on the actual activities/concentrations) with the solubility products (representing the equilibrium activities/concentrations). A detailed discussion of the assessment of the saturation state of solutions can be found in Chapter 8 (Section 8.4). The comparison of the different precipitation behavior of calcium and magnesium at high pH values is subject of Problem 8.4 in Chapter 8. This aspect is particularly relevant for water treatment processes, especially lime softening (Worch 2019), but it is not considered in the common definition of carbonate hardness.

Calcium and hydrogencarbonate are components of an equilibrium (calco-carbonic equilibrium, Chapter 9) which can be described by:

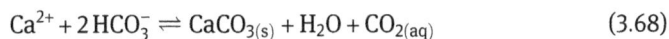

$$Ca^{2+} + 2\,HCO_3^- \rightleftharpoons CaCO_{3(s)} + H_2O + CO_{2(aq)} \tag{3.68}$$

If this equilibrium is disturbed by heating connected with partial removal of $CO_2$ and water, calcium carbonate (scale) precipitates. As can be seen from equation (3.68), the precipitation of $Ca^{2+}$ as $CaCO_3$ requires an equivalent concentration of hydrogencarbonate. The equivalence condition, derived from the reaction stoichiometry, can be expressed by molar concentrations or equivalent concentrations as:

$$2\,c(Ca^{2+}) = c(HCO_3^-) \tag{3.69}$$

$$c\left(\frac{1}{2}Ca^{2+}\right) = c\left(\frac{1}{1}HCO_3^-\right) = c(HCO_3^-) \tag{3.70}$$

In a more general form (including magnesium and carbonate), this equivalence condition is the theoretical basis of the definition of the carbonate hardness.

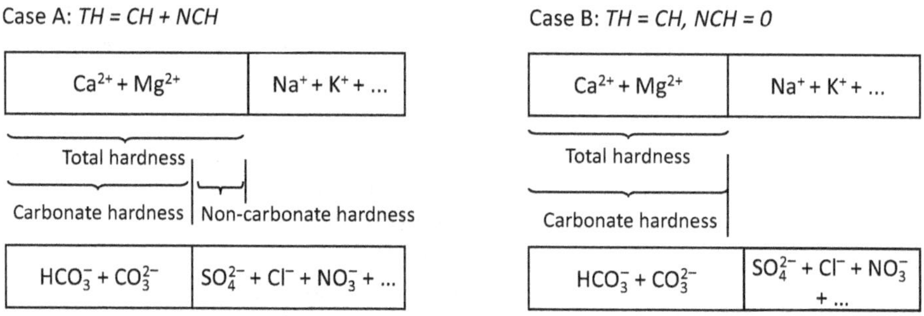

Case A: *TH = CH + NCH*                                          Case B: *TH = CH, NCH = 0*

**Figure 3.3:** Definition of carbonate hardness for different ratios of the equivalent concentrations $c(1/2\ Ca^{2+} + 1/2\ Mg^{2+})$ and $c(HCO_3^- + 1/2\ CO_3^{2-})$. Case A: $c(1/2\ Ca^{2+} + 1/2\ Mg^{2+}) > c(HCO_3^- + 1/2\ CO_3^{2-})$. Case B: $c(1/2\ Ca^{2+} + 1/2\ Mg^{2+}) < c(HCO_3^- + 1/2\ CO_3^{2-})$.

If there are more calcium and magnesium ions in the system than required by the equivalence condition, these excess ions cannot take part in the carbonate precipitation reaction and form the non-carbonate (permanent) hardness.

The definitions of carbonate and non-carbonate hardness are depicted in Figure 3.3. In principle, two different cases can be found with respect to the water composition. In the case A, the equivalent concentrations of calcium and magnesium are higher than those of hydrogencarbonate and carbonate. Therefore, a part of calcium and magnesium is equivalent to hydrogencarbonate/carbonate (carbonate hardness, *CH*) and another part is equivalent to other anions (non-carbonate hardness, *NCH*). Accordingly, the balance equation is:

$$TH = CH + NCH \tag{3.71}$$

where *TH* is the total hardness, the sum of the total concentrations of calcium and magnesium. *TH* can be calculated from the analytical data. *CH* has to be calculated on the basis of the definition. Since carbonate hardness is the concentration of calcium and magnesium ions that is equivalent to the concentrations of hydrogencarbonate and carbonate ions, the following material balance holds:

$$c\left(\frac{1}{2}Ca^{2+} + \frac{1}{2}Mg^{2+}\right)_{CH} = c\left(\frac{1}{1}HCO_3^-\right) + c\left(\frac{1}{2}CO_3^{2-}\right) \approx c\left(HCO_3^-\right) \qquad (3.72)$$

As already mentioned, the carbonate concentration in the medium pH range is very low in comparison to the hydrogencarbonate concentration and can therefore be neglected. If the carbonate hardness should be given in molar concentrations instead of equivalent concentrations, equation (3.72) has to be transformed under consideration of equation (3.6) to:

$$2c\left(Ca^{2+} + Mg^{2+}\right)_{CH} \approx c\left(HCO_3^-\right) \qquad (3.73)$$

$$c\left(Ca^{2+} + Mg^{2+}\right)_{CH} \approx \frac{1}{2}c\left(HCO_3^-\right) \qquad (3.74)$$

Accordingly, the carbonate hardness can be calculated from the hydrogencarbonate concentration. However, it has to be noted that equation (3.74) only gives feasible results if the equivalent concentration of hydrogencarbonate (and carbonate) is lower than or exactly equal to the sum of the equivalent concentrations of calcium and magnesium ions. In other words, the carbonate hardness can be at maximum as high as the total hardness but not higher.

Case B shows a situation where the equivalent concentration of hydrogencarbonate (and carbonate) is higher than the sum of the equivalent concentrations of calcium and magnesium ions. In this case, the total amount of calcium and magnesium is equivalent to only a part of the hydrogencarbonate/carbonate. The excess of hydrogencarbonate/carbonate ions is here equivalent to other cations. In this case, non-carbonate hardness does not exist and the total hardness equals the carbonate hardness ($TH = CH$). Here, a formal application of equation (3.74) would lead to a carbonate hardness that is higher than the total hardness, which is not possible. In this case, the calculation result has to be rejected and carbonate hardness has to be set equal to the total hardness. This specific aspect is often not considered.

---

**Example 3.6**
In water sample 1, the molar concentrations of $Ca^{2+}$ and $Mg^{2+}$ are 1.8 mmol/L and 0.4 mmol/L, respectively, and the hydrogencarbonate concentration was measured to be 2.6 mmol/L. In water sample 2, the molar concentrations of $Ca^{2+}$ and $Mg^{2+}$ are 1.0 mmol/L and 0.3 mmol/L, respectively, and the hydrogencarbonate concentration amounts to 3.0 mmol/L. Calculate the total hardness, the carbonate hardness, and the non-carbonate hardness (in mmol/L) for both water samples.

**Solution:**
Water sample 1:

$$TH = c\left(Ca^{2+} + Mg^{2+}\right) = 1.8 \text{ mmol/L} + 0.4 \text{ mmol/L} = 2.2 \text{ mmol/L}$$

$$CH = c\left(Ca^{2+} + Mg^{2+}\right)_{CH} = \frac{1}{2}c(HCO_3^-) = \frac{2.6 \text{ mmol/L}}{2} = 1.3 \text{ mmol/L}$$

Since $CH < TH$, the result for $CH$ is plausible and $NCH$ can be calculated as difference between $TH$ and $CH$:

$$NCH = c\left(Ca^{2+} + Mg^{2+}\right)_{NCH} = TH - CH = 2.2 \text{ mmol/L} - 1.3 \text{ mmol/L} = 0.9 \text{ mmol/L}$$

Water sample 2:

$$TH = c\left(Ca^{2+} + Mg^{2+}\right) = 1.0 \text{ mmol/L} + 0.3 \text{ mmol/L} = 1.3 \text{ mmol/L}$$

$$CH = c\left(Ca^{2+} + Mg^{2+}\right)_{CH} = \frac{1}{2}c(HCO_3^-) = \frac{3.0 \text{ mmol/L}}{2} = 1.5 \text{ mmol/L} (!)$$

The formally calculated $CH$ is higher than $TH$ and is therefore not plausible and the result has to be rejected. Instead of this, $CH$ and $NCH$ are given as:

$$CH = TH = 1.3 \text{ mmol/L}, \quad NCH = 0 \text{ mmol/L}$$

---

Instead of the unit mmol/L that conforms to the international system of units (SI), in practice traditional units are often in use. They are typically defined with $CaO$ or $CaCO_3$ as a reference substance. Conversions can be carried out by stoichiometric calculations (Table 3.1). The magnesium concentration has to be converted into units related to $CaO$ or $CaCO_3$ under consideration of the ratio of the molecular weights $M(Ca) : M(Mg) = 40.08 \text{ g/mol} : 24.305 \text{ g/mol} = 1.649$.

**Table 3.1:** Units of water hardness.

| Unit | Definition/reference | 1 mmol/L $Ca^{2+}$ equals | 1 mg/L $Ca^{2+}$ equals |
|---|---|---|---|
| mmol/L (conform to SI) | mmol/L $Ca^{2+}$ and $Mg^{2+}$ | 1 mmol/L $Ca^{2+}$ | 0.025 mmol/L $Ca^{2+}$ |
| mg/L (or ppm) | mg/L $CaCO_3$ | 100.1 mg/L $CaCO_3$ | 2.5 mg/L $CaCO_3$ |
| °dH (German degree) | 10 mg/L CaO | 5.608 °dH | 0.14 °dH |
| °Clark (Clark degree, English degree) | 1 grain/imperial gallon (= 14.254 mg/L) $CaCO_3$ | 7.023 °Clark | 0.175 °Clark |
| °f (French degree) | 10 mg/L $CaCO_3$ | 10.01 °f | 0.25 °f |

From the equivalence factor Ca/Mg and the conversion factor given in Table 3.1, the following equation for conversion of mass concentrations of $Ca^{2+}$ and $Mg^{2+}$ into hardness in mg/L $CaCO_3$ can be derived:

$$\text{Hardness } [mg/L\, CaCO_3] = 2.5\, \rho^*(Ca^{2+})[mg/L] + 4.12\, \rho^*(Mg^{2+})[mg/L] \qquad (3.75)$$

**Example 3.7**
Calculate the hardness caused by 10 mg/L $Mg^{2+}$ in mg/L $CaCO_3$ and in °dH.

**Solution:**
According to the ratio of the molecular weights, 10 mg/L $Mg^{2+}$ is equivalent to:

$$(1.649)\, (10\ mg/L\ Mg^{2+}) = 16.49\ mg/L\ Ca^{2+}$$

With the conversion factors given in Table 3.1, we find:

$$16.49\ mg/L\ Ca^{2+}\ \frac{2.5\ mg/L\ CaCO_3}{mg/L\ Ca^{2+}} = 41.23\ mg/L\ CaCO_3$$

and

$$16.49\ mg/L\ Ca^{2+}\ \frac{0.14\ °dH}{mg/L\ Ca^{2+}} = 2.31\ °dH$$

## 3.8 Activities and Activity Coefficients

In real aqueous solutions, the dissolved species are not isolated from each other but are subject to interactions. The strength of the interactions between the solute species depends on their chemical nature and their concentration. In particular, the relatively strong electrostatic interactions between ionic species are of practical relevance. According to Coulomb's law, the strength of electrostatic interactions increases with increasing charge and decreasing distance between the oppositely charged ions (equation (2.12) in Chapter 2). With increasing concentration, the distance between the species decreases in the statistical mean. Therefore, an increase of the strength of interactions with increasing ion concentration is to be expected. On the other hand, the lower the concentration is, the larger is the mean distance between the species and the weaker are the interactions. If the concentration is small enough, the interactions can be neglected. For this limiting case, the terms ideal solution (in contrast to real solution) or ideal dilute solution are applied.

The occurrence of interactions in real solutions has consequences for the behavior of the dissolved species in reactions. Due to the interactions, the species cannot act in reactions as strongly as can be expected from their concentration. Therefore, there is a need to distinguish between the measured concentration that is completely active in reactions only under ideal conditions and a (reduced) effective concentration that acts under real conditions. This effective concentration is referred to as activity. Consequently, exact equilibrium calculations have to be done on the basis of activities rather than concentrations.

The activity, $a$, can be found by multiplying the molar concentration with a correction factor that is referred to as the activity coefficient:

$$a = \gamma \frac{c}{c_{std}} \tag{3.76}$$

where $\gamma$ is the activity coefficient and $c_{std}$ is the concentration of the standard state.

It has to be noted that equation (3.76) is the rigorous thermodynamic definition from which it follows that the activity is dimensionless. The commonly used standard concentration for aqueous systems is $c_{std} = 1$ mol/L. In analogy to equation (3.76), dimensionless activities for other concentration measures can also be defined, for instance for the mole fraction or the molality. As a consequence of the fact that the activities are dimensionless, all thermodynamic equilibrium constants in laws of mass action must also be dimensionless (Chapter 5). However, dimensionless activities and equilibrium constants may cause confusion because the direct relation to the underlying concentration measures and the kind of formulation of the law of mass action gets lost. Therefore, a more convenient approach is frequently used in practice, which is based on a simplified definition of the activity:

$$a = \gamma c \tag{3.77}$$

Here, the activity formally has the same unit as the concentration. Consequently, the equilibrium constants also have units in this case (Chapter 5). In this book, we will follow this approach.

In real aqueous solutions, where the concentrations are not very high, the activity coefficient, $\gamma$, is lower than 1 and approaches 1 with decreasing concentration. In ideal dilute solutions without any interactions, the activity coefficient is 1 and the activity equals the concentration:

$$c \rightarrow 0, \quad \gamma \rightarrow 1, \quad a \rightarrow c \tag{3.78}$$

As mentioned previously, the interactions between ions (and the resulting deviations from ideal behavior) are determined by the charges and the concentrations of the ions. The quantity ionic strength allows quantifying both factors by means of only one parameter. The ionic strength is defined as:

$$I = 0.5 \sum_i c_i z_i^2 \tag{3.79}$$

where $z_i$ is the charge number of the ion $i$.

To determine the ionic strength exactly, the concentrations of all ionic water constituents (at least of the major ions) have to be known. As an alternative, an empirical correlation between the ionic strength and the electrical conductivity, which depends on the same factors (charge and concentration) as the ionic strength, can be used:

$$I \text{ (in mol/L)} = \frac{\kappa_{25} \text{ (in mS/m)}}{6\,200} \tag{3.80}$$

where $\kappa_{25}$ is the electrical conductivity at 25 °C.

The ionic strength can be used to estimate activity coefficients. The following equation for the activity coefficient of an ion with the charge $z$ results from the Debye–Hückel theory (see also Chapter 2, Section 2.2.6) and is valid for aqueous solutions and very low ionic strengths (<5 mmol/L). It is also known as Debye–Hückel limiting law:

$$\log \gamma_z = -A\, z^2 \sqrt{\frac{I}{\text{mol/L}}} \tag{3.81}$$

where $I$ is the ionic strength in mol/L. To make the expression under the root sign dimensionless, the ionic strength is divided by the unit mol/L. The parameter $A$ depends on the relative permittivity, $\varepsilon_r$, and the absolute temperature, $T$:

$$A = \frac{1.8246 \times 10^6}{\left(\varepsilon_r \dfrac{T}{K}\right)^{3/2}} \tag{3.82}$$

Note that the temperature $T$ is divided by the unit K to make the term dimensionless.

Values of $A$ for different temperatures calculated with the permittivity data given in Table 2.3 (Chapter 2, Section 2.2.6) are listed in Table 3.2. As can be seen, a value of 0.5 can be used as a good approximation in the practically relevant temperature range.

**Table 3.2:** Parameter $A$ of the Debye–Hückel equation at different temperatures.

| Temperature (°C) | A |
|---|---|
| 0 | 0.4896 |
| 5 | 0.4898 |
| 10 | 0.4964 |
| 15 | 0.5003 |
| 20 | 0.5046 |
| 25 | 0.5091 |
| 30 | 0.5141 |

To extend the application range to higher concentrations, several modifications of the Debye–Hückel equation have been proposed. The Güntelberg equation (for $I < 0.1$ mol/L) is frequently used for hydrochemical calculations:

$$\log \gamma_z = -A \, z^2 \frac{\sqrt{\dfrac{I}{mol/L}}}{1+1.4\sqrt{\dfrac{I}{mol/L}}} \tag{3.83}$$

The Davies equation (for $I < 0.5$ mol/L) is another modification of the Debye–Hückel equation:

$$\log \gamma_z = -A \, z^2 \left( \frac{\sqrt{\dfrac{I}{mol/L}}}{1+\sqrt{\dfrac{I}{mol/L}}} - 0.3 \frac{I}{mol/L} \right) \tag{3.84}$$

Note that the factor in the linear term was originally 0.2 but later changed to 0.3.

The parameter $A$ in equations (3.83) and (3.84) is the same as in equation (3.81) and can therefore also be calculated using equation (3.82) or can be set approximately to 0.5.

For a univalent ion, equations (3.81), (3.83), and (3.84) simplify to:

$$\log \gamma_1 = -A\sqrt{\frac{I}{mol/L}} \tag{3.85}$$

$$\log \gamma_1 = -A \frac{\sqrt{\dfrac{I}{mol/L}}}{1+1.4\sqrt{\dfrac{I}{mol/L}}} \tag{3.86}$$

and

$$\log \gamma_1 = -A \left( \frac{\sqrt{\dfrac{I}{mol/L}}}{1+\sqrt{\dfrac{I}{mol/L}}} - 0.3 \frac{I}{mol/L} \right) \tag{3.87}$$

Accordingly, the relationship between the activity coefficients of univalent and polyvalent ions in all cases is given by:

$$\log \gamma_z = z^2 \log \gamma_1 \qquad \gamma_z = \gamma_1^{z^2} \tag{3.88}$$

In the range of typical ionic strengths of freshwater, the Güntelberg equation and the Davies equation give comparable results (Example 3.8).

**Example 3.8**
Calculate the activity coefficient of a bivalent ion for an ionic strength of 50 mmol/L by means of the Güntelberg equation and the Davies equation ($A = 0.5$).

**Solution:**
To apply the equations for estimating the activity coefficients, we need the ionic strength in mol/L and the square root of the ionic strength:

$$I = 50 \, \text{mmol/L} = 0.05 \, \text{mol/L}$$

$$\sqrt{\frac{I}{\text{mol/L}}} = \sqrt{0.05} = 0.224$$

Güntelberg equation:

$$\log \gamma_2 = -A \, z^2 \, \frac{\sqrt{\dfrac{I}{\text{mol/L}}}}{1 + 1.4 \sqrt{\dfrac{I}{\text{mol/L}}}} = (-0.5)(4) \frac{0.224}{1 + (1.4)(0.224)} = -0.341$$

$$\gamma_2 = 10^{-0.341} = 0.456$$

Davies equation:

$$\log \gamma_2 = -A \, z^2 \left( \frac{\sqrt{\dfrac{I}{\text{mol/L}}}}{1 + \sqrt{\dfrac{I}{\text{mol/L}}}} - 0.3 \, \frac{I}{\text{mol/L}} \right) = (-0.5)(4) \left( \frac{0.224}{1 + 0.224} - (0.3)(0.05) \right) = -0.336$$

$$\gamma_2 = 10^{-0.336} = 0.461$$

Table 3.3 shows activity coefficients of univalent and polyvalent ions for different ionic strengths, calculated by means of the Güntelberg equation. The deviations from $y = 1$ (ideal solution) increase with increasing charge (at given ionic strength) and with increasing ionic strength (at given charge).

**Table 3.3:** Activity coefficients calculated by means of the Güntelberg equation.

| Ionic strength (mol/L) | Activity coefficient of univalent ions, $\gamma_1$ | Activity coefficient of bivalent ions, $\gamma_2$ | Activity coefficient of trivalent ions, $\gamma_3$ |
|---|---|---|---|
| 0.001 | 0.966 | 0.870 | 0.731 |
| 0.01 | 0.904 | 0.668 | 0.403 |
| 0.1 | 0.777 | 0.364 | 0.103 |

It has to be noted that the Debye–Hückel equation (equation (3.81)), the Güntelberg equation (equation (3.83)), and the Davies equation (equation (3.84)) consider only the charge of the ions but not their individual properties (in particular their sizes).

This means that for all ions with the same charge, the same activity coefficients are calculated. By contrast, the extended Debye–Hückel equation:

$$\log \gamma_z = -A\,z^2 \frac{\sqrt{\dfrac{I}{\mathrm{mol/L}}}}{1 + B\,a\,\sqrt{\dfrac{I}{\mathrm{mol/L}}}} \tag{3.89}$$

provides the opportunity to estimate individual activity coefficients for the different ions. Here, $a$ is an ion size parameter and $B$ is given by:

$$B = \frac{50.3}{\sqrt{\varepsilon_r\,\dfrac{T}{K}}} \tag{3.90}$$

where $\varepsilon_r$ is the relative permittivity ("dielectric constant") of water and $T$ is the absolute temperature. Since $\varepsilon_r$ decreases with increasing $T$, the value of $B$ shows only very weak temperature dependence ($B = 0.324$ at 0 °C, $B = 0.326$ at 10 °C, $B = 0.329$ at 25 °C). The ion size parameters for the major ions are given in Table 3.4. The extended Debye–Hückel equation is applicable for ionic strengths $I < 0.1$ mol/L.

**Table 3.4:** Ion size parameters to be used in the extended Debye–Hückel equation.

| Ion size parameter | Ions |
| --- | --- |
| 3 | $K^+$, $NH_4^+$, $Cl^-$, $NO_3^-$, $OH^-$ |
| 4 | $Na^+$, $HCO_3^-$, $H_2PO_4^-$, $HPO_4^{2-}$, $PO_4^{3-}$, $SO_4^{2-}$ |
| 5 | $CO_3^{2-}$ |
| 6 | $Ca^{2+}$, $Fe^{2+}$, $Mn^{2+}$ |
| 8 | $Mg^+$ |
| 9 | $H^+$ |

For the most practical cases, however, the simpler Güntelberg equation is exact enough. In the typical range of the ionic strengths found for natural freshwater, the results of the Güntelberg equation and the extended Debye–Hückel equation differ only slightly. For instance, the activity coefficient of $Ca^{2+}$ at $I = 0.02$ mol/L and 25 °C, calculated by the extended Debye–Hückel equation, is found to be 0.601, whereas the Güntelberg equation gives 0.581.

## 3.9 Problems

Molecular weights necessary for the solution of the problems:

$M(Na) = 23$ g/mol, $\quad M(K) = 39.1$ g/mol, $\quad M(Ca) = 40.1$ g/mol,

$M(Mg) = 24.3$ g/mol, $\quad M(Cl) = 35.5$ g/mol, $\quad M(H) = 1$ g/mol,

$M(C) = 12$ g/mol, $\quad M(S) = 32.1$ g/mol, $\quad M(O) = 16$ g/mol.

**3.1.** A water sample contains 50 mg/L $Ca^{2+}$ and 20 mg/L $SO_4^{2-}$. Calculate the molar concentrations and the equivalent concentrations of both ions.

**3.2.** 1 kg of an aqueous solution contains 60 mg $Na^+$. What is the mass fraction of $Na^+$ in g/g, mass percent, and ppm?

**3.3.** The molar concentration of $Ca^{2+}$ in a water sample was found to be 2 mmol/L. What is the mole fraction? You can use the approximate conversion equation for dilute aqueous solutions.

**3.4.** Atmospheric air contains 78 vol% nitrogen. What is the molar concentration of nitrogen in the air? The temperature is assumed to be 25 °C and the gas constant is $R = 0.083145$ L·bar/(mol·K). The total pressure is 1 bar.

**3.5.** A mineral water contains the following concentrations of major ions: $\rho^*(Na^+) =$ 118 mg/L, $\rho^*(K^+) = 11$ mg/L, $\rho^*(Ca^{2+}) = 348$ mg/L, $\rho^*(Mg^{2+}) = 108$ mg/L, $\rho^*(Cl^-) =$ 40 mg/L, $\rho^*(HCO_3^-) = 1\,816$ mg/L, $\rho^*(SO_4^{2-}) = 38$ mg/L. Establish the ion balance and calculate the balance error. Is this analysis plausible?

**3.6.** For a water, the following concentrations were determined: $c(Ca^{2+}) = 1.25$ mmol/L, $c(Mg^{2+}) = 0.35$ mmol/L, $c(HCO_3^-) = 2.46$ mmol/L. Calculate the total, the carbonate and the non-carbonate hardness in mmol/L as well as the total hardness in mg/L $CaCO_3$.

**3.7.** What is the ionic strength of a water that contains $c(Na^+) = 0.58$ mmol/L, $c(K^+) =$ 0.05 mmol/L, $c(Ca^{2+}) = 1.71$ mmol/L, $c(Mg^{2+}) = 0.51$ mmol/L, $c(Cl^-) = 0.65$ mmol/L, $c(HCO_3^-) = 2.51$ mmol/L, and $c(SO_4^{2-}) = 0.94$ mmol/L? Calculate the activity coefficients for univalent and bivalent ions by means of the Güntelberg equation.

# 4 Colligative Properties

## 4.1 Introduction

Colligative properties are solution properties that depend on the amount of dissolved species in the solution but not on their chemical identity. This means that equal amounts of dissolved species always cause the same effects independent of their chemical nature. To express the amount of species present in the solution, the mole fraction, the molality, or the molarity have to be used depending on the type of the colligative property.

There are four colligative properties: vapor pressure lowering, boiling point elevation, freezing point depression, and osmotic pressure. This means that a solution shows a decreased vapor pressure, an increased boiling point, and a decreased freezing point in comparison to the pure solvent (water in our case). Furthermore, each solution causes a concentration-dependent osmotic pressure. In the following sections, at first, the calculation of colligative properties in ideal solutions is considered. After that, the deviations in real systems are discussed separately in Section 4.5.

## 4.2 Vapor Pressure Lowering

The vapor pressure of a liquid is the result of the transfer of molecules from the liquid phase to the gas phase. The transferred molecules cause a certain pressure in the gas phase, the vapor pressure. Since the likelihood of such a transfer increases with increasing temperature, the vapor pressure increases with increasing temperature. If a substance is dissolved in the liquid, a certain number of solvent molecules are replaced from the surface by the solute species as schematically shown in Figure 4.1. Consequently, at the same temperature, less solvent molecules can be transferred to the gas phase and the vapor pressure is reduced in comparison to the pure solvent. Furthermore, additional interactions between the solvent and the solute (solvation) reduce the number of solvent molecules that are able to leave the liquid phase.

The vapor pressure of the solvent in the solution, $p_1$, is given by:

$$p_1 = x_1 p_{01} \tag{4.1}$$

where $p_{01}$ is the vapor pressure of the pure solvent (water in our case) and $x_1$ is the mole fraction of the solvent in the solution. Given that the sum of mole fractions equals 1 and assuming that there is only a single nondissociating solute in the solution, we can write:

$$p_1 = x_1 p_{01} = (1 - x_2) p_{01} = p_{01} - x_2 p_{01} \tag{4.2}$$

where $x_2$ is the mole fraction of the solute. Accordingly, the relative lowering of the vapor pressure is given by the mole fraction of the solute:

https://doi.org/10.1515/9783110758788-004

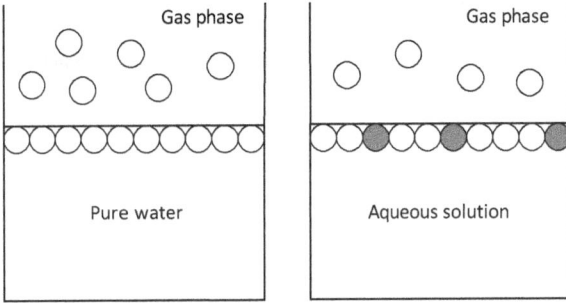

**Figure 4.1:** Vapor pressure lowering as a consequence of reduced transfer of solvent molecules from a solution to the gas phase. Gray circles: solute molecules or ions.

$$\frac{p_{01} - p_1}{p_{01}} = x_2 = \frac{n_{\text{solute}}}{n_{\text{solvent}} + n_{\text{solute}}} \tag{4.3}$$

If a dissociating compound is added to water, the number of species increases due to the dissociation. This can be considered by introducing the van't Hoff factor, $i$:

$$i = 1 + (v - 1)\alpha \tag{4.4}$$

where $v$ is the number of the potentially releasable species in a formula unit (e.g., number of ions in a formula unit of a salt) and $\alpha$ is the degree of dissociation. For NaCl, for instance, we find $v = 2$ (NaCl dissociates into the ions $Na^+$ and $Cl^-$) and $\alpha = 1$ (complete dissociation). Accordingly, the van't Hoff factor for NaCl is $i = 2$. The respective equation for the relative vapor pressure lowering is:

$$\frac{p_{01} - p_1}{p_{01}} = x_2 = \frac{i\,n_{\text{solute}}}{n_{\text{solvent}} + i\,n_{\text{solute}}} \tag{4.5}$$

In a multisolute solution with $N$ components (solvent + $(N - 1)$ solutes), where we know the mole fractions of all dissolved species $(x_2 \ldots x_N)$, we can replace the mole fraction of the solvent, $x_1$, in equation (4.1) by:

$$x_1 = 1 - \sum_{i=2}^{N} x_i \tag{4.6}$$

and the relative lowering of the vapor pressure is then given by:

$$\frac{p_{01} - p_1}{p_{01}} = \sum_{i=2}^{N} x_i \tag{4.7}$$

**Example 4.1**

What is the relative lowering of the water vapor pressure if 1 mol $CaCl_2$ is added to 1 L water? The molecular weight of water is 18 g/mol and the density is approximately 1 g/cm$^3$.

**Solution:**

With the density of water ($\rho = 1$ g/cm$^3 = 1$ kg/L) and the molecular weight ($M = 18$ g/mol), we can compute the moles of water in 1 L:

$$n(H_2O) = \frac{m(H_2O)}{M(H_2O)} = \frac{\rho(H_2O)\,V(H_2O)}{M(H_2O)} = \frac{(1\text{ kg/L})\,(1\text{ L})}{18\text{ g/mol}} = \frac{1000\text{ g}}{18\text{ g/mol}} = 55.56\text{ mol}$$

The amount of $CaCl_2$ in 1 L water is $n(CaCl_2) = 1$ mol and the van't Hoff factor is $i = 3$ (due to the dissociation $CaCl_2 \rightarrow Ca^{2+} + 2\ Cl^-$). Accordingly, we obtain:

$$x_2 = \frac{i\,n(CaCl_2)}{i\,n(CaCl_2) + n(H_2O)} = \frac{3\text{ mol}}{3\text{ mol} + 55.56\text{ mol}} = 0.051$$

which equals the relative lowering of the vapor pressure:

$$\frac{p_{01} - p_1}{p_{01}} = x_2 = 0.051$$

If 1 mol calcium chloride is added to 1 L water, the vapor pressure is decreased by 5.1% in comparison to the vapor pressure of pure water.

Alternatively, we can also apply equation (4.7) to solve the problem. As the result of the dissociation, there are 1 mol $Ca^{2+}$ and 2 mol $Cl^-$ in the solution. The total number of moles is:

$$n_{total} = n(Ca^{2+}) + n(Cl^-) + n(H_2O) = 1\text{ mol} + 2\text{ mol} + 55.56\text{ mol} = 58.56\text{ mol}$$

and the respective mole fractions are:

$$x(Ca^{2+}) = \frac{n(Ca^{2+})}{n_{total}} = \frac{1\text{ mol}}{58.56\text{ mol}} = 0.017$$

$$x(Cl^-) = \frac{n(Cl^-)}{n_{total}} = \frac{2\text{ mol}}{58.56\text{ mol}} = 0.034$$

Finally, we can compute the relative vapor pressure lowering by:

$$\frac{p_{01} - p_1}{p_{01}} = \sum_{i=2}^{N} x_i = x(Ca^{2+}) + x(Cl^-) = 0.017 + 0.034 = 0.051$$

## 4.3 Boiling Point Elevation and Freezing Point Depression

Boiling point elevation and freezing point depression can be considered secondary effects of the vapor pressure lowering. This can be derived from Figure 4.2 which shows vapor pressure curves for pure water, an aqueous solution, and ice. Let us first look at the vapor pressure curve of the pure solvent water. As already discussed, the vapor pressure increases with increasing temperature. The curve ends at the boiling point which is the point where the vapor pressure equals the external pressure (here assumed to be the atmospheric pressure of about 1 bar, exactly 1.013 bar). At the lower end, the vapor pressure curve of the liquid water meets the vapor pressure curve of ice. That point, where liquid and solid have the same vapor pressure, is the freezing point. As we have seen in the previous section, the vapor pressure of a solution is lower than that of the pure solvent. Accordingly, the vapor pressure curve of an aqueous solution is located below the curve of pure water. Therefore, this curve meets the line of the external pressure at a higher temperature. The difference to the boiling point of the solvent is referred to as boiling point elevation. Since during freezing of a solution a separation of solute and solvent occurs, the vapor pressure curve of ice remains unchanged and the freezing point of the solution equals the intersection point of the vapor pressure curve of the solution and the vapor pressure curve of ice. This point is shifted to lower temperatures in comparison to the intersection point of the vapor pressure curves of pure water and ice. This difference is referred to as freezing point depression. Both effects are directly proportional to the molality of the solute and the respective equations have an equal mathematical form but with different proportionality factors.

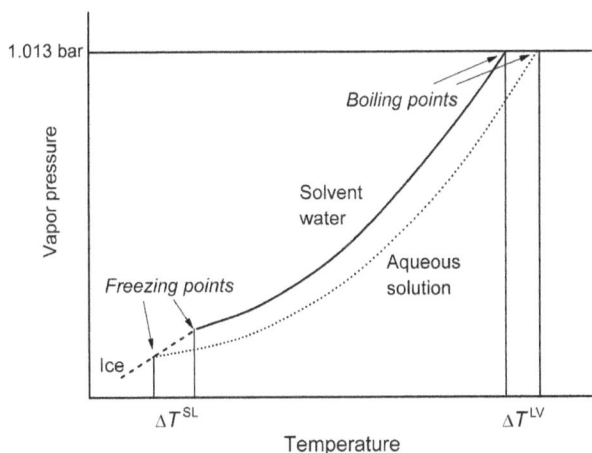

**Figure 4.2:** Boiling point elevation and freezing point depression as secondary effects of vapor pressure lowering.

The boiling point elevation, $\Delta T^{LV}$, is given as:

$$\Delta T^{LV} = b\,K_B \tag{4.8}$$

where $b$ is the molality of the solute (mol solute per kg solvent) and $K_B$ is the ebullioscopic constant (the molal boiling point elevation). The value of the ebullioscopic constant depends on the type of the solvent. For water, the ebullioscopic constant is 0.513 K·kg/mol.

For dissociating compounds, we can extend the equation with the van't Hoff factor, $i$:

$$\Delta T^{LV} = i\, b\, K_B \tag{4.9}$$

and for multisolute systems of known composition, such as natural waters, we can also write:

$$\Delta T^{LV} = \sum_i b_i\, K_B \tag{4.10}$$

The respective equations for the freezing point depression, $\Delta T^{SL}$, are:

$$\Delta T^{SL} = b\, K_F \tag{4.11}$$

$$\Delta T^{SL} = i\, b\, K_F \tag{4.12}$$

and

$$\Delta T^{SL} = \sum_i b_i\, K_F \tag{4.13}$$

for single solutes, for dissociating solutes, and multicomponent systems, respectively. The proportionality factor is referred to as the cryoscopic constant and amounts to $K_F = -1.86$ K·kg/mol for water.

---

**Example 4.2**

What are the boiling point elevation and the freezing point depression of an aqueous NaCl solution with the molality of 2 mol/kg? $K_B = 0.513$ K·kg/mol, $K_F = -1.86$ K·kg/mol.

**Solution:**

With $i = 2$ (NaCl $\rightarrow$ Na$^+$ + Cl$^-$), we find for the boiling point elevation and the freezing point depression:

$$\Delta T^{LV} = i\, b\, K_B = 2\,(2\text{ mol/kg})\,(0.513\text{ K} \cdot \text{kg/mol}) = 2.05\text{ K}$$

$$\Delta T^{SL} = i\, b\, K_F = 2\,(2\text{ mol/kg})\,(-1.86\text{ K} \cdot \text{kg/mol}) = -7.44\text{ K}$$

Accordingly, the aqueous NaCl solution boils at 102.05 °C and freezes at −7.44 °C.

---

## 4.4 Osmotic Pressure

Osmosis refers to the flow of solvent molecules to a solution through a semipermeable membrane that stops the flow of the solutes only (Figure 4.3). As a result, the pressure on the solution side increases, which is illustrated as hydrostatic pressure in Figure 4.3.

The flow can be stopped by applying a counter pressure. The pressure that is needed to stop the flow is referred to as the osmotic pressure, $\pi$. If an external pressure higher than the osmotic pressure acts on the solution side, a solvent flow out of the solution is initiated. This reversed process is referred to as reverse osmosis. Reverse osmosis can be used in water treatment, for instance for sea water desalination or for production of ultrapure water.

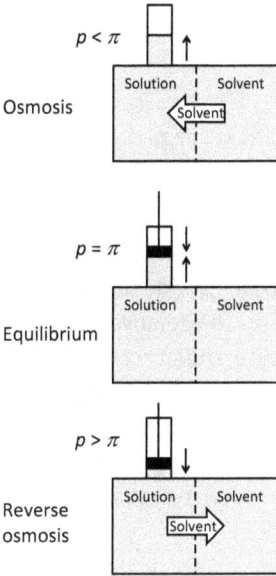

**Figure 4.3:** Osmosis and osmotic pressure.

The osmotic pressure depends on the molar concentration of the solutes. For single solutes, the osmotic pressure is given by the van't Hoff equation:

$$\pi = cRT \tag{4.14}$$

where $R$ is the gas constant ($R = 8.3145$ J/(mol · K) = 0.083145 bar · L/(mol · K)) and $T$ is the absolute temperature. As in the cases of vapor pressure lowering, boiling point elevation and freezing point depression, equations for dissociating solutes and multicomponent systems can be formulated as follows:

$$\pi = icRT \tag{4.15}$$

$$\pi = \sum_i c_i RT \tag{4.16}$$

---

**Example 4.3**

What is the osmotic pressure of seawater at $\vartheta = 25\ °C$? For simplification, only the four major ions, $Na^+$, $Mg^{2+}$, $Cl^-$, and $SO_4^{2-}$, should be considered in the calculation. The concentrations of the major ions in seawater are $c\,(Na^+) = 479.7$ mmol/L, $c\,(Mg^{2+}) = 54.4$ mmol/L, $c\,(Cl^-) = 559.4$ mmol/L, and $c\,(SO_4^{2-}) = 28.9$ mmol/L (see also Table 1.2 in Chapter 1). The universal gas constant is $R = 0.083145$ bar·L/(mol·K).

**Solution:**

The sum of the molar concentrations is:

$$\sum_i c_i = (479.7 + 54.4 + 559.4 + 28.9)\ \text{mmol/L} = 1\,122.4\ \text{mmol/L} = 1.12\ \text{mol/L}$$

For this total concentration, we find the osmotic pressure to be:

$$\pi = \sum_i c_i\,R\,T = (1.12\ \text{mol/L})\,(0.083145\ \text{bar·L/(mol·K)})\,(298.15\ \text{K}) = 27.76\ \text{bar}$$

---

The osmotic pressure of seawater calculated in Example 4.3 is the pressure that has to be overcome for desalination of seawater by reverse osmosis. For the practical process of desalination, it has to be considered that due to the removal of water from the solution, the concentrations in the solution increase and accordingly also the osmotic pressure increases with increasing removal of water. Since the process stops when the osmotic pressure reaches the external pressure, in practice, much higher pressures than the calculated initial osmotic pressure are applied (typically 60–80 bar) to ensure a long-lasting driving force for the water removal from the concentrated solution.

## 4.5 Colligative Properties of Real Solutions

The equations for calculating the colligative properties given in the previous sections are strictly valid only for ideal solutions. The deviations found for real solutions are typically expressed by the osmotic coefficient, $\varphi$, that relates the (measured) osmotic pressure of a real solution to the osmotic pressure of an ideal solution which can be calculated by one of equations (4.14)–(4.16):

$$\varphi = \frac{\pi_{\text{real}}}{\pi_{\text{ideal}}} \tag{4.17}$$

The osmotic coefficient can also be used as a correction factor for the other colligative properties.

The osmotic coefficient depends on the concentrations of the solutes and on the temperature of the solution. The osmotic coefficients show larger deviations from 1 (ideal solution) only at higher concentrations. For seawater (salinity 35 g/kg), for instance, the osmotic coefficient within the relevant temperature range is between 0.90 (0 °C) and 0.91 (30 °C). For dilute solutions, such as freshwater, $\varphi$ approaches 1 and the deviations from the state of an ideal solution can be neglected.

## 4.6 Problems

**4.1.** The vapor pressure of water at 20 °C is 23.4 mbar. What is the vapor pressure of a NaCl solution with the molality of $b = 1$ mol/kg? The molecular weight of water is 18 g/mol.

**4.2.** An aqueous salt solution freezes at −4 °C. What is the osmotic pressure of this solution at 20 °C? Note that for dilute aqueous solutions the numerical value of the molarity is approximately the same as the value of the molality. The cryoscopic constant is $K_F = -1.86$ K·kg/mol and the gas constant is $R = 0.083145$ bar·L/(mol·K).

**4.3.** What is the freezing point of seawater? For simplification, only the major ions Cl⁻ (19 350 ppm), Na⁺ (10 760 ppm), $SO_4^{2-}$ (2 710 ppm), and $Mg^{2+}$ (1 290 ppm) should be considered in the calculation and the solution is assumed to be ideal. The total salinity of seawater amounts to 3.5% and the cryoscopic constant of water is $K_F = -1.86$ K·kg/mol. The molecular weights of the dissolved ions are $M(Cl^-) = 35.5$ g/mol, $M(Na^+) = 23$ g/mol, $M(SO_4^{2-}) = 96.1$ g/mol, and $M(Mg^{2+}) = 24.3$ g/mol.

**4.4.** To avoid freezing of water in cooling systems, antifreezing agents, such as ethylene glycol, are added to the water. What volume of ethylene glycol (density: 1.11 g/cm³, molecular weight: 62 g/mol) has to be added to 1 L water (density: 1 g/cm³) to prevent it from freezing down to −20 °C? Note that ethylene glycol does not dissociate in water. $K_F = -1.86$ K·kg/mol.

**4.5.** What is the osmotic pressure of a freshwater ($\vartheta = 10$ °C) that contains the following concentrations of major ions: $c(Na^+) = 0.50$ mmol/L, $c(K^+) = 0.03$ mmol/L, $c(Ca^{2+})$ $= 1.25$ mmol/L, $c(Mg^{2+}) = 0.40$ mmol/L, $c(HCO_3^-) = 3.00$ mmol/L, $c(Cl^-) = 0.23$ mmol/L, $c(SO_4^{2-}) = 0.30$ mmol/L? The gas constant is $R = 0.083145$ bar·L/(mol·K).

**4.6.** The osmotic pressure of seawater at 25 °C is found to be 27.8 bar under the assumption that the total molar concentration of the major ions is 1.12 mol/L and seawater behaves as an ideal solution (see Example 4.3). What is the theoretical volume of pure water that can be produced from 1 m³ seawater by reverse osmosis at 25 °C if an external pressure of 80 bar is applied in the process? The gas constant is $R = 0.083145$ bar·L/(mol·K).

# 5 The Chemical Equilibrium: Some General Aspects

## 5.1 Introduction

The fate of water constituents in aqueous systems is determined by various phase transfer processes and chemical reactions, such as partitioning between aqueous and gas phase, acid/base reactions, precipitation/dissolution, redox reactions, complex formation, and sorption. These processes will be discussed in the following chapters with focus on the equilibrium state, which is the endpoint of the processes where the involved compounds no longer change their concentrations.

Before starting the discussion on the most important types of equilibria in aqueous systems, it is necessary to present some important basics of the chemical equilibrium and its mathematical description. This will be done in this chapter in a compacted form without going into excessive detail.

## 5.2 Law of Mass Action and Equilibrium Constants

The law of mass action is the fundamental law of the chemical equilibrium. Its relevance arises from the fact that most chemical reactions are not completed, which means that the reactants are not totally converted to products. Instead, the reactions end in an equilibrium state characterized by a specific ratio of the concentrations (or more exactly activities) of the chemicals involved in the reaction. That is also true for the reactions that occur in aqueous systems. By means of the law of mass action, this concentration (or activity) ratio can be expressed by a characteristic parameter, the equilibrium constant. The respective laws of mass action together with material balance equations provide the basis for hydrochemical calculations as will be shown in the following chapters.

For a general reaction equation with the reactants A and B and the products C and D:

$$\nu_A A + \nu_B B \rightleftharpoons \nu_C C + \nu_D D \tag{5.1}$$

the law of mass action reads:

$$K^* = \frac{(a_C)_{eq}^{\nu_C} (a_D)_{eq}^{\nu_D}}{(a_A)_{eq}^{\nu_A} (a_B)_{eq}^{\nu_B}} \tag{5.2}$$

In the law of mass action, the equilibrium activities of the products are written in the numerator whereas the equilibrium activities of the reactants are written in the denominator. The stoichiometric factors, $\nu$, occur as exponents in the law of mass action. The equilibrium constant $K^*$, defined according to equation (5.2), is referred to as the thermodynamic equilibrium constant. In general, the thermodynamic

https://doi.org/10.1515/9783110758788-005

equilibrium constant depends on pressure and temperature. For aqueous systems, however, the pressure dependence can be neglected and only the temperature dependence has to be considered.

Given that the activities are related to the molar concentrations by $a = \gamma c$ (Chapter 3) with $\gamma$ as activity coefficient, equation (5.2) can be rewritten in the form:

$$K* \frac{(\gamma_A)^{\nu_A} (\gamma_B)^{\nu_B}}{(\gamma_C)^{\nu_C} (\gamma_D)^{\nu_D}} = K = \frac{(c_C)_{eq}^{\nu_C} (c_D)_{eq}^{\nu_D}}{(c_A)_{eq}^{\nu_A} (c_B)_{eq}^{\nu_B}} \tag{5.3}$$

We can derive from equation (5.3) that the thermodynamic constant has to be corrected by a term that includes the activity coefficients if we want to use measurable concentrations in the law of mass action. The resulting constant $K$ is referred to as the conditional constant, because it is valid only for a specific condition characterized by specific values of the activity coefficients or, due to the relationship between ionic strength and activity coefficients, by a specific ionic strength (see Section 3.8 in Chapter 3). Consequently, the conditional equilibrium constant $K$ depends on the temperature and the ionic strength of the aqueous solution.

For dilute solutions, the difference between $K*$ and $K$ vanishes, because the strength of interactions between the species decreases and the real state approaches the ideal state ($\gamma = 1$ for all species):

$$c \to 0, \quad \gamma \to 1, \quad a \to c, \quad K \to K* \tag{5.4}$$

If $K$ is assumed to be equal to $K*$ (ideal dilute solution), concentrations can be used in the law of mass action together with the thermodynamic constant which considerably simplifies equilibrium calculations. As long as only an approximate solution is desired, this simplification can be used for most practical cases, in particular if the concentrations of the water constituents are not very high (e.g., in surface water, groundwater, or drinking water). In this book, we will make use of this simplification for most theoretical considerations and examples. However, for exact calculations, activities have to be used, even for waters with relatively low concentrations. For typical values of activity coefficients with dependence on ionic strength, see Table 3.3 in Chapter 3.

At this point, we should recall the fact that there is a difference between the definition of the activity used here (with the same unit as the concentration) and the strict thermodynamic definition as a dimensionless quantity (Section 3.8 in Chapter 3). The second approach would lead to dimensionless constants. In this case, each concentration on the right-hand side of equation (5.3) must be normalized with the standard concentration 1 mol/L to obtain a dimensionless constant. However, independent of the applied definition, the numerical values for the activities and constants remain the same. The disadvantage of dimensionless constants consists in a loss of information regarding the concentration measures used for the formulation of the law of mass action.

## 5.3 Conventions on the Use of Concentration Measures in the Law of Mass Action

In Section 5.2, the molar concentration or the related activity was used to formulate the law of mass action. In principle, the laws of mass action can also be formulated with other types of concentration measures, such as molalities, mol fractions, or partial pressures (for gas components). It is also possible to use different measures for different species in the same law of mass action. This underlines the statement regarding the disadvantage of dimensionless activities and equilibrium constants given in the previous section.

Although there is no strict rule as to which concentration or activity measures have to be used, the following conventions are typically made to express the laws of mass action for hydrochemical reactions and to define the related equilibrium constants:

- The concentrations of dissolved species are given as molar concentration in mol/L or as respective activities.
- For pure solid phases in contact with water, the mole fraction, $x$, or the related activity, $a_x$, is used and assumed to be 1. As a consequence, solids should not be considered in the formulation of the law of the mass action.
- For the solvent $H_2O$, which is the main component in aqueous systems and exists in excess in comparison to the dissolved components, there are two possibilities that are equivalent: to use the mol fraction, $x$, or the related activity, $a_x$, and to assume for both a value of 1 (pure component) or to use the molar concentration and consider this concentration as constant, because it will not change considerably during the reaction with water constituents of much lower concentrations. In the latter case, this constant $H_2O$ concentration is included in the equilibrium constant as a constant factor. In both cases, water concentration or activity does not occur on the right-hand side of the law of mass action.
- For gas components, the partial pressure over the liquid phase (in bar) is used unless otherwise stated.

To avoid confusion about the aggregate state of the species occurring in a given reaction equation and the related convention to be used, the species formulae can be extended by the indices "(s)," "(g)," and "(aq)" to indicate the state of the species (solid state, gas state, and dissolved state, respectively). In clear cases (e.g., ions), "(aq)" can also be omitted. Example 5.1 demonstrates the application of the conventions. Note that in the example the conditional constant is used.

---

**Example 5.1**
What is the law of mass action (expressed by the conditional constant) for the following reaction equation that includes a gas, a solid, and water? What is the unit of the constant?

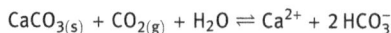

$$CaCO_{3(s)} + CO_{2(g)} + H_2O \rightleftharpoons Ca^{2+} + 2\,HCO_3^-$$

**Solution:**
According to the abovementioned conventions, the law of mass action reads:

$$K = \frac{c(Ca^{2+})\, c^2(HCO_3^-)}{p(CO_2)}$$

The $CO_2$ concentration in the gas phase is expressed by the partial pressure, water is not considered, the solid $CaCO_3$ is also not considered, and the concentrations of the hydrogencarbonate and calcium ions are expressed as molar concentrations. Accordingly, the unit of the constant is $mol^3/(L^3 \cdot bar)$.

## 5.4 Gibbs Energy of Reaction, Equilibrium Constants, and Reaction Quotients

The Gibbs energy, $\Delta G$, is a thermodynamic function that gives the information in which direction a process proceeds spontaneously under constant temperature and pressure. The change of the Gibbs energy comprises two quantities, the enthalpy change and the entropy change. For a chemical reaction, the Gibbs energy can be expressed as:

$$\Delta_R G = \Delta_R H - T\,\Delta_R S \tag{5.5}$$

where $\Delta_R G$ is the change of the Gibbs energy of the reaction, $\Delta_R H$ is the enthalpy change of the reaction, $\Delta_R S$ is the entropy change of the reaction, and $T$ is the absolute temperature.

The change of the reaction enthalpy indicates whether energy is released (exothermic process, $\Delta_R H$ negative) or consumed (endothermic process, $\Delta_R H$ positive) during the reaction. The change of the entropy describes the change in the state of order of the system during the reaction. A positive value of $\Delta_R S$ indicates an increase in disorder in the considered system. Conversely, a negative value is related to an increase in the degree of order.

Depending on the values of the enthalpy and the entropy term, $\Delta_R G$ can be positive or negative. The sign indicates if a reaction proceeds spontaneously in the direction as written in the reaction equation or not. A reaction proceeds spontaneously in the considered direction if $\Delta_R G$ is negative (exergonic process). If $\Delta_R G$ is positive (endergonic process), the process does not proceed spontaneously in the considered direction but in the reverse direction. If $\Delta_R G = 0$, no reaction occurs; the system is in the state of equilibrium.

The change of the Gibbs energy of reaction is related to the ratio of the activities (or concentrations) of the components involved in the reaction by:

$$\Delta_R G = \Delta_R G^0 + R\,T \ln Q \tag{5.6}$$

where $\Delta_R G^0$ is the Gibbs energy for a defined standard state, $R$ is the gas constant, and $Q$ is the reaction quotient. The reaction quotient is defined formally in the same way as the equilibrium constant, but in contrast to the equilibrium constant, it is more general and not restricted to equilibrium activities (or concentrations). Instead, it includes the actual activities or concentrations. For the general reaction given in equation (5.1), the reaction quotient reads:

$$Q = \frac{(a_C)^{\nu_C} (a_D)^{\nu_D}}{(a_A)^{\nu_A} (a_B)^{\nu_B}} \tag{5.7}$$

In nonequilibrium states, $Q$ is higher or lower than $K^*$. In the state of equilibrium, the activities equal the equilibrium activities and $Q$ equals $K^*$ (equation (5.2)). Furthermore, $\Delta_R G$ becomes zero in the state of equilibrium as mentioned above and equation (5.6) therefore simplifies to:

$$\Delta_R G^0 = -R\,T \ln K^* \tag{5.8}$$

Equation (5.8) can be introduced into equation (5.6) to receive the general relationship:

$$\Delta_R G = -R\,T \ln K^* + R\,T \ln Q = R\,T \ln \frac{Q}{K^*} \tag{5.9}$$

Since the quotient $Q/K^*$ is directly related to $\Delta_R G$, it is a useful tool to predict the direction in which a reaction will proceed. The different cases are listed in Table 5.1.

**Table 5.1:** Assessment of aqueous systems with respect to the establishment of equilibrium by comparison of $Q$ and $K^*$.

| Condition | Change of Gibbs energy | Direction of reaction | Change of activities necessary to reach the state of equilibrium |
|---|---|---|---|
| $Q < K^*$ | $\Delta_R G < 0$ | Reaction proceeds spontaneously in the direction as written in the reaction equation (from left to right) | Activities of A and B in equation (5.7) decrease and activities of C and D increase until $Q = K^*$ |
| $Q = K^*$ | $\Delta_R G = 0$ | Equilibrium state, no reaction | No change of the activities |
| $Q > K^*$ | $\Delta_R G > 0$ | Reaction proceeds spontaneously in the reverse direction (from right to left) | Activities of C and D in equation (5.7) decrease and activities of A and B increase until $Q = K^*$ |

## 5.5 Estimation of Equilibrium Constants

The standard Gibbs energy of reaction can be estimated from the standard Gibbs energies of formation of the reactants and the products, $G^0_{f,i}$, according to the following equation:

$$\Delta_R G^0 = \sum_i \nu_i\, G^0_{f,i}(\text{products}) - \sum_i \nu_i\, G^0_{f,i}(\text{reactants}) \qquad (5.10)$$

Standard Gibbs energies of formation can be found for a large number of compounds in respective handbooks or databases. If $\Delta_R G^0$ has been computed, $K^*$ is available from equation (5.8). It has to be noted that all reactants and products have to be considered in the calculation, independent of their occurrence in the law of mass action.

Since the standard Gibbs energies of formation are typically given in databases for the standard temperature of 25 °C, the calculated equilibrium constants are also valid for 25 °C. On the other hand, it has to be considered that equilibrium constants depend on the temperature. This temperature dependence is discussed in more detail in the following section.

---

**Example 5.2**
What is the equilibrium constant for the reaction:

$$HCO_3^- \rightleftharpoons H^+ + CO_3^{2-}$$

at 25 °C? The following standard Gibbs energies of formation are given: $G^0_f\,(HCO_3^-) = -587$ kJ/mol, $G^0_f\,(H^+) = 0$ kJ/mol, $G^0_f\,(CO_3^{2-}) = -528$ kJ/mol. The universal gas constant is $R = 8.3145 \times 10^{-3}$ kJ/(mol·K).

**Solution:**
By using equation (5.10), we find:

$$\Delta_R G^0 = 0\ \text{kJ/mol} - 528\ \text{kJ/mol} - (-587\ \text{kJ/mol}) = 59\ \text{kJ/mol}$$

$\log K^*$ can be calculated by means of equation (5.8):

$$\log K^* = \frac{\ln K^*}{2.303} = -\frac{\Delta_R G^0}{2.303\ R\,T} = -\frac{59\ \text{kJ/mol}}{(2.303)\,(8.3145 \times 10^{-3}\ \text{kJ/(mol·K)})\,(298.15\ \text{K})}$$

$$\log K^* = -10.33$$

and the thermodynamic constant is:

$$K^* = \frac{a(H^+)\,a(CO_3^{2-})}{a(HCO_3^-)} = 4.68 \times 10^{-11}\ \text{mol/L}$$

The unit follows from the law of mass action.

---

## 5.6 Temperature Dependence of Equilibrium Constants

The derivative of $\ln K^*$ with respect to the temperature $T$ can be found by combining equations (5.8) and (5.5):

$$\frac{\partial \ln K^*}{\partial T} = \frac{\Delta_R H^0}{R\,T^2} \qquad (5.11)$$

where $\Delta_R H^0$ is the molar standard enthalpy of reaction, which can be found from the molar standard enthalpies of formation $\Delta H_{f,i}^0$ in an analogous manner as shown for the standard Gibbs energy of reaction (equation (5.10)).

Separation of the variables and integration gives:

$$\ln \frac{K_{T_2}^*}{K_{T_1}^*} = \frac{\Delta_R H^0}{R} \left( \frac{1}{T_1} - \frac{1}{T_2} \right) \tag{5.12}$$

If the equilibrium constant at the temperature $T_1$ and the molar standard enthalpy of reaction are known, equation (5.12) can be used to find the equilibrium constant at any other temperature $T_2$. Typically, the required data ($K_{T_1}^*$ and $\Delta_R H^0$) are given for the standard temperature 25 °C (Section 5.5). Therefore, we can write equation (5.12) also in a modified form:

$$\ln \frac{K^*}{K_0^*} = \frac{\Delta_R H^0}{R} \left( \frac{1}{T_0} - \frac{1}{T} \right) \tag{5.13}$$

where $T_0$ is the standard temperature (298.15 K = 25 °C), $K_0^*$ is the equilibrium constant at 25 °C, $\Delta_R H^0$ is the molar standard enthalpy of reaction determined for 25 °C, and $K^*$ is the equilibrium constant at any other temperature $T$.

It has to be noted that equations (5.12) and (5.13) are only valid if the molar enthalpy of reaction is assumed to be constant (independent of temperature). For more exact calculations, the temperature dependence of $\Delta_R H^0$ can be considered in the following manner:

$$\ln \frac{K^*}{K_0^*} = \frac{\Delta_R H^0}{R} \left( \frac{1}{T_0} - \frac{1}{T} \right) + \frac{c_p}{R} \left( \ln \frac{T}{T_0} + \frac{T_0}{T} - 1 \right) \tag{5.14}$$

where $c_p$ is the molar heat capacity of the reaction.

To apply equation (5.14), $c_p$ of the reaction has to be known. An alternative approach to describe the temperature dependence of equilibrium constants in the case of a nonconstant standard enthalpy of reaction is to measure constants at different temperatures and to describe the temperature dependence by means of an empirical polynom, for instance:

$$\ln K^* = a + \frac{b}{T} + \frac{c}{T^2} \tag{5.15}$$

where $a$, $b$, and $c$ are characteristic constants for the considered reaction.

**Example 5.3**

Water dissociates according to:

$$H_2O \rightleftharpoons H^+ + OH^-$$

The respective dissociation constant for 25 °C is $K_w^* = 1 \times 10^{-14}$ mol$^2$/L$^2$. What is the dissociation constant for 5 °C? Calculate the dissociation constant (a) without and (b) with consideration of the temperature dependence of $\Delta_R H^\circ$. $\Delta_R H^\circ = 56\,530$ J/mol, $c_p = -197$ J/(mol · K). The universal gas constant is $R = 8.3145$ J/(mol · K).

**Solution:**

(a) If we set $T_0 = 298.15$ K (25 °C), $T = 278.15$ K (5 °C), and $K_0^* = 1 \times 10^{-14}$ mol$^2$/L$^2$ and introduce the data into equation (5.13), we get:

$$\ln K_T^* - \ln(1 \times 10^{-14}) = \frac{56\,530 \text{ J/mol}}{8.3145 \text{ J/(mol · K)}} \left( \frac{1}{298.15 \text{ K}} - \frac{1}{278.15 \text{ K}} \right)$$

$$\ln K_T^* = \ln(1 \times 10^{-14}) + \frac{56\,530 \text{ J/mol}}{8.3145 \text{ J/(mol · K)}} \left( \frac{1}{298.15 \text{ K}} - \frac{1}{278.15 \text{ K}} \right)$$

$$\ln K_T^* = -32.24 + 6\,798.97 \left( 3.354 \times 10^{-3} - 3.595 \times 10^{-3} \right) = -32.24 - 1.64 = -33.88$$

$$K_T^* = K_{5\,°C}^* = 1.93 \times 10^{-15} \text{ mol}^2/\text{L}^2$$

(b) If we want to consider the temperature dependence of the reaction enthalpy, we have to apply equation (5.14):

$$\ln K_T^* = \ln(1 \times 10^{-14}) + \frac{56\,530 \text{ J/mol}}{8.3145 \text{ J/(mol · K)}} \left( \frac{1}{298.15 \text{ K}} - \frac{1}{278.15 \text{ K}} \right)$$

$$+ \frac{-197 \text{ J/(mol · K)}}{8.3145 \text{ J/(mol · K)}} \left[ \ln \left( \frac{278.15 \text{ K}}{298.15 \text{ K}} \right) + \frac{298.15 \text{ K}}{278.15 \text{ K}} - 1 \right]$$

$$\ln K_T^* = -32.24 + 6\,798.97 \left( 3.354 \times 10^{-3} - 3.595 \times 10^{-3} \right) - 23.69 (-0.069 + 1.072 - 1)$$

$$\ln K_T^* = -32.24 - 1.64 - 0.07 = -33.95$$

$$K_T^* = K_{5\,°C}^* = 1.80 \times 10^{-15} \text{ mol}^2/\text{L}^2$$

There is only a small difference in the calculated dissociation constants, which shows that equation 5.13 is a good approximation for narrow temperature ranges.

## 5.7 Equilibrium Constants of Reverse and Overall Reactions

The general reaction equation given in Section 5.1 describes the process of approaching the equilibrium starting with A and B as reactants. However, the same equilibrium state can also be reached from the other side, starting with C and D as reactants. For this reverse reaction, the reaction equation is:

$$v_C \, C + v_D \, D \rightleftharpoons v_A \, A + v_B \, B \qquad (5.16)$$

And, leaving out the indices "eq," the related law of mass action reads:

$$K^*_{\text{reverse}} = \frac{(a_A)^{\nu_A} (a_B)^{\nu_B}}{(a_C)^{\nu_C} (a_D)^{\nu_D}} \tag{5.17}$$

As can be derived from a comparison of equations (5.2) and (5.17), the equilibrium constant of the reverse reaction is the reciprocal of the constant for the forward reaction. Generally, it holds:

$$K^*_{\text{reverse}} = \frac{1}{K^*_{\text{forward}}} \tag{5.18}$$

If reactions are coupled, which is often the case in aqueous systems, the elementary reaction equations can be added to an overall reaction equation.

Taking the reactions:

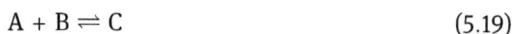

$$A + B \rightleftharpoons C \tag{5.19}$$

and

$$C \rightleftharpoons D + E \tag{5.20}$$

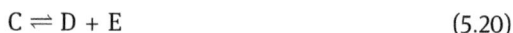

as examples (stoichiometric factors are assumed to be 1 for simplification), the overall reaction equation, found by adding the equations (5.19) and (5.20) is:

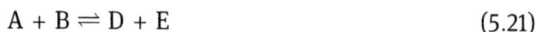

$$A + B \rightleftharpoons D + E \tag{5.21}$$

The law of mass action for the overall reaction can then be expressed as:

$$K^*_{\text{overall}} = \frac{a_D \, a_E}{a_A \, a_B} \tag{5.22}$$

The same quotient results from the multiplication of the laws of mass action of both elementary reactions:

$$K^*_1 \, K^*_2 = \frac{a_C}{a_A \, a_B} \frac{a_D \, a_E}{a_C} = \frac{a_D \, a_E}{a_A \, a_B} = K^*_{\text{overall}} \tag{5.23}$$

We can derive from this example that in the case of addition of reaction equations to an overall reaction equation, the constant of the overall reaction can be found by multiplying the constants of the elementary reactions.

## 5.8 Problems

**5.1.** Calculate the equilibrium constant for the dissolution of solid calcium carbonate according to:

$$CaCO_{3(s)} \rightleftharpoons Ca^{2+} + CO_3^{2-} ?$$

at 25 °C. The standard Gibbs energies of formation are $G_f^0(CaCO_3) = -1\,129.2$ kJ/mol, $G_f^0(Ca^{2+}) = -552.8$ kJ/mol, and $G_f^0(CO_3^{2-}) = -528.0$ kJ/mol. The universal gas constant is $R = 8.3145$ J/(mol · K).

**5.2.** It is known that in aqueous systems $Mn^{2+}$ and $Fe^{2+}$ are oxidized by dissolved oxygen to form $MnO_2$ and $Fe(OH)_3$, respectively. Is a comparable reaction for $Pb^{2+}$ to be expected under the following conditions: $pH = 7$, $c(Pb^{2+}) = 1 \times 10^{-8}$ mol/L, $c(O_2) = 0.261$ mmol/L (saturation concentration of atmospheric $O_2$ in water at 25 °C)? The equilibrium constant for the oxidation reaction:

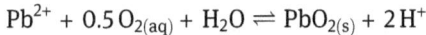

$$Pb^{2+} + 0.5\,O_{2(aq)} + H_2O \rightleftharpoons PbO_{2(s)} + 2\,H^+$$

is $K^* = 4 \times 10^{-7}$ $mol^{0.5}/L^{0.5}$ (25 °C).

**5.3.** For the dissolution of calcium carbonate:

$$CaCO_{3(s)} \rightleftharpoons Ca^{2+} + CO_3^{2-}$$

the following equilibrium constants (solubility products) were found: $3.93 \times 10^{-9}$ $mol^2/L^2$ at 10 °C and $3.30 \times 10^{-9}$ $mol^2/L^2$ at 25 °C. Calculate the standard enthalpy of reaction for this temperature interval. Is the dissolution of $CaCO_3$ an endothermic or an exothermic process? A possible further reaction of carbonate to hydrogencarbonate should not be considered here. The universal gas constant is $R = 8.3145$ J/(mol · K).

**5.4.** The equilibrium constants (at 25 °C) of the reactions:

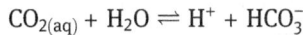

$$CO_{2(aq)} + H_2O \rightleftharpoons H^+ + HCO_3^-$$

and

$$HCO_3^- \rightleftharpoons H^+ + CO_3^{2-}$$

are $K_1^* = 4.4 \times 10^{-7}$ mol/L and $K_2^* = 4.7 \times 10^{-11}$ mol/L, respectively. What is the equilibrium constant of the reaction:

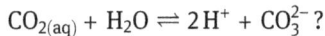

$$CO_{2(aq)} + H_2O \rightleftharpoons 2\,H^+ + CO_3^{2-} ?$$

# 6 Gas–Water Partitioning

## 6.1 Introduction

Aqueous systems are often in contact with an adjacent gas phase. Typical examples are rain or surface water, which are in contact with the atmospheric air, or seepage water in the vadose (unsaturated) zone, which is in contact with the soil atmosphere. As a result of such a liquid/gas phase contact, a distribution of gas phase components between the liquid and the gas phase occurs. The distribution of a gas or a vapor between the gas phase and the aqueous phase is also referred to as gas–water partitioning. Gas–water partitioning is relevant for many natural processes in water bodies. It is, for instance, one mechanism of oxygen input into aquatic systems. Dissolved oxygen is the most important oxidant in water bodies. Gas–water partitioning of carbon dioxide, as a further example, controls the carbonic acid system, which is involved in many other processes (biological processes, precipitation/dissolution processes, buffer reactions). Furthermore, gas–water partitioning is also relevant for a number of water treatment processes, where gases have to be introduced into the water, such as aeration, ozonation or chlorination of drinking water or where dissolved gases or volatile substances have to be removed from water, such as stripping of carbon dioxide, methane, or volatile organics.

If water comes into contact with a gas phase, partitioning of gaseous components between both phases takes place until an equilibrium state is established. In principle, the approach to the equilibrium state can proceed from both sides. The transfer of a gas component to the aqueous phase is referred to as absorption, whereas the transfer of a dissolved gas or vapor from the aqueous phase to the gas phase is referred to as desorption (Figure 6.1). If the rate of absorption equals the rate of desorption, a dynamic equilibrium is established that is characterized by a defined ratio between the concentrations of the considered component in the gas phase and in the aqueous phase.

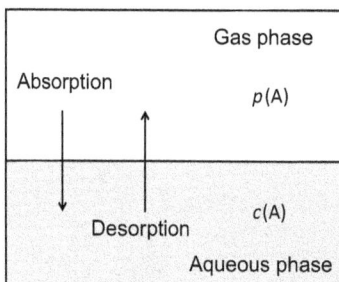

**Figure 6.1:** Gas–water partitioning.

After absorption, which is a physical process, some dissolved gases take part in chemical reactions within the aqueous phase. The main part of this chapter deals

https://doi.org/10.1515/9783110758788-006

with the basics of physical partitioning. Coupled equilibria consisting of absorption and chemical reaction are discussed separately in Section 6.8.

## 6.2 Henry's Law

Although gas–water partitioning is not a real chemical reaction but a phase transfer process, it can be formally described as reaction equilibrium by using the law of mass action. Taking an arbitrary gas A as an example, the dissolution of A in the aqueous phase can be written as:

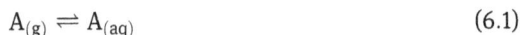

$$A_{(g)} \rightleftharpoons A_{(aq)} \tag{6.1}$$

If the content of the gas A in the gas phase is expressed by its partial pressure (see Chapter 5, Section 5.3), and the concentration of the dissolved gas is given as molar concentration, the formal law of mass action reads:

$$H(A) = \frac{c(A)}{p(A)} \tag{6.2}$$

This law is also known as Henry's law and, accordingly, the equilibrium constant in this special case is referred to as the Henry's law constant or simply Henry constant and abbreviated with $H$ instead of the common $K$ for equilibrium constants. Typically, Henry's law is written in the form:

$$c(A) = H(A)\, p(A) \tag{6.3}$$

which shows that the equilibrium concentration of a dissolved gaseous component in the aqueous phase (or, in other words, its solubility) is proportional to its partial pressure in the gas phase that is in contact with the aqueous phase. The unit of $H$ in this case is mol/(L·bar). If Henry's law is written in the form of equation (6.3), increasing Henry constants indicate increasing solubilities. Table 6.1 lists some Henry constants in the order of increasing solubility.

**Table 6.1:** Selected Henry constants at 25 °C.

| Compound | Henry constant, $H$ [mol/(L·bar)] | Compound | Henry constant, $H$ [mol/(L·bar)] |
|---|---|---|---|
| Nitrogen | 0.0006 | Dihydrogen sulfide | 0.102 |
| Oxygen | 0.0012 | Benzene | 0.182 |
| Methane | 0.0016 | Dichloromethane | 0.385 |
| Carbon dioxide | 0.033 | Methyl-*tert*-butyl ether | 1.80 |
| 1,1,1-Trichloroethane | 0.066 | Ammonia | 59.88 |

As with other equilibrium constants, the Henry constant depends on the temperature. Generally, the solubility of gases increases with decreasing temperature. Table 6.2 shows by example the Henry constants of the major gases in the atmosphere for different temperatures.

The temperature dependence of $H$ can be described in the same manner as the temperature dependence of equilibrium constants of chemical reactions (Chapter 5, Section 5.6). According to equation (5.13) in Chapter 5 we can write:

$$\ln \frac{H}{H_0} = \frac{\Delta H_{sol}^0}{R} \left( \frac{1}{T_0} - \frac{1}{T} \right) \tag{6.4}$$

where $T_0$ is the standard temperature 298.15 K (25 °C) and $H_0$ is the Henry constant at 298.15 K. Note that in equation (6.4) the standard enthalpy of solution, $\Delta H_{sol}^0$, takes the place of the standard enthalpy of reaction, $\Delta_R H^0$.

**Table 6.2:** Henry constants of the major gas components in the atmosphere.

| Gas | Temperature (°C) | Henry constant, $H$ [$10^{-3}$ mol/(L·bar)] |
|---|---|---|
| Nitrogen, $N_2$ | 10 | 0.828 |
| Nitrogen, $N_2$ | 25 | 0.646 |
| Oxygen, $O_2$ | 10 | 1.674 |
| Oxygen, $O_2$ | 25 | 1.247 |
| Carbon dioxide, $CO_2$ | 10 | 52.47 |
| Carbon dioxide, $CO_2$ | 25 | 33.42 |

As an alternative to equation (6.1), the reaction equation can be formulated for the reverse process (desorption) in the form:

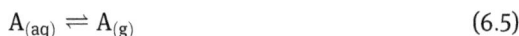

$$A_{(aq)} \rightleftharpoons A_{(g)} \tag{6.5}$$

In this case, the law of mass action has to be written as:

$$H_{inv}(A) = \frac{p(A)}{c(A)} \tag{6.6}$$

or

$$p(A) = H_{inv}(A)\, c(A) \tag{6.7}$$

where $H_{inv}(A)$ is the inverse of the equilibrium constant $H(A)$:

$$H_{inv}(A) = \frac{1}{H(A)} \tag{6.8}$$

In contrast to $H$, increasing $H_{inv}$ indicates increasing volatility. It has to be noted that both methods of writing Henry's law can be found in the literature and often the same abbreviation $H$ (without qualifier) is used, which may cause confusion. It is therefore important to notice the unit when Henry constants are taken from the literature or from databases.

Another point to keep in mind when using Henry constants concerns the fact that a number of modified formulations of Henry's law can be found in the literature where other concentration measures are used to describe the composition of the gas and/or the aqueous phase. Some of these modifications will be summarized in the next section.

## 6.3 Alternative Formulations of Henry's Law

The partial pressure of a gas A in a gas mixture is related to its molar concentration in the gas phase, $c_g$, by the following expression that can be derived from the state equation of an ideal gas (Section 3.5 in Chapter 3):

$$p(A) = \frac{n(A)}{V} R\,T = c_g(A)\,R\,T \qquad (6.9)$$

Substituting the partial pressure in equation (6.3) by equation (6.9) gives:

$$c_{aq}(A) = H(A)\,R\,T\,c_g(A) \qquad (6.10)$$

Here, the subscripts "aq" and "g" were introduced to distinguish between the aqueous-phase and the gas-phase concentration. Equation (6.10) can be simplified to:

$$c_{aq}(A) = K_c(A)\,c_g(A) \qquad (6.11)$$

where $K_c$ is a dimensionless distribution constant that is related to $H$ by:

$$K_c(A) = H(A)\,R\,T \qquad (6.12)$$

With the molecular weight, $M$, of the dissolved gas, we can write:

$$K_c(A) = \frac{c_{aq}(A)}{c_g(A)} = \frac{c_{aq}(A)\,M(A)}{c_g(A)\,M(A)} = \frac{\rho_{aq}^*(A)}{\rho_g^*(A)} \qquad (6.13)$$

Accordingly, the value of $K_c$ is not only valid for the ratio of the molar concentrations but also for the ratio of the mass concentrations.

If the mole fraction, $x$, is used to describe the liquid-phase composition, Henry's law can be written as:

$$x(A) = H_x(A)\,p(A) \qquad (6.14)$$

where the Henry constant $H_x$ has the unit 1/bar. $H_x$ can be related to $H$ by using the approximation:

$$x(A) = \frac{n(A)}{\sum n} \approx \frac{n(A)}{n(H_2O)} = \frac{c_{aq}(A)}{c(H_2O)} = \frac{c_{aq}(A)}{55.56 \text{ mol/L}} \tag{6.15}$$

$$H_x(A) \approx \frac{H(A)}{55.56 \text{ mol/L}} \tag{6.16}$$

In equation (6.15), it is assumed that the amount of the dissolved gas is much lower than the amount of the solvent water ($n(A) \ll n(H_2O)$) and that the concentration of water can be approximated by the quotient of its density ($\rho = 1\,000$ g/L) and its molecular weight ($M = 18$ g/mol). For these approximations, see also Section 3.3 in Chapter 3.

Since at constant total pressure, $p_{total}$, the partial pressure, $p$, is proportional to the gas-phase mole fraction, $y$:

$$p = y\, p_{total} \tag{6.17}$$

the composition of both phases can be expressed by mole fractions. In this case, Henry's law reads:

$$x(A) = H_x(A)\, p_{total}\, y(A) \tag{6.18}$$

or

$$x(A) = K_{xy}(A)\, y(A) \tag{6.19}$$

where the dimensionless distribution constant, $K_{xy}$ ($= H_x\, p_{total}$), gives the ratio of the mole fractions.

It has to be noted that all modified versions of Henry's law shown here can also be written in inverse form as demonstrated in Section 6.2 for the basic form.

---

**Example 6.1**
The Henry constant of oxygen at 25 °C is $H = 1.247 \times 10^{-3}$ mol/(L · bar). What are the values of the alternative constants $K_c$ and $H_x$? $R = 0.083145$ bar · L/(mol · K).

**Solution:**
The relationship between $H$ and $K_c$ is given by $K_c = H\,R\,T$. After introducing $T = 298.15$ K ($= 25$ °C) and $R = 0.083145$ bar · L/(mol · K), we get:

$$K_c = H\,R\,T = \left(1.247 \times 10^{-3} \frac{\text{mol}}{\text{L} \cdot \text{bar}}\right)\left(0.083145 \frac{\text{bar} \cdot \text{L}}{\text{mol} \cdot \text{K}}\right)(298.15 \text{ K}) = 3.09 \times 10^{-2}$$

For the conversion of $H$ into $H_x$, we have to apply the approximation:

$$H_x \approx \frac{H}{55.56 \text{ mol/L}}$$

Introducing $H$ gives:

$$H_x \approx \frac{1.247 \times 10^{-3} \text{ mol}/(\text{L} \cdot \text{bar})}{55.56 \text{ mol}/\text{L}} = 2.24 \times 10^{-5} \text{ bar}^{-1}$$

## 6.4 Estimation of Henry's Law Constants for Volatile Substances

For numerous volatile substances, Henry's law constants can be found in respective databases. If the required data are not available, these constants can also be estimated approximately from other substance properties. Given that the relationship between the partial pressure and the concentration in the aqueous phase is linear and the saturation limits in both phases are given by the vapor pressure and the aqueous solubility of the substance, the Henry's law constant can be found from the ratio:

$$H = \frac{c_s}{p_v} \tag{6.20}$$

where $c_s$ is the aqueous solubility in mol/L and $p_v$ is the vapor pressure in bar.

---

**Example 6.2**

The vapor pressure of benzene at 25 °C is $p_v = 0.127$ bar and its aqueous solubility at the same temperature is $\rho_s^* = 1790$ mg/L. What is the Henry's law constant at 25 °C? The molecular weight of benzene is $M = 78.1$ g/mol.

**Solution:**

The solubility in mol/L can be found by dividing the mass concentration by the molecular weight:

$$c_s = \frac{\rho_s^*}{M} = \frac{1.79 \text{ g}/\text{L}}{78.1 \text{ g}/\text{mol}} = 0.023 \text{ mol}/\text{L}$$

With this value of $c_s$, the Henry's law constant can be found according to:

$$H = \frac{c_s}{p_v} = \frac{0.023 \text{ mol}/\text{L}}{0.127 \text{ bar}} = 0.181 \text{ mol}/(\text{L} \cdot \text{bar})$$

---

## 6.5 Open and Closed Systems

With respect to the application of Henry's law, it is necessary to distinguish between open and closed systems (Figure 6.2). Open systems are characterized by a nearly infinitely large volume of the gas phase, whereas the volume of the aqueous phase is limited and comparatively small. Surface water in contact with the atmosphere is a typical example of an open system. Due to the large volume of the gas phase and the limited capacity of the liquid phase, the amounts of gases that can be transferred are not large enough to change the partial pressures in the gas phase considerably. Such open systems are therefore characterized by practically constant partial pressures of the gas-phase constituents. Accordingly, the equilibrium solubilities of the atmospheric gases (e.g., oxygen, nitrogen, carbon dioxide) can be easily computed by applying a constant partial pressure in Henry's law (Section 6.6).

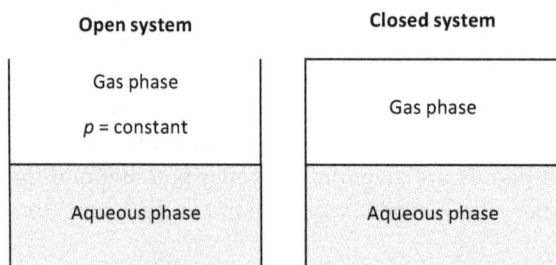

**Figure 6.2:** Gas–water partitioning: open and closed systems.

For gases or vapors that are not natural constituents of the atmosphere or occur only in negligible concentrations, the partial pressure is zero or practically zero. If such components are desorbed in limited amounts from the aqueous phase, the partial pressure does not change considerably and remains practically zero due to the large volume of the gas phase. Since the equilibrium concentration that has to be reached by desorption at a partial pressure of $p = 0$ bar is $c = 0$ mol/L, there is a steady driving force for the transfer of the component to the gas phase as long as the actual concentration in the aqueous phase is not zero. As a result, the total amount of the dissolved gas or vapor can be expected to be transferred to the atmosphere without measurable increase of its partial pressure. Often such compounds (e.g., volatile organic solvents) are not stable in the atmosphere and subject to chemical or photochemical degradation processes with the consequence that the very small amounts transferred disappear after a sufficiently long time.

The situation in a closed system is quite different from the situation in open systems discussed previously. In closed systems (e.g., reactors, tubes, bottles), the gas phase has a limited volume that is in the same order of magnitude as the volume of the aqueous phase. Consequently, each transfer of gases or vapors from the liquid phase to the gas phase or from the gas phase to the liquid phase results not only in a change of the concentration in the liquid phase but also in a change of the partial pressure (or concentration) in the gas phase. The equilibrium that is reached after the transfer is also described by Henry's law. However, to estimate the concentration changes in both phases, Henry's law has to be combined with an appropriate material balance equation (Section 6.7).

## 6.6 Solubilities of Atmospheric Gases in Water

The major constituents of the atmosphere are nitrogen (78.1 vol%), oxygen (20.9 vol%), and carbon dioxide (0.0415 vol% = 415 ppm$_v$). Given that the volume percentages divided by 100% give the volume fractions and that, for an ideal gas mixture, the volume fractions equal the mole fractions (see Section 3.5 in Chapter 3), the mole fractions of the atmospheric gases are $y(N_2) = 0.781$, $y(O_2) = 0.209$, and $y(CO_2) = 0.000415$. If we assume that the total pressure in the atmosphere is about 1 bar, the related partial pressures are $p(N_2) = 0.781$ bar, $p(O_2) = 0.209$ bar, and $p(CO_2) = 0.000415$ bar (equation (6.17)). With these partial pressures and the Henry constants given in Table 6.2, the equilibrium concentrations in the aqueous phase can be calculated by Henry's law (equation (6.3)) as shown in Example 6.3.

---

**Example 6.3**

What is the solubility (in mg/L) of atmospheric oxygen ($M = 32$ g/mol) at 25 °C? The atmosphere contains 20.9 vol% oxygen, the total atmospheric pressure is assumed to be 1 bar, and the Henry constant at 25 °C is $1.247 \times 10^{-3}$ mol/(L · bar).

**Solution:**

The mole fraction of oxygen in the gas phase equals the volume fraction and is therefore $y = 0.209$. The partial pressure is given by:

$$p(O_2) = y(O_2)\, p_{total} = (0.209)\,(1\,\text{bar}) = 0.209\,\text{bar}$$

and the molar solubility can be estimated from Henry's law according to:

$$c(O_2) = H(O_2)p(O_2) = (1.247 \times 10^{-3}\,\text{mol}/(\text{L} \cdot \text{bar}))(0.209\,\text{bar})$$

$$c(O_2) = 2.61 \times 10^{-4}\,\text{mol/L} = 0.261\,\text{mmol/L}$$

Finally, the mass concentration is:

$$\rho^*(O_2) = c(O_2)\, M(O_2) = (0.261\,\text{mmol/L})\,(32\,\text{mg/mmol}) = 8.35\,\text{mg/L}$$

---

Table 6.3 lists the solubilities of atmospheric gases for two different temperatures. As mentioned previously, the solubilities decrease with increasing temperature. The solubility of oxygen is only slightly lower than that of nitrogen, because the lower partial pressure is mostly compensated for by the higher value of the Henry constant. It can further be seen from the data that, despite the high value of the Henry constant, the solubility of carbon dioxide is considerably lower in comparison to the solubilities of oxygen and nitrogen. This is a result of the much lower partial pressure of $CO_2$ in the atmospheric air.

**Table 6.3:** Solubilities of the major gases in the atmosphere at different temperatures.

| Gas | Partial pressure (bar) | Temperature (°C) | Equilibrium concentration (solubility) | |
|---|---|---|---|---|
| | | | mmol/L | mg/L |
| $N_2$ | 0.781 | 10 | 0.647 | 18.11 |
| $N_2$ | 0.781 | 25 | 0.505 | 14.13 |
| $O_2$ | 0.209 | 10 | 0.350 | 11.20 |
| $O_2$ | 0.209 | 25 | 0.261 | 8.35 |
| $CO_2$ | 0.000415 | 10 | 0.0218 | 0.959 |
| $CO_2$ | 0.000415 | 25 | 0.0139 | 0.612 |

## 6.7 Calculation of Equilibrium Concentrations in Closed Systems

Basic parameters and relationship used for closed systems are illustrated in Figure 6.3. In order to calculate the partitioning of a gas or vapor between the gas phase and the aqueous phase in a closed system, the material balance has to be combined with Henry's law. In this specific case, it is reasonable to use the same concentration unit for both phases, for instance the molar concentration or the mass concentration. In the following considerations, we want to use the mass concentration.

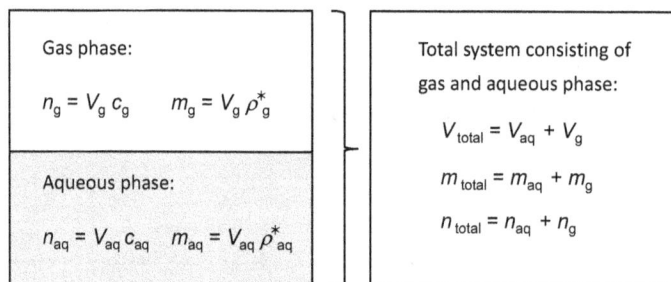

**Figure 6.3:** Closed system: basic parameters and relationships.

The material balance equation for a substance distributed between gas and aqueous phase in a closed system reads:

$$m_{total} = m_g + m_{aq} = \rho_g^* V_g + \rho_{aq}^* V_{aq} \tag{6.21}$$

where $m_{total}$ is the total mass of the considered gas or vapor in the closed system, $m_g$ and $m_{aq}$ are the masses in the gas phase and in the aqueous phase, respectively, $V_g$

and $V_{aq}$ are the volumes of the gas and the aqueous phase, respectively, and $\rho_g^*$ and $\rho_{aq}^*$ are the respective mass concentrations.

If we factor out the mass concentration in the gas phase and consider equation (6.13), we receive:

$$m_{total} = \rho_g^* \left( V_g + \frac{\rho_{aq}^*}{\rho_g^*} V_{aq} \right) = \rho_g^* \left( V_g + K_c\, V_{aq} \right) \tag{6.22}$$

where the distribution constant, $K_c$, is given by:

$$K_c = \frac{c_{aq}}{c_g} = \frac{c_{aq}\, M}{c_g\, M} = \frac{\rho_{aq}^*}{\rho_g^*} \tag{6.23}$$

with

$$K_c = H\,R\,T \tag{6.24}$$

After rearranging equation (6.22), the gas-phase concentration, $\rho_g^*$, in the state of equilibrium is found to be:

$$\rho_g^* = \frac{m_{total}}{\left( V_g + K_c\, V_{aq} \right)} \tag{6.25}$$

and the equilibrium concentration in the aqueous phase is given by:

$$\rho_{aq}^* = K_c\, \rho_g^* \tag{6.26}$$

Alternatively, $\rho_{aq}^*$ can also be calculated by:

$$\rho_{aq}^* = \frac{m_{total}}{\left( \dfrac{V_g}{K_c} + V_{aq} \right)} \tag{6.27}$$

Equation (6.27) can be derived in the same way as shown for equation (6.25) but with factoring out $\rho_{aq}^*$ instead of $\rho_g^*$ on the right-hand side of equation (6.21).

---

**Example 6.4**

A bottle with the total volume of 1 L was filled halfway with a solution of 1,1,1-trichloroethane (initial concentration: $\rho_{aq,\,in}^* = 100$ mg/L). After that, it was closed and kept for a long time at 20 °C so that establishment of equilibrium can be assumed. What are the equilibrium concentrations of 1,1,1-trichloroethane in the gas phase and in the aqueous phase? The Henry constant of 1,1,1-trichloroethane is $H = 0.078$ mol/(L·bar) at 20 °C and the gas constant is $R = 0.083145$ bar·L/(mol·K).

**Solution:**

At first, the dimensionless distribution constant, necessary for application of the material balance equation, has to be calculated:

$$K_c = HRT = \left(0.078\,\frac{\text{mol}}{\text{L}\cdot\text{bar}}\right)\left(0.083145\,\frac{\text{bar}\cdot\text{L}}{\text{mol}\cdot\text{K}}\right)(293.15\ \text{K}) = 1.9$$

At the beginning, the total amount of 1,1,1-trichloroethane is dissolved in the aqueous phase. The volume of the solution is $V_{aq} = 0.5$ L. Consequently, the total mass of 1,1,1-trichloroethane in the system is:

$$m_{total} = \rho^*_{aq,\,in}\, V_{aq} = (100\ \text{mg/L})(0.5\ \text{L}) = 50\ \text{mg}$$

Now, the concentration in the gas phase after the establishment of equilibrium can be calculated by:

$$\rho^*_g = \frac{m_{total}}{(V_g + K_c\, V_{aq})} = \frac{50\ \text{mg}}{0.5\ \text{L} + (1.9)(0.5\ \text{L})} = 34.48\ \text{mg/L}$$

and the aqueous-phase concentration can be computed by means of the distribution constant:

$$\rho^*_{aq} = K_c\, \rho^*_g = (1.9)(34.48\ \text{mg/L}) = 65.51\ \text{mg/L}$$

To validate the results, the masses in both phases can be calculated. Their sum must equal the total mass of 50 mg:

$$m_{aq} = V_{aq}\, \rho^*_{aq} = (0.5\ \text{L})(34.48\ \text{mg/L}) = 17.24\ \text{mg}$$

$$m_g = V_g\, \rho^*_g = (0.5\ \text{L})(65.51\ \text{mg/L}) = 32.76\ \text{mg}$$

$$m_{total} = m_{aq} + m_g = 17.24\ \text{mg} + 32.76\ \text{mg} = 50\ \text{mg}$$

## 6.8 Coupling of Gas–Water Partitioning and Chemical Reaction

In the previous sections, we have only considered the physical partitioning between the gas and the liquid phase. However, some dissolved gases are also subject to chemical reactions within the aqueous phase. In this context, acid/base reactions are particularly relevant, because a number of gases act as acids or bases in aqueous solutions. The atmospheric gas carbon dioxide, $CO_2$, is a typical example. In aqueous solutions, $CO_2$ acts as an acid ("carbonic acid," Figure 6.4).

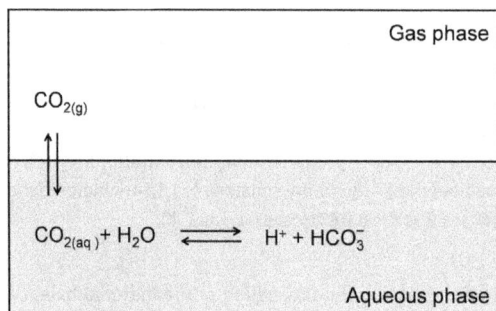

**Figure 6.4:** Dissolution of carbon dioxide: coupling of gas–water partitioning and acid reaction.

The coupled process consisting of gas–water partitioning and acid reaction in aqueous solution can be written as:

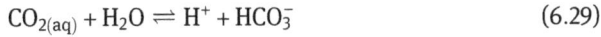

$$CO_{2(g)} \rightleftharpoons CO_{2(aq)} \tag{6.28}$$

$$CO_{2(aq)} + H_2O \rightleftharpoons H^+ + HCO_3^- \tag{6.29}$$

$$\overline{CO_{2(g)} + H_2O \rightleftharpoons H^+ + HCO_3^-} \tag{6.30}$$

Note that under most natural conditions (medium pH range), the possible second dissociation step of the carbonic acid is of minor relevance. Therefore, it will be neglected here.

The laws of the mass action for the elementary processes gas–water partitioning and acid reaction (first dissociation step of the carbonic acid) are:

$$H = \frac{c(CO_2)}{p(CO_2)} \tag{6.31}$$

$$K_a = \frac{c(H^+)\, c(HCO_3^-)}{c(CO_2)} \tag{6.32}$$

where $K_a$ is the conditional acidity constant (for definition and further details see Chapter 7, Section 7.4) and $c$ refers to the aqueous-phase concentration.

According to the combination principle demonstrated in Chapter 5 (Section 5.7), we get for the overall reaction:

$$K_{overall} = H\, K_a = \frac{c(CO_2)}{p(CO_2)} \cdot \frac{c(H^+)\, c(HCO_3^-)}{c(CO_2)} = \frac{c(H^+)\, c(HCO_3^-)}{p(CO_2)} \tag{6.33}$$

Equation (6.33) describes the simultaneous equilibrium and can be used, for instance, to find the pH value resulting from the dissolution of carbon dioxide (Chapter 7, Section 7.7.7).

For more general calculations, additional reactions has to be taken into account, for instance the dissociation of hydrogencarbonate ($HCO_3^- \rightleftharpoons H^+ + CO_3^{2-}$) and the precipitation/dissolution of carbonates (in particular calcium carbonate). This coupling of equilibria will be demonstrated in the context of the calco-carbonic equilibrium in Chapter 9 (Section 9.6).

Other examples of gases that are subject to acid/base reactions in the aqueous phase are hydrogen sulfide ($H_2S$, also referred to as dihydrogen sulfide), and ammonia ($NH_3$). Dissolved $H_2S$ acts as an acid, whereas dissolved $NH_3$ acts as a base. The respective reaction equations are:

$$H_2S_{(g)} \rightleftharpoons H_2S_{(aq)} \tag{6.34}$$

$$H_2S_{(aq)} \rightleftharpoons H^+ + HS^- \tag{6.35}$$

$$\overline{\phantom{H_2S_{(aq)} \rightleftharpoons H^+ + HS^-}}$$

$$H_2S_{(g)} \rightleftharpoons H^+ + HS^- \tag{6.36}$$

and

$$NH_{3(g)} \rightleftharpoons NH_{3(aq)} \tag{6.37}$$

$$NH_{3(aq)} + H_2O \rightleftharpoons NH_4^+ + OH^- \tag{6.38}$$

$$\overline{\phantom{NH_{3(aq)} + H_2O \rightleftharpoons NH_4^+ + OH^-}}$$

$$NH_{3(g)} + H_2O \rightleftharpoons NH_4^+ + OH^- \tag{6.39}$$

## 6.9 Problems

**6.1** In a database, the following Henry's law constant of benzene at 25 °C was found: 5.495 L · bar/mol. What is the corresponding formulation of Henry's law and what is the value of $K_c$ if $K_c$ is defined by $K_c = c_{aq}/c_g$? $R = 0.083145$ bar · L/(mol · K).

**6.2** What is the aqueous solubility (in mg/L) of pure oxygen at 25 °C and at a total pressure of 10 bar? The Henry's law constant amounts to $H(O_2) = 1.247$ mol/(m³· bar). The molecular weight of molecular oxygen is $M = 32$ g/mol.

**6.3** The Henry's law constant of $CO_2$ at 25 °C is $3.342 \times 10^{-2}$ mol/(L · bar), the enthalpy of solution is $\Delta H_{sol} = -20.79$ kJ/mol. What is the Henry's law constant of $CO_2$ at 20 °C? $R = 8.3145$ J/(mol · K).

**6.4.** Due to biodegradation of organic material, the $CO_2$ content of the soil atmosphere (gas phase between the soil particles) is much higher than the $CO_2$ content of the normal atmosphere, which is $p(CO_2) = 0.000415$ bar. Calculate the $CO_2$ concentration (in mmol/L) that would be reached in seepage water (at 10 °C) in equilibrium with the soil atmosphere when the $CO_2$ content in the soil atmosphere would be 50 times higher than in the normal atmosphere. $H(CO_2) = 0.0525$ mol/(L·bar) at 10 °C.

**6.5.** A 1 L flask contains 0.8 L of an aqueous solution of trichloroethylene ($H = 0.085$ mol/(L·bar) at 25 °C) with the initial concentration of $\rho^*_{aq,\, in} = 100$ mg/L. A second 1 L flask contains 0.2 L of the same solution. Calculate the concentrations remaining in the aqueous phase after establishment of the equilibrium at 25 °C for both cases. The universal gas constant is $R = 0.083145$ bar·L/(mol·K).

**6.6.** In drinking water treatment, it is often necessary to remove dissolved $CO_2$ from raw water due to its aggressive (corrosive) character. Stripping by air is one of the possible processes that can be applied to remove $CO_2$ from water. What is the minimum $CO_2$ mass concentration in the water that can be reached by this process at 10 °C? The molecular weight of $CO_2$ is $M = 44$ g/mol and the Henry's law constant at 10 °C amounts to $H = 0.0525$ mol/(L·bar). The $CO_2$ content in the process air (total pressure: 1 bar) is assumed to be 415 $ppm_v$.

# 7 Acid/Base Equilibria

## 7.1 Introduction

The relevance of acid/base equilibria for aqueous systems results from the fact that numerous water constituents are acids or bases according to Brønsted's acid/base theory (Section 7.2). Moreover, water itself is also involved in acid/base reactions (Section 7.3). Since, after Brønsted's acid/base theory, the definition of acids and bases is based on the capability to donate or accept protons ($H^+$), the pH as a measure of the proton concentration plays an important role in acid/base systems. The degree of proton release or acceptance determines the strength of acids and bases (Section 7.4). Furthermore, the pH determines the concentration distribution of conjugate acid/base pairs. This is in particular true for weak acids and bases which are not completely deprotonated or protonated over the whole pH range. Typical examples for such acid/base systems with pH-dependent species distribution are $CO_2/HCO_3^-/CO_3^{2-}$, $HPO_4^{2-}/H_2PO_4^-$, and $NH_3/NH_4^+$. Another important aspect is that the solubility of many solids and gases depends on pH due to the coupling of dissolution with acid/base reactions.

Acids and bases can also be found within the group of organic water constituents. Humic and fulvic acids are frequently occurring natural acids. Basic amines and acidic phenols originating from natural and anthropogenic sources are further relevant substance groups.

From the practical point of view, there are two types of problems with respect to acid/base systems. The first is to answer the question of what pH results after dissolution of a defined amount of an acid, a base, or a salt (Section 7.5). The second problem concerns the mathematical description of the pH-dependent speciation (Section 7.6).

A special section of this chapter is dedicated to the carbonic acid system due to its high relevance for natural water bodies (Section 7.7).

## 7.2 Brønsted's Acid/Base Theory

Amongst the different acid/base theories existing in chemistry, Brønsted's acid/base theory (also referred to as the Brønsted–Lowry acid/base theory) is the most important theory for describing acid/base reactions in aquatic systems. After Brønsted's acid/base theory, acids are defined as species that are able to donate protons, whereas bases are defined as species that are able to accept protons. Accordingly, acids are proton donors and bases are proton acceptors. The process of proton release is also referred to as deprotonation, whereas the process of proton acceptance is referred to as protonation.

https://doi.org/10.1515/9783110758788-007

According to the general equation:

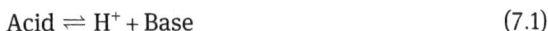

$$\text{Acid} \rightleftharpoons \text{H}^+ + \text{Base} \qquad (7.1)$$

an acid is always related to a base and vice versa. Such acid/base pairs are referred to as conjugate acid/base pairs. In other words, the conjugate base is the deprotonated form of the related acid, and the conjugate acid is the protonated form of the related base. The reaction described by equation (7.1) is also referred to as acid dissociation.

Since isolated protons do not exist in aqueous solutions, the released proton is transferred (donated) to another base according to:

$$\text{Acid 1} \rightleftharpoons \text{H}^+ + \text{Base 1} \qquad (7.2)$$

$$\text{H}^+ + \text{Base 2} \rightleftharpoons \text{Acid 2} \qquad (7.3)$$

$$\overline{\text{Acid 1} + \text{Base 2} \rightleftharpoons \text{Acid 2} + \text{Base 1}} \qquad (7.4)$$

Here, the proton from acid 1 is donated to base 2. Acid 1, which has lost its proton, is transformed to base 1, the conjugate base of acid 1. Base 2, which has accepted the proton from acid 1, is transformed to acid 2, the conjugate acid of base 2. Accordingly, two conjugate acid/base pairs are always involved in a complete acid/base reaction. As can be derived from the previous discussion, proton transfer is the characteristic property of an acid/base reaction. Such a proton transfer reaction is also referred to as protolysis.

In aquatic systems, the solvent water itself forms two acid/base pairs according to:

$$\text{H}_2\text{O} + \text{H}_2\text{O} \rightleftharpoons \text{H}_3\text{O}^+ + \text{OH}^- \qquad (7.5)$$

Here, water acts both as an acid and as a base. This process is also referred to as autoprotolysis. Each of the conjugate acid/base pairs ($\text{H}_2\text{O}/\text{OH}^-$ and $\text{H}_3\text{O}^+/\text{H}_2\text{O}$) can also react with any other acid or base, here shown by example for the acid HCl and the base $\text{NH}_3$:

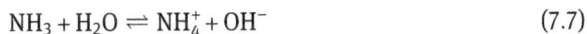

$$\text{HCl} + \text{H}_2\text{O} \rightleftharpoons \text{H}_3\text{O}^+ + \text{Cl}^- \qquad (7.6)$$

$$\text{NH}_3 + \text{H}_2\text{O} \rightleftharpoons \text{NH}_4^+ + \text{OH}^- \qquad (7.7)$$

In the first case, water acts as a base and accepts the proton from HCl, whereas in the second case, water is the acid that gives its proton to $\text{NH}_3$.

According to equations (7.2)–(7.4), equations (7.6) and (7.7) can be considered overall equations consisting of two elementary reactions, one describing the loss and the other the acceptance of the proton. Accordingly, we can write for the acid HCl:

$$\text{HCl} \rightleftharpoons \text{H}^+ + \text{Cl}^- \qquad (7.8)$$

$$\text{H}^+ + \text{H}_2\text{O} \rightleftharpoons \text{H}_3\text{O}^+ \qquad (7.9)$$

$$\overline{\text{HCl} + \text{H}_2\text{O} \rightleftharpoons \text{H}_3\text{O}^+ + \text{Cl}^-} \qquad (7.10)$$

If we define $H^+$ as shorthand notation of $H_3O^+$ (since all protons in water occur as hydrated protons) and take into account that water does not appear in the law of mass action (Chapter 5, Section 5.3), equations (7.8) and (7.10) and the related laws of mass actions become equivalent:

$$K^*(HCl) = \frac{a(H^+)\,a(Cl^-)}{a(HCl)} = \frac{a(H_3O^+)\,a(Cl^-)}{a(HCl)} \tag{7.11}$$

Therefore, we are free to use equation (7.8) or equation (7.10) to express the reaction of the acid HCl in water. This is also true for all other acids. In this book, as a rule, we will use the simpler notation according to equation (7.8).

For the base $NH_3$, we can write the elementary reactions and the overall reaction as:

$$H_2O \rightleftharpoons H^+ + OH^- \tag{7.12}$$

$$NH_3 + H^+ \rightleftharpoons NH_4^+ \tag{7.13}$$

$$\overline{NH_3 + H_2O \rightleftharpoons NH_4^+ + OH^-} \tag{7.14}$$

In principle, we can use either equation (7.13) or equation (7.14) to express the protonation of the base in water. However, in contrast to the case of acids discussed before, the different formulations lead to different laws of mass action. For equation (7.13), which is the reverse reaction of the deprotonation of $NH_4^+$, we obtain:

$$K^*(NH_3) = \frac{1}{K^*(NH_4^+)} = \frac{a(NH_4^+)}{a(NH_3)\,a(H^+)} \tag{7.15}$$

whereas the law of mass action corresponding to equation (7.14) reads:

$$K^*(NH_3) = \frac{a(NH_4^+)\,a(OH^-)}{a(NH_3)} \tag{7.16}$$

In principle, both types of writing the base reaction are equivalent. It depends on the specific context which form is more advantageous.

The equilibrium constants that describe the deprotonation of acids and the protonation of bases are referred to as acidity and basicity constants, respectively. The definitions of these constants and the relationship between them will be discussed in more detail in Section 7.4.

It has to be noted that the definition of acids and bases is strictly based on the proton transfer only and is therefore independent of the charge of the species. Neutral species but also cations and anions can act as Brønsted acids, and the same is true for Brønsted bases as will be shown by the following examples:

$$H_2SO_4 \rightleftharpoons H^+ + HSO_4^- \tag{7.17}$$

$$NH_4^+ \rightleftharpoons H^+ + NH_3 \tag{7.18}$$

$$HCO_3^- \rightleftharpoons H^+ + CO_3^{2-} \tag{7.19}$$

$$\left[Al(H_2O)_6\right]^{3+} \rightleftharpoons H^+ + \left[Al(H_2O)_5OH\right]^{2+} \tag{7.20}$$

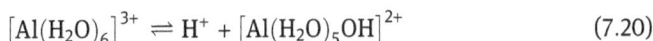

Moreover, certain species can accept and release protons. In particular, partly deprotonated species belong to this group. These species are formed during the incomplete dissociation of acids with more than one proton (multiprotic acids). Taking sulfuric acid as an example, two steps of dissociation can be written as follows:

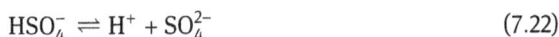

$$H_2SO_4 \rightleftharpoons H^+ + HSO_4^- \tag{7.21}$$

$$HSO_4^- \rightleftharpoons H^+ + SO_4^{2-} \tag{7.22}$$

In the first reaction, hydrogensulfate, $HSO_4^-$, is the base, whereas it acts as the acid in the second dissociation step. Such species, which can act as acid and base, are referred to as amphoteric species or ampholytes. Other examples are $H_2PO_4^-$, $HPO_4^{2-}$, or $HCO_3^-$.

Neutral species can also exhibit amphoteric character. As we have seen before, water ($H_2O$) can accept or release a proton (equation (7.5)). Another type of amphoteric compounds comprises organic substances with basic and acidic functional groups, such as amino acids. For instance, the amino acid glycine ($NH_2-CH_2-COOH$) occurs as zwitterion $NH_3^+-CH_2-COO^-$ that can accept or release protons depending on pH:

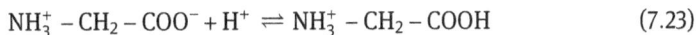

$$NH_3^+ - CH_2 - COO^- + H^+ \rightleftharpoons NH_3^+ - CH_2 - COOH \tag{7.23}$$

$$NH_3^+ - CH_2 - COO^- \rightleftharpoons NH_2 - CH_2 - COO^- + H^+ \tag{7.24}$$

## 7.3 Water as an Acid/Base System

Before starting the discussion on other acid/base equilibria, at first, water, the solvent in aquatic systems, will be considered from the viewpoint of acid/base equilibria. Liquid water dissociates to a small extent into protons and hydroxide ions according to equation (7.5). Using the shorthand notation $H^+$ instead of $H_3O^+$, we can write:

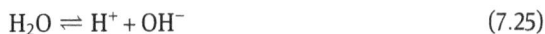

$$H_2O \rightleftharpoons H^+ + OH^- \tag{7.25}$$

According to the conventions mentioned in Chapter 5, the law of mass action is given by:

$$K_w^* = a(H^+)\,a(OH^-) \tag{7.26}$$

where $K_w^*$ is the dissociation (or autoprotolysis) constant of water which has a value of $1 \times 10^{-14}\ \text{mol}^2/\text{L}^2$ at 25 °C. The constant $K_w^*$, also referred to as the ion product of water, is frequently given in logarithmic form as:

$$pK_w^* = -\log K_w^* \tag{7.27}$$

The dissociation reaction described by equation (7.25) and the related constant constitute the basis for the definition of the conventional pH scale and its neutral point. If the activities (or simply the concentrations) of the protons and the hydroxide ions are described by the parameters pH and pOH according to:

$$pH = -\log a(H^+) \approx -\log c(H^+) \tag{7.28}$$

$$pOH = -\log a(OH^-) \approx -\log c(OH^-) \tag{7.29}$$

equation (7.26) can also be written in the form:

$$pK_w^* = pH + pOH \tag{7.30}$$

Given that $pK_w^*$ equals 14, it can be derived from equation (7.30) that both pH and pOH can cover a range from 0 to 14. Therefore, the conventional pH scale ranges from 0 to 14. At each pH, the related pOH is given as difference to 14. It has to be noted that the very complex problem of negative pH and pOH values is not considered here because it is irrelevant for aquatic systems.

Since neutrality is defined as that state where the positive charges equal the negative charges, for pure water, the condition:

$$a(H^+) = a(OH^-) = 1 \times 10^{-7}\ \text{mol/L} \tag{7.31}$$

can be derived, which is equivalent to:

$$pH = pOH = 7 \tag{7.32}$$

Equation (7.32) defines pH = 7 as the neutral point. However, it has to be noted that $K_w^*$ (or $pK_w^*$), like all other equilibrium constants, depends on the temperature. Therefore, $pK_w^* = 14$ is strictly valid only for 25 °C. At other temperatures, $pK_w^*$ has slightly different values. The temperature dependence of $pK_w^*$ is shown in Figure 7.1 and selected values are given in Table 7.1. It follows from Table 7.1 that, for instance, the $pK_w^*$ is 13.54 at 40 °C, and the neutral point is then pH = 6.77.

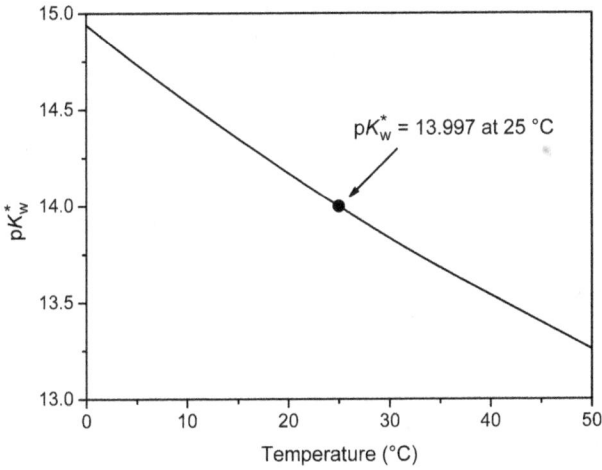

**Figure 7.1:** Temperature dependence of $pK_w^*$.

**Table 7.1:** Dissociation constant
of water at different temperatures.

| Temperature (°C) | $pK_w^*$ |
|---|---|
| 0 | 14.94 |
| 5 | 14.73 |
| 10 | 14.54 |
| 15 | 14.35 |
| 20 | 14.17 |
| 25 | 14.00 |
| 30 | 13.83 |
| 35 | 13.68 |
| 40 | 13.54 |
| 45 | 13.40 |
| 50 | 13.26 |

## 7.4 Protolysis of Acids and Bases

The protolysis of an acid HA (where A stands for an arbitrary anion) can be written as:

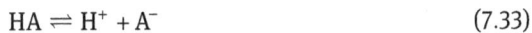

$$HA \rightleftharpoons H^+ + A^-$$ (7.33)

The respective law of mass action reads:

$$K_a^* = \frac{a(H^+)\, a(A^-)}{a(HA)}$$ (7.34)

where $K_a^*$ is the (thermodynamic) acidity constant. Introducing the conditional acidity constant, $K_a$, the law of mass action can be written by using concentrations instead of activities:

$$K_a = \frac{c(H^+)\, c(A^-)}{c(HA)} \tag{7.35}$$

Note that for dilute solutions, the difference between $K_a^*$ and $K_a$ diminishes. Since the acidity constants expand over broad ranges, a logarithmic notation (p notation) is frequently used according to the definition:

$$pK_a^* = -\log K_a^* \tag{7.36}$$

It follows from equation (7.36) that a high value of the acidity constant $K_a^*$ corresponds to a low value of $pK_a^*$ and vice versa. The acidity constant quantifies the degree of dissociation. The higher the value of $K_a^*$ is, the higher is the degree of dissociation or, in other words, the more the equilibrium described by equation (7.33) is shifted to the right and the more protons are released. Depending on the value of $K_a^*$, it can be distinguished between strong and weak acids. Strong acids are characterized by very high values of the acidity constant, $K_a^*$, or low values of $pK_a^*$. Consequently, if such acids are dissolved in water, the degree of dissociation is very high. Very strong acids, such as HCl or $H_2SO_4$, dissociate completely. On the contrary, weak acids with low values of $K_a^*$ (high values of $pK_a^*$) show only a partial dissociation during dissolution. As a consequence, the undissociated acid and the anion (the dissociation product) coexist in the solution.

In principle, the protonation of bases can be considered the reverse reaction of the acid dissociation given by equation (7.33). The definition of the basicity constant, however, is not based on this reverse reaction but on an overall reaction that additionally includes the dissociation of water (see Section 7.2):

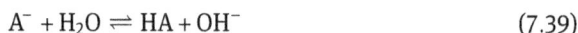

$$A^- + H^+ \rightleftharpoons HA \tag{7.37}$$

$$H_2O \rightleftharpoons H^+ + OH^- \tag{7.38}$$

$$\overline{A^- + H_2O \rightleftharpoons HA + OH^-} \tag{7.39}$$

The advantage of using equation (7.39) as the basis of the definition of the basicity constant is that it makes clear that the protolysis of bases is coupled with the formation of hydroxide ions.

To indicate that $A^-$ is the base, we will use the symbol B and write equation (7.39) in the alternative form:

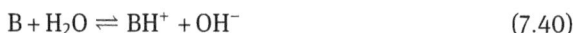

$$B + H_2O \rightleftharpoons BH^+ + OH^- \tag{7.40}$$

where $BH^+$ stands for the protonated base.

The law of mass action related to equation (7.40) reads:

$$K_b^* = \frac{a(BH^+)\, a(OH^-)}{a(B)}$$

(7.41)

or

$$K_b = \frac{c(BH^+)\, c(OH^-)}{c(B)}$$

(7.42)

where $K_b^*$ is the thermodynamic basicity constant and $K_b$ is the conditional basicity constant.

Analogous to the acidity constant, a logarithmic form of the basicity constant can be defined by:

$$pK_b^* = -\log K_b^*$$

(7.43)

A relationship between the acidity constant and the basicity constant of the conjugate base can be derived from equations (7.37) to (7.39). The constant related to the overall reaction (equation (7.39)) is $K_b^*$. It is given by the product of the constants of the elementary reactions (equations (7.37) and (7.38)), where the constant for the first reaction is $1/K_a^*$ (reverse reaction to the reaction that defines $K_a^*$ of the conjugate acid) and the constant for the second reaction is the dissociation constant of water, $K_w^*$. Hence, we can write for the overall reaction:

$$K_b^* = \frac{K_w^*}{K_a^*} = \frac{a(H^+)\, a(OH^-)\, a(HA)}{a(A^-)\, a(H^+)} = \frac{a(HA)\, a(OH^-)}{a(A^-)}$$

(7.44)

Rearranging equation (7.44) gives:

$$K_a^*\, K_b^* = K_w^*$$

(7.45)

which can be expressed in logarithmic form as:

$$pK_a^* + pK_b^* = pK_w^*$$

(7.46)

Table 7.2 lists acidity constants as $pK_a^*$ for 25 °C in the order of decreasing acid strength for selected acids frequently occurring in aqueous systems. The related $pK_b^*$ values for the conjugate bases can be found from equation (7.46). The formal acidity constant given for $H_2O$ is valid for the formulation of the law of mass action according to the definition of the acidity constant (equation (7.34)), that is:

$$K_a^*(H_2O) = \frac{a(H^+)\, a(OH^-)}{a(H_2O)}$$

(7.47)

The value of the constant $K_a^*$ is therefore different from that of the constant $K_w^*$, which includes the activity of water (equation (7.26)). If we substitute the activity of water by its concentration (approximately 55.56 mol/L, see Section 3.3.1 in Chapter 3), we can take this concentration as a conversion factor. Thus, the relationship between the formal acidity constant, $K_a^*$, and the dissociation constant of water, $K_w^*$, is given by:

$$K_a^* \ (H_2O) \times (55.56 \, mol/L) = K_w^* \tag{7.48}$$

It has to be further noted that the $pK_a^*$ for the carbonic acid given in Table 7.2 is valid for the sum of true carbonic acid, $H_2CO_3$, and dissolved $CO_2$ (see also Section 7.7).

The acidic character of the trivalent ions $Fe^{3+}$ and $Al^{3+}$ results from the protolysis of water molecules in the hydration shell according to:

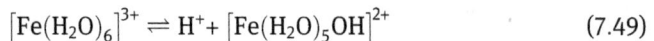

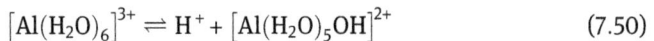

$$\left[Fe(H_2O)_6\right]^{3+} \rightleftharpoons H^+ + \left[Fe(H_2O)_5OH\right]^{2+} \tag{7.49}$$

$$\left[Al(H_2O)_6\right]^{3+} \rightleftharpoons H^+ + \left[Al(H_2O)_5OH\right]^{2+} \tag{7.50}$$

Equations (7.49) and (7.50) describe the first step of a series of protolysis reactions that can be alternatively considered a specific form of complex formation (see Chapter 11).

**Table 7.2:** $pK_a^*$ values of selected acids (25 °C).

| Acid | Formula | $pK_a^*$ |
|---|---|---|
| Hydrochloric acid | HCl | −6 |
| Sulfuric acid | $H_2SO_4$ | −3 |
| Nitric acid | $HNO_3$ | −1 |
| Hydrogensulfate ion | $HSO_4^-$ | 1.9 |
| Phosphoric acid | $H_3PO_4$ | 2.1 |
| Hexaaquairon(III) ion | $[Fe(H_2O)_6]^{3+}$ | 2.2 |
| Arsenic acid | $H_3AsO_4$ | 2.2 |
| Formic acid | HCOOH | 3.8 |
| Acetic acid | $CH_3COOH$ | 4.8 |
| Hexaaquaaluminum ion | $[Al(H_2O)_6]^{3+}$ | 4.9 |
| Carbonic acid | $CO_{2(aq)}$ | 6.4 |
| Dihydrogenarsenate ion | $H_2AsO_4^-$ | 7.0 |
| Hydrosulfuric acid | $H_2S$ | 7.1 |
| Dihydrogenphosphate ion | $H_2PO_4^-$ | 7.2 |
| Boric acid | $H_3BO_3$ | 9.3 |
| Ammonium ion | $NH_4^+$ | 9.3 |
| Silicic acid | $H_4SiO_4$ | 9.5 |
| Phenol | $C_6H_5OH$ | 9.9 |
| Hydrogencarbonate ion | $HCO_3^-$ | 10.3 |
| Hydrogenarsenate ion | $HAsO_4^{2-}$ | 11.5 |
| Hydrogenphosphate ion | $HPO_4^{2-}$ | 12.4 |
| Silicate ion | $H_3SiO_4^-$ | 12.6 |
| Water | $H_2O$ | 15.7 |
| Hydrogensulfide ion | $HS^-$ | 17.0 |

It has to be noted that Table 7.2 lists the thermodynamic constants. As mentioned previously, for ideal dilute solutions, the activities can be replaced by the concentrations and $pK_a^*$ equals $pK_a$. The same is true for the corresponding basicity constants that can be found by using equation (7.46). In the following sections of this chapter, as a rule, we will apply the assumption of an ideal dilute solution.

## 7.5 pH of Aqueous Solutions of Acids, Bases, and Salts

### 7.5.1 pH of Acid Solutions

To calculate the pH that arises after dissolution of an acid, different approaches are possible depending on the strength of the acid. In the most general case, a set of four equations is necessary: the law of mass action for the acid dissociation (equation (7.51)), the law of mass action for the dissociation of water (equation (7.52)), the material balance equation (equation (7.53)), and the charge balance equation (equation (7.54)):

$$K_a = \frac{c(H^+)\,c(A^-)}{c(HA)} \qquad (7.51)$$

$$K_w = c(H^+)\,c(OH^-) \qquad (7.52)$$

$$c_0(\text{acid}) = c(HA) + c(A^-) \qquad (7.53)$$

$$c(H^+) = c(A^-) + c(OH^-) \qquad (7.54)$$

The material balance equation results from the fact that the initially existing acid concentration $c_0$ must be equal to the sum of the concentrations of the remaining undissociated acid and the formed dissociation product that exist simultaneously in the solution after the equilibrium has been established. The charge balance equation is based on the electroneutrality condition that requires that the total concentration of the positive charges must equal the total concentration of the negative charges. It has to be noted that the concentration of protons on the left-hand side of equation (7.54) is the sum of the $H^+$ concentrations resulting from the dissociation of water and from the dissociation of the acid, whereas the $OH^-$ concentration on the right-hand side results only from the water dissociation.

Starting with equation (7.53) and substituting $c(HA)$, $c(A^-)$, and $c(OH^-)$ by the respective equilibrium relationships and the charge balance equation gives after some rearrangements:

$$c^3(H^+) + K_a\,c^2(H^+) - (K_a\,c_0(\text{acid}) + K_w)\,c(H^+) - K_w\,K_a = 0 \qquad (7.55)$$

The cubic equation has to be solved for $c(H^+)$ by an iteration procedure. After that, the pH can be found by using equation (7.28). However, the application of the cubic equation is only necessary if the concentration of $A^-$ is in the same order of magnitude as the concentration of $OH^-$ resulting from the water dissociation. This is typically the case when the acid is a very weak acid (low value of $K_a$) and the initial acid concentration is low.

If $c(A^-) \gg c(OH^-)$ (higher initial acid concentration and/or higher value of $K_a$), equation (7.54) simplifies to:

$$c(H^+) = c(A^-) \tag{7.56}$$

where only the dissociation products of the acid are considered and the $H^+$ and $OH^-$ concentrations from the water dissociation are neglected. Introducing equation (7.56) into the material balance gives:

$$c(HA) = c_0(\text{acid}) - c(H^+) \tag{7.57}$$

With equations (7.56) and (7.57), the law of mass action for the acid dissociation can be written as:

$$K_a = \frac{c^2(H^+)}{c_0(\text{acid}) - c(H^+)} \tag{7.58}$$

or

$$c^2(H^+) + K_a\, c(H^+) - K_a\, c_0(\text{acid}) = 0 \tag{7.59}$$

Applying the general formula for solving a quadratic equation to equation (7.59), we find:

$$c(H^+) = -\frac{K_a}{2} \pm \sqrt{\frac{K_a^2}{4} + K_a\, c_0(\text{acid})} \tag{7.60}$$

Of the two possible solutions, the negative one can be canceled because the concentration of the protons cannot be negative.

A further simplification can be made for weak acids (low degree of dissociation) where the condition:

$$c(H^+) \ll c_0(\text{acid}) \tag{7.61}$$

holds. In this case, the law of mass action (equation (7.58)) simplifies to:

$$K_a = \frac{c^2(H^+)}{c_0(\text{acid})} \tag{7.62}$$

and the proton concentration can be found by:

$$c(H^+) = \sqrt{K_a\, c_0(\text{acid})} \tag{7.63}$$

Alternatively, equation (7.63) can be written in logarithmic form as:

$$pH = \frac{1}{2}(pK_a - \log c_0(\text{acid}))\qquad(7.64)$$

The simplest version of pH calculation results for very strong (completely dissociated) acids. Here, the concentration of the protons is the same as the initial concentration of the acid:

$$c_0(\text{acid}) \approx c(A^-) \approx c(H^+) \qquad c(HA) \to 0 \qquad(7.65)$$

and the pH can be found from:

$$pH = -\log c_0(\text{acid})\qquad(7.66)$$

It has to be noted that at high concentrations (strong deviation from ideal conditions), the activity has to be used instead of the concentration according to the exact definition of pH (see equation (7.28)).

---

**Example 7.1**

What is the pH of an acetic acid solution ($pK_a = 4.76$) with $c_0 = 0.1$ mol/L?

**Solution:**

We start with the quadratic equation:

$$c(H^+) = -\frac{K_a}{2} \pm \sqrt{\frac{K_a^2}{4} + K_a\, c_0(\text{acid})}$$

With

$$K_a = 10^{-pK_a} = 1.74 \times 10^{-5}\ \text{mol/L}$$

we obtain:

$$c(H^+) = -8.7 \times 10^{-6}\ \text{mol/L} \pm \sqrt{7.57 \times 10^{-11}\ \text{mol}^2/L^2 + (1.74 \times 10^{-5}\ \text{mol/L})\,(0.1\,\text{mol/L})}$$

$$c(H^+) = 1.31 \times 10^{-3}\ \text{mol/L (negative result canceled)}$$

and finally, we find the pH to be:

$$pH = -\log c(H^+) = 2.88$$

In this case, the simplified equation:

$$pH = \frac{1}{2}(pK_a - \log c_0(\text{acid})) = \frac{1}{2}(4.76 + 1) = 2.88$$

gives the same result.

---

We can conclude from the example that the use of the quadratic equation is only necessary for stronger but not completely dissociated acids. Those are acids with $pK_a^*$ in the range of $4.5 > pK_a^* > -1.74$. The latter is the $pK_a^*$ of $H_3O^+$ that assigns the limit between strong and very strong (completely dissociated) acids.

## 7.5.2 pH of Base Solutions

The equations that can be used to calculate the pOH values of base solutions have the same mathematical form as the equations derived for the calculation of the pH values of acid solutions. The only difference is that $K_a$ has to be replaced by $K_b$, $c_0$(acid) by $c_0$(base), and $c(H^+)$ by $c(OH^-)$. Accordingly, the general solution equation is:

$$c^3(OH^-) + K_b\,c^2(OH^-) - (K_b\,c_0(\text{base}) + K_w)\,c(OH^-) - K_w\,K_b = 0 \qquad (7.67)$$

and the simplified quadratic solution for $c(HB^+) \gg c(H^+)$ is:

$$c(OH^-) = -\frac{K_b}{2} \pm \sqrt{\frac{K_b^2}{4} + K_b\,c_0(\text{base})} \qquad (7.68)$$

The equations for weak bases read:

$$c(OH^-) = \sqrt{K_b\,c_0(\text{base})} \qquad (7.69)$$

and

$$pOH = \frac{1}{2}(pK_b - \log c_0(\text{base})) \qquad (7.70)$$

and finally, the equation:

$$pOH = -\log c_0(\text{base}) \qquad (7.71)$$

can be used for very strong bases.

Equations (7.67)–(7.71) yield at first the hydroxide ion concentration or the pOH. The proton concentration and the related pH are available from equations (7.52) or (7.30).

## 7.5.3 pH of Salt Solutions

Independent of its true origin, each salt can be formally considered to be a product of the neutralization of an acid with a base. Ammonium chloride, for instance, can be considered to be the product of the reaction of the base ammonia with hydrochloric acid:

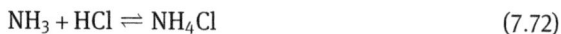

$$NH_3 + HCl \rightleftharpoons NH_4Cl \qquad (7.72)$$

As can be derived from equation (7.72), the cation of the salt originates from the base and the anion originates from the acid. Regarding the strength of the acid and the base, different combinations are possible: strong acid + strong base, strong acid + weak base, weak acid + strong base, and weak acid + weak base. The kind of combination determines the pH that will arise in the salt solution.

In aqueous solutions, the salts dissociate into cations and anions, for instance:

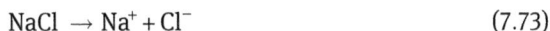

$$NaCl \rightarrow Na^+ + Cl^- \tag{7.73}$$

If the cation originates from a strong base and the anion from a strong acid as in the given example (strong base NaOH and strong acid HCl), no reaction with water will occur because both strong acid and strong base are completely protolyzed. This means that in our example neither the $Na^+$ ions react with $OH^-$ ions from water to restitute NaOH nor the $Cl^-$ ions react with $H^+$ ions from water to restitute HCl. Consequently, the solution remains neutral.

A shift in the pH will occur when one partner in the neutralization is weak and the corresponding ion reacts with $H^+$ or $OH^-$ originating from water dissociation, under partial restitution of the conjugate acid or base compound, whereas the ion from the strong partner remains unchanged. In this case, either $H^+$ or $OH^-$ ions are consumed which results in a change of pH.

Let us take $NH_4Cl$ (salt of a strong acid and a weak base) as an example. In aqueous solution we have:

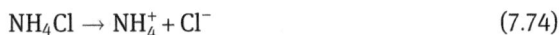

$$NH_4Cl \rightarrow NH_4^+ + Cl^- \tag{7.74}$$

The cation $NH_4^+$ (from the weak base $NH_3$) partially reacts with $OH^-$ to restitute $NH_3$:

$$NH_4^+ + OH^- \rightleftharpoons NH_3 + H_2O \tag{7.75}$$

whereas $Cl^-$, the anion of the strong acid HCl shows no reaction with $H^+$. Therefore, only the $OH^-$ concentration decreases (due to the consumption by $NH_4^+$) with the consequence that the pH of the solution decreases. The overall reaction that determines the pH of the salt solution can be derived from a combination of equation (7.75) with the water dissociation:

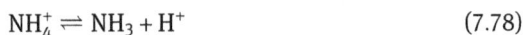

$$H_2O \rightleftharpoons H^+ + OH^- \tag{7.76}$$

$$NH_4^+ + OH^- \rightleftharpoons NH_3 + H_2O \tag{7.77}$$

$$\overline{\phantom{xxxxxxxxxxxxxxxxxxxxx}}$$

$$NH_4^+ \rightleftharpoons NH_3 + H^+ \tag{7.78}$$

Thus, we can generalize that a solution of a salt, which consists of the cation of a weak base and the anion of a strong acid, reacts acidic.

Next, we will consider a salt consisting of the cation of a strong base and the anion of a weak acid, for instance $Na_2CO_3$. Here, the anion reacts with protons from water, whereas the cation (from the strong base) remains unchanged:

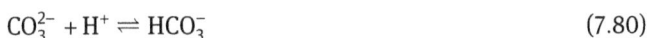

$$Na_2CO_3 \rightarrow 2\,Na^+ + CO_3^{2-} \tag{7.79}$$

$$CO_3^{2-} + H^+ \rightleftharpoons HCO_3^- \tag{7.80}$$

$$HCO_3^- + H^+ \rightleftharpoons (H_2CO_3) \rightleftharpoons H_2O + CO_{2(aq)} \tag{7.81}$$

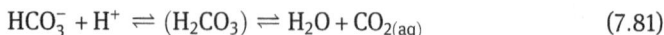

As a consequence, there is a consumption of $H^+$ leading to an increase of pH. It has to be noted that the contribution of the last reaction to the proton consumption is small (see also Section 7.7). Thus, the overall reaction can be written in a simplified form as:

$$H_2O \rightleftharpoons H^+ + OH^- \tag{7.82}$$

$$CO_3^{2-} + H^+ \rightleftharpoons HCO_3^- \tag{7.83}$$

$$\overline{CO_3^{2-} + H_2O \rightleftharpoons HCO_3^- + OH^-} \tag{7.84}$$

Generally, a solution of a salt formed by a weak acid and a strong base reacts basic.

As can be seen from the previous discussion, the pH-determining reactions are identical with the protolysis of an acid (equation (7.78)) or of a base (equation (7.84)). Consequently, the calculation of the pH that will arise in the salt solution does not differ from the calculation of the pH of an acid or base solution as shown in Sections 7.5.1 and 7.5.2, respectively. After identification of the pH-determining reaction, one of the equations for weak acids or bases given in these sections can be used to find the resulting pH. The initial concentration of the acid (cation) or the base (anion) can be derived from the initial concentration of the salt under the assumption that the salt is completely dissociated.

---

**Example 7.2**
What is the pH of an $NH_4Cl$ solution with $c_0 = 0.1$ mol/L? $pK_a$ ($NH_4^+$) = 9.25.

**Solution:**
Ammonium chloride dissociates during dissolution according to:

$$NH_4Cl \rightarrow NH_4^+ + Cl^-$$

Since $NH_4^+$ comes from a weak base ($NH_3$) and $Cl^-$ from a strong acid (HCl), the pH-determining reaction is the protolysis of the cation $NH_4^+$ that reacts as an acid according to:

$$NH_4^+ \rightleftharpoons NH_3 + H^+$$

The relatively high $pK_a$ allows calculating the pH by the simplified equation (7.64) (see also Example 7.1):

$$pH = \frac{1}{2}(pK_a - \log c_0(acid))$$

Given that the salt is completely dissociated, the initial concentration of the acid $NH_4^+$ is the same as the salt concentration (0.1 mol/L). Consequently, the pH is:

$$pH = \frac{1}{2}(9.25 + 1) = 5.13$$

If a salt is formed from a weak acid and a weak base, both protons and hydroxide ions are consumed by the partial restitution of the related acid and base, respectively. Therefore, we can expect that the total pH change is smaller than in the cases strong acid + weak base or weak acid + strong base due to the compensation of the opposite effects. It depends on the relative strengths of the partners if slightly more protons or slightly more OH⁻ ions are consumed and in which direction the pH is shifted. Let us take ammonium formiate $(HCOONH_4)$ as an example. After dissolution of the salt and dissociation into $NH_4^+$ and $HCOO^-$, we have to consider the following reactions:

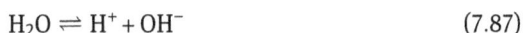

$$NH_4^+ \rightleftharpoons NH_3 + H^+ \tag{7.85}$$

$$HCOO^- + H_2O \rightleftharpoons HCOOH + OH^- \tag{7.86}$$

$$H_2O \rightleftharpoons H^+ + OH^- \tag{7.87}$$

The respective laws of mass action are:

$$K_a = \frac{c(NH_3)\,c(H^+)}{c(NH_4^+)} \tag{7.88}$$

$$K_b = \frac{c(HCOOH)\,c(OH^-)}{c(HCOO^-)} \tag{7.89}$$

$$K_w = c(H^+)\,c(OH^-) \tag{7.90}$$

If we consider the fact that only portions of the cations and anions are transformed to the corresponding base or acid and residuals of the ions remain unchanged, the material balances have to be written as:

$$c_0(NH_4^+) = c(NH_4^+) + c(NH_3) \tag{7.91}$$

and

$$c_0(HCOO^-) = c(HCOO^-) + c(HCOOH) \tag{7.92}$$

where

$$c_0(NH_4^+) = c_0(HCOO^-) = c_0(salt) \tag{7.93}$$

Finally, the charge balance reads:

$$c(NH_4^+) + c(H^+) = c(HCOO^-) + c(OH^-) \tag{7.94}$$

Substituting $c(NH_3)$ and $c(HCOOH)$ in equations (7.91) and (7.92) by the respective laws of mass action and inserting the resulting expressions into the charge balance gives:

$$\frac{c_0(\text{salt})}{1 + \dfrac{K_a}{c(H^+)}} + c(H^+) = \frac{c_0(\text{salt})}{1 + \dfrac{K_b}{c(OH^-)}} + c(OH^-) \tag{7.95}$$

Rearranging under consideration of equation (7.90) leads to:

$$\frac{c_0(\text{salt})\, c(H^+)}{c(H^+) + K_a} + c(H^+) = \frac{c_0(\text{salt})\, \dfrac{K_w}{c(H^+)}}{\dfrac{K_w}{c(H^+)} + K_b} + \frac{K_w}{c(H^+)} \tag{7.96}$$

This equation has to be solved by an iterative method.

Frequently, the second terms on both sides are considerably smaller than the first terms and can therefore be neglected. This is particularly true when the salt concentration is not too low and/or the pH difference to 7 ($c(H^+) = c(OH^-) = 1 \times 10^{-7}$ mol/L) is small. The latter can be expected if the strength of the acid and the base are only slightly different. This simplification leads, after some rearrangements, to the equation:

$$c(H^+) \approx \sqrt{\frac{K_a\, K_w}{K_b}} \tag{7.97}$$

Under this condition, the pH is no longer influenced by the salt concentration.

From equation (7.97), the following conclusions can be drawn. If $K_a = K_b$, the opposite protolysis effects compensate for each other and the pH remains constant ($c(H^+) = \sqrt{K_w} = 1 \times 10^{-7}$ mol/L, pH = 7). If $K_a > K_b$, then the solution will become acidic with $c(H^+) > \sqrt{K_w}$, $c(H^+) > 1 \times 10^{-7}$ mol/L, pH < 7. On the contrary, if $K_b > K_a$, then the solution becomes alkaline with $c(H^+) < \sqrt{K_w}$, $c(H^+) < 1 \times 10^{-7}$ mol/L, pH > 7.

---

**Example 7.3**

Calculate the pH of an ammonium formiate ($HCOONH_4$) solution with $c_0 = 0.1$ mol/L. $pK_a(NH_4^+) = 9.25$, $pK_a(HCOOH) = 3.77$, $pK_w = 14$.

**Solution:**

The pH is determined by both protolysis reactions:

$$NH_4^+ \rightleftharpoons H^+ + NH_3$$

and

$$HCOO^- + H_2O \rightleftharpoons HCOOH + OH^-$$

For the second reaction, we need the $pK_b$. We can compute it from the $pK_a$ by:

$$pK_b = pK_w - pK_a = 14 - 3.77 = 10.23$$

Since the $pK_b$ of $HCOO^-$ is greater than the $pK_a$ of $NH_4^+$ ($K_b < K_a$), we expect the solution to be acidic. First, we want to apply the simplified equation (7.97). Later, we can take the result to prove the validity of the simplifying assumptions by using the exact equation (7.96).

The required constants are:

$$NH_4^+ : K_a = 10^{-pK_a} \text{ mol/L} = 5.62 \times 10^{-10} \text{ mol/L}$$

$$HCOO^- : K_b = 10^{-pK_b} \text{ mol/L} = 5.89 \times 10^{-11} \text{ mol/L}$$

$$H_2O : K_w = 10^{-pK_w} \text{ mol}^2/L^2 = 1 \times 10^{-14} \text{ mol}^2/L^2$$

Introducing the constants into equation (7.97) gives:

$$c(H^+) = \sqrt{\frac{K_a \, K_w}{K_b}} = \sqrt{\frac{(5.62 \times 10^{-10} \text{ mol/L}) \, (1 \times 10^{-14} \text{ mol}^2/L^2)}{5.89 \times 10^{-11} \text{ mol/L}}}$$

$$c(H^+) = \sqrt{9.54 \times 10^{-14} \text{ mol}^2/L^2} = 3.09 \times 10^{-7} \text{ mol/L}$$

$$pH = -\log(3.09 \times 10^{-7}) = 6.51$$

As expected, we find for this salt, originating from a weak base and a weak acid, only a small shift of pH from 7 to 6.51 into the acidic range. To verify the approximate solution, we can now substitute the found proton concentration and the constants into equation (7.96):

$$\frac{c_0(\text{salt}) \, c(H^+)}{c(H^+) + K_a} + c(H^+) = \frac{c_0(\text{salt}) \dfrac{K_w}{c(H^+)}}{\dfrac{K_w}{c(H^+)} + K_b} + \frac{K_w}{c(H^+)}$$

$$\frac{(0.1 \text{ mol/L}) \, (3.09 \times 10^{-7} \text{ mol/L})}{3.09 \times 10^{-7} \text{ mol/L} + 5.62 \times 10^{-10} \text{ mol/L}} + 3.09 \times 10^{-7} \text{ mol/L}$$

$$= \frac{0.1 \text{ mol/L} \dfrac{1 \times 10^{-14} \text{ mol}^2/L^2}{3.09 \times 10^{-7} \text{ mol/L}}}{\dfrac{1 \times 10^{-14} \text{ mol}^2/L^2}{3.09 \times 10^{-7} \text{ mol/L}} + 5.89 \times 10^{-11} \text{ mol/L}} + \frac{1 \times 10^{-14} \text{ mol}^2/L^2}{3.09 \times 10^{-7} \text{ mol/L}}$$

$$\frac{3.09 \times 10^{-8} \text{ mol}^2/L^2}{3.096 \times 10^{-7} \text{ mol/L}} + 3.09 \times 10^{-7} \text{ mol/L} = \frac{3.24 \times 10^{-9} \text{ mol}^2/L^2}{3.24 \times 10^{-8} \text{ mol/L}} + 3.24 \times 10^{-8} \text{ mol/L}$$

$$0.1 \text{ mol/L} + 3.09 \times 10^{-7} \text{ mol/L} = 0.1 \text{ mol/L} + 3.24 \times 10^{-8} \text{ mol/L}$$

We can see that, indeed, the second terms of both sides are negligible in comparison to the first terms which was the assumption made for deriving the simplified equation (7.97). Accordingly, the small difference between the second terms of both sides has no impact on the total balance.

---

## 7.5.4 Buffer Systems

A buffer system is a specific acid/base system that is able to keep the pH value of the aqueous solution nearly constant. A typical buffer system consists of the components of a conjugate acid/base pair in comparable concentrations. The buffer effect is based on the ability of the buffer base and the buffer acid to bind protons or hydroxide ions originating from strong acids or bases (e.g., HCl or NaOH) introduced into the buffered solution from outside. The pH that is kept approximately constant despite the introduction of $H^+$ or $OH^-$ depends on the acidity constant of the buffer acid and the concentration ratio of the buffer base and the buffer acid and is given by:

$$pH = pK_a + \log \frac{c(\text{buffer base})}{c(\text{buffer acid})} \tag{7.98}$$

For the buffer system HAc/Ac⁻ (HAc stands here for the acetic acid, $CH_3COOH$, and Ac⁻ for the acetate anion, $CH_3COO^-$), the buffer reactions can be described as follows:

Addition of a strong acid (addition of $H^+$):

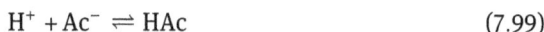

$$H^+ + Ac^- \rightleftharpoons HAc \tag{7.99}$$

Addition of a strong base (addition of $OH^-$):

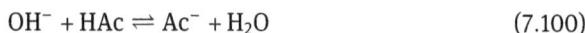

$$OH^- + HAc \rightleftharpoons Ac^- + H_2O \tag{7.100}$$

If a strong acid is added to the buffered solution, the buffer base (Ac⁻) binds the introduced protons under formation of the buffer acid (HAc). On the contrary, if a strong base is added to the buffered solution, the buffer acid (HAc) binds the introduced hydroxide ions under formation of the buffer base (Ac⁻). The higher the concentration of the buffer system is, the more protons or hydroxide ions can be bound, and the higher the buffer capacity.

Considering the conversion reactions and the related changes in the concentrations of the buffer acid and buffer base, the change in pH can be calculated by:

$$pH = pK_a + \log \frac{c(\text{buffer base}) - c(H^+)}{c(\text{buffer acid}) + c(H^+)} \tag{7.101}$$

$$pH = pK_a + \log \frac{c(\text{buffer base}) + c(OH^-)}{c(\text{buffer acid}) - c(OH^-)} \tag{7.102}$$

where $c(H^+)$ and $c(OH^-)$ are the concentrations of the protons and hydroxide ions from the added strong acid and base, respectively.

---

**Example 7.4**
What is the pH value in a buffer system consisting of 0.4 mol/L HAc ($pK_a$ = 4.76) and 0.4 mol/L NaAc? How does the pH change after addition of 0.1 mol/L HCl and what would be the pH of a 0.1 molar solution of HCl without any buffer?

**Solution:**
The pH of the buffer system can be calculated from equation (7.98):

$$pH = pK_a + \log \frac{c(Ac^-)}{c(HAc)} = 4.76 + \log \frac{0.4 \text{ mol/L}}{0.4 \text{ mol/L}} = 4.76$$

Since HCl is a strong (completely dissociated) acid, the proton concentration is $c(H^+) = c(HCl) = 0.1$ mol/L. The pH after addition of 0.1 mol/L HCl is:

$$pH = pK_a + \log\frac{c(Ac^-) - c(H^+)}{c(HAc) + c(H^+)} = 4.76 + \log\frac{0.4\,\text{mol/L} - 0.1\,\text{mol/L}}{0.4\,\text{mol/L} + 0.1\,\text{mol/L}}$$

$$pH = 4.76 + \log\frac{0.3}{0.5} = 4.76 - 0.22 = 4.54$$

Despite the addition of a relative high concentration of a very strong acid, there is only a small decrease in pH. For comparison, the pH of an unbuffered HCl solution with $c = 0.1$ mol/L would be:

$$pH = -\log c(H^+) = -\log(c_0(HCl)) = -\log(0.1) = 1$$

according to equation (7.66).

---

Amphoteric anions can also act as buffers because they are able to bind protons as well as hydroxide ions. The most relevant amphoteric buffer ion in natural waters is hydrogencarbonate. Amongst all potential buffers in natural waters, it typically occurs in the highest concentration and therefore determines the buffer capacity of the water. Waters with high hydrogencarbonate concentrations are therefore also referred to as well-buffered waters. The respective buffer reactions are:

$$H^+ + HCO_3^- \rightleftharpoons CO_2 + H_2O \tag{7.103}$$

$$OH^- + HCO_3^- \rightleftharpoons CO_3^{2-} + H_2O \tag{7.104}$$

The carbonic acid system is discussed in more detail in Section 7.7.

## 7.6  Degree of Protolysis and Acid/Base Speciation

### 7.6.1 Monoprotic Acids

Monoprotic acids are acids that can donate only one proton. The degree of protolysis of a monoprotic acid, $\alpha$, describes the ratio of the concentration of the dissociation product and the total initial concentration of the acid, $c_0(acid)$. It follows from the material balance that the latter must be equal to the sum of the concentrations of the residual undissociated acid, $c(HA)$, and the concentration of the formed anion, $c(A^-)$, after establishment of the equilibrium:

$$\alpha = \frac{c(A^-)}{c_0(acid)} = \frac{c(A^-)}{c(HA) + c(A^-)} \tag{7.105}$$

The degree of protolysis depends on the pH as can be derived by combining the definition equation given above with the law of mass action. It follows from equation (7.35) that:

$$\frac{c(HA)}{c(A^-)} = \frac{c(H^+)}{K_a} \tag{7.106}$$

After introducing pH and p$K_a$, we can write equation (7.106) in the form:

$$\log \frac{c(HA)}{c(A^-)} = pK_a - pH \tag{7.107}$$

or

$$\frac{c(HA)}{c(A^-)} = 10^{pK_a - pH} \tag{7.108}$$

Combining equations (7.105) and (7.108) gives:

$$\frac{1}{\alpha} = \frac{c(HA)}{c(A^-)} + 1 = 10^{pK_a - pH} + 1 \tag{7.109}$$

and

$$\alpha = \frac{1}{10^{pK_a - pH} + 1} \tag{7.110}$$

For the considered acid HA, $\alpha$ represents the fraction of the anion, $f(A^-)$, which is the conjugate base of HA and, consequently, $1-\alpha$ represents the fraction of the un-dissociated acid, $f(HA)$. More generally speaking, $\alpha$ is the fraction of the base and $1-\alpha$ is the fraction of the conjugate acid:

$$f(\text{acid}) = 1 - \alpha \tag{7.111}$$

$$f(\text{base}) = \alpha \tag{7.112}$$

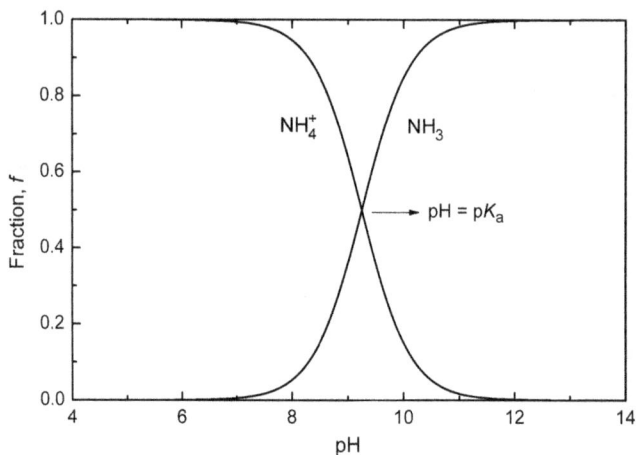

**Figure 7.2:** Speciation in the system $NH_3/NH_4^+$ (25 °C).

Consequently, equation (7.110) can be used to find the species distribution of an acid/base pair as a function of pH, also referred to as acid/base speciation. It has to be noted that, although derived for the neutral acid HA, these relationships hold independently of the charge of the monoprotic acid and its conjugate base. An acid/base speciation example is shown in Figure 7.2 for the ammonium ion ($pK_a = 9.25$ at 25 °C) and its conjugate base ammonia.

---

**Example 7.5**

What is the fraction of phenolate in the system phenol/phenolate at pH = 8? $pK_a$ (phenol) = 9.9.

**Solution:**

Phenol is a weak monoprotic acid that dissociates according to:

$$C_6H_5OH \rightleftharpoons H^+ + C_6H_5O^-$$

where phenolate is the conjugate base. The degree of protolysis at pH = 8 can be found from:

$$\alpha = \frac{1}{10^{pK_a - pH} + 1} = \frac{1}{10^{9.9-8} + 1} = \frac{1}{10^{1.9} + 1} = 0.012$$

According to equation (7.112), the degree of protolysis equals the fraction of the base. Therefore, the result is:

$$f(\text{phenolate}) = \alpha = 0.012$$

---

To get a quick insight into the speciation of a given acid/base system, it is helpful to consider characteristic pH values in relation to the $pK_a$. For instance, it follows from equation (7.110) that for pH = $pK_a$, the degree of protolysis is 0.5. Table 7.3 summarizes the degrees of protolysis and the related speciation for five selected pH values. It can be derived from Table 7.3 that at pH < $pK_a$ − 2, the deprotonation is negligible and the acid is the only relevant species in the acid/base system. By contrast, at pH > $pK_a$ +2, the deprotonation is nearly complete and the base dominates the composition of the acid/base system.

**Table 7.3:** Degrees of protolysis and acid/base speciation at selected pH values.

| pH | Degree of protolysis, $\alpha$ | $f$ (acid) | $f$ (base) | Concentration ratio $c$(acid) : $c$(base) |
|---|---|---|---|---|
| pH = $pK_a$ − 2 | 0.01 | 0.99 | 0.01 | 100:1 |
| pH = $pK_a$ − 1 | 0.09 | 0.91 | 0.09 | 10:1 |
| pH = $pK_a$ | 0.5 | 0.5 | 0.5 | 1:1 |
| pH = $pK_a$ + 1 | 0.91 | 0.09 | 0.91 | 1:10 |
| pH = $pK_a$ + 2 | 0.99 | 0.01 | 0.99 | 1:100 |

## 7.6.2 Polyprotic Acids

The calculation of the speciation of polyprotic acids (acids with more than one proton) as a function of pH is a bit more complicated than that of monoprotic acids. A general way to find the species distribution as a function of pH is to start with a material balance of the acid and the dissociation products and to subsequently substitute the unknown concentrations by the equilibrium relationships in such a manner that only one unknown concentration is left. This concentration is then related to the total concentration to find the fraction, $f$, of the considered species in the system. The fractions of the other species can be found in the same manner. Generally, the fractions, $f$, of all species only depend on pH and on the acidity constants for the different protolysis steps.

The elementary dissociation steps and the overall reaction equation of a diprotic acid $H_2A$ as well as the related laws of mass action are:

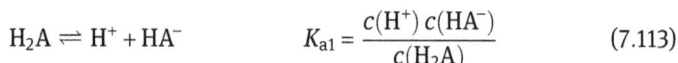

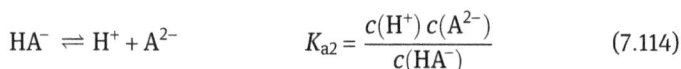

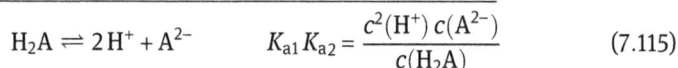

$$H_2A \rightleftharpoons H^+ + HA^- \qquad K_{a1} = \frac{c(H^+)\,c(HA^-)}{c(H_2A)} \qquad (7.113)$$

$$HA^- \rightleftharpoons H^+ + A^{2-} \qquad K_{a2} = \frac{c(H^+)\,c(A^{2-})}{c(HA^-)} \qquad (7.114)$$

$$H_2A \rightleftharpoons 2H^+ + A^{2-} \qquad K_{a1}K_{a2} = \frac{c^2(H^+)\,c(A^{2-})}{c(H_2A)} \qquad (7.115)$$

The material balance equation reads:

$$c_{total} = c(H_2A) + c(HA^-) + c(A^{2-}) \qquad (7.116)$$

where $c_{total}$ is the total concentration of all species in the acid/base system.

To find the fraction of the undissociated acid $H_2A$, at first, the concentrations of $HA^-$ and $A^{2-}$ are substituted by means of the equations (7.113) and (7.115):

$$c_{total} = c(H_2A) + \frac{c(H_2A)\,K_{a1}}{c(H^+)} + \frac{c(H_2A)\,K_{a1}\,K_{a2}}{c^2(H^+)} \qquad (7.117)$$

After factoring out $c(H_2A)$ and rearranging the equation, an expression for the fraction $f(H_2A)$ can be found:

$$c_{total} = c(H_2A)\left[1 + \frac{K_{a1}}{c(H^+)} + \frac{K_{a1}\,K_{a2}}{c^2(H^+)}\right] \qquad (7.118)$$

$$f(H_2A) = \frac{c(H_2A)}{c_{total}} = \frac{1}{1 + \dfrac{K_{a1}}{c(H^+)} + \dfrac{K_{a1}\,K_{a2}}{c^2(H^+)}} \qquad (7.119)$$

The fractions $f(HA^-)$ and $f(A^{2-})$ can be derived in an analogous manner:

$$f(HA^-) = \frac{c(HA^-)}{c_{total}} = \frac{1}{\dfrac{c(H^+)}{K_{a1}} + 1 + \dfrac{K_{a2}}{c(H^+)}} \qquad (7.120)$$

$$f(A^{2-}) = \frac{c(A^{2-})}{c_{total}} = \frac{1}{\dfrac{c^2(H^+)}{K_{a1}\,K_{a2}} + \dfrac{c(H^+)}{K_{a2}} + 1} \qquad (7.121)$$

In the same way, the speciation of acids with more than two protons can also be found. The speciation of a triprotic acid will be shown here by considering the example of phosphoric acid that dissociates in three steps according to:

$$H_3PO_4 \rightleftharpoons H^+ + H_2PO_4^- \qquad pK_a = 2.1 \qquad (7.122)$$

$$H_2PO_4^- \rightleftharpoons H^+ + HPO_4^{2-} \qquad pK_a = 7.2 \qquad (7.123)$$

$$HPO_4^{2-} \rightleftharpoons H^+ + PO_4^{3-} \qquad pK_a = 12.4 \qquad (7.124)$$

The given $pK_a$ values are valid for 25 °C. The distribution of the phosphoric acid species as a function of pH is depicted in Figure 7.3. It can be seen that in the medium pH range, dihydrogenphosphate and hydrogenphosphate are the dominant species, whereas phosphate, $PO_4^{3-}$, dominates only at very high pH values. Nevertheless, phosphate concentrations in natural waters are often reported as $c(PO_4^{3-})$ or $\rho^*(PO_4^{3-})$.

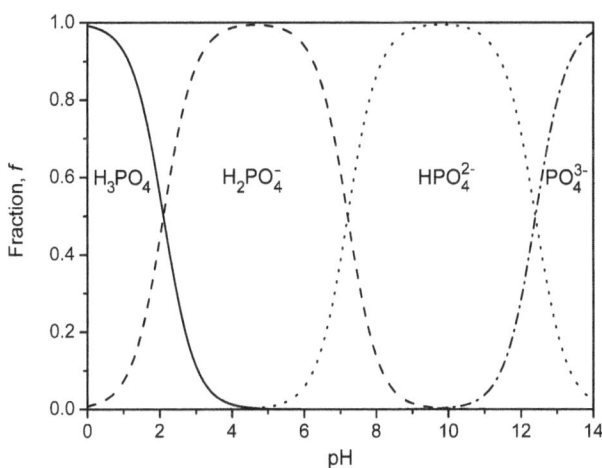

**Figure 7.3:** Speciation of phosphoric acid (25 °C).

## 7.7 Carbonic Acid

### 7.7.1 Relevance

Carbonic acid and its dissociation products play an important role in aquatic systems. Carbonic acid is mostly dissolved carbon dioxide, and only a very small part occurs as $H_2CO_3$. However, $[CO_2 + H_2O]$ reacts in the same way as $H_2CO_3$ and therefore the name carbonic acid is also used for dissolved carbon dioxide. Carbon dioxide is an essential part of the biogeochemical carbon cycle. Biomass is produced from carbon dioxide and water by algae and plants through photosynthesis. On the other hand, degradation (mineralization) of biomass leads to carbon dioxide. Furthermore, the gas carbon dioxide is a constituent of atmospheric air and is introduced into aquatic systems according to Henry's law (Chapter 6). Its solubility is further enhanced by the subsequent protolysis of the dissolved $CO_2$. $CO_{2(aq)}$ is also related to the dissolution/precipitation of the frequently occurring mineral calcite ($CaCO_3$) within the calco–carbonic equilibrium (Chapter 9). The first dissociation product of carbonic acid, hydrogencarbonate (bicarbonate), belongs to the major ions in natural waters. It is often the anion with the highest concentration. Carbonic acid and its dissociation products act as buffer in aquatic systems. As demonstrated in Section 7.5.4, hydrogencarbonate can bind protons (under formation of carbon dioxide) and hydroxide ions (under formation of carbonate) and thus buffers against strong acids and bases.

Due to the relevance of the carbonic acid system for natural waters, it is worthwhile to discuss it in more detail.

### 7.7.2 Speciation of Carbonic Acid

The dissociation of the carbonic acid proceeds according to:

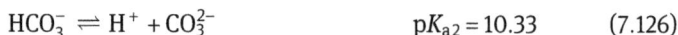

$$CO_{2(aq)} + H_2O \rightleftharpoons (H_2CO_3) \rightleftharpoons H^+ + HCO_3^- \qquad pK_{a1} = 6.36 \qquad (7.125)$$

$$HCO_3^- \rightleftharpoons H^+ + CO_3^{2-} \qquad pK_{a2} = 10.33 \qquad (7.126)$$

The given $pK_a$ values are valid for 25 °C.

For the sake of simplicity, the index (aq) is omitted in the following sections. Accordingly, if not otherwise stated, $c(CO_2)$ means concentration of dissolved $CO_2$. It has to be noted that $K_{a1}$, in a rigorous sense, is the acidity constant of the analytical sum of $H_2CO_3$ and $CO_{2(aq)}$. However, the fraction of $H_2CO_3$ is very small (about 0.3% at 25 °C) and its contribution can therefore be neglected.

The speciation can be calculated following the scheme given in Section 7.6.2. The material balance reads:

$$c_{total} = c(DIC) = c(CO_2) + c(HCO_3^-) + c(CO_3^{2-}) \qquad (7.127)$$

where the total concentration of the carbonic acid species is expressed by the frequently used term dissolved inorganic carbon (DIC).

According to equations (7.119)–(7.121), the fractions of the carbonic acid species are given by:

$$f(CO_2) = \frac{c(CO_2)}{c(DIC)} = \frac{1}{1 + \dfrac{K_{a1}}{c(H^+)} + \dfrac{K_{a1}\, K_{a2}}{c^2(H^+)}} \qquad (7.128)$$

$$f(HCO_3^-) = \frac{c(HCO_3^-)}{c(DIC)} = \frac{1}{\dfrac{c(H^+)}{K_{a1}} + 1 + \dfrac{K_{a2}}{c(H^+)}} \qquad (7.129)$$

$$f(CO_3^{2-}) = \frac{c(CO_3^{2-})}{c(DIC)} = \frac{1}{\dfrac{c^2(H^+)}{K_{a1}\, K_{a2}} + \dfrac{c(H^+)}{K_{a2}} + 1} \qquad (7.130)$$

The resulting speciation is shown in Figure 7.4.

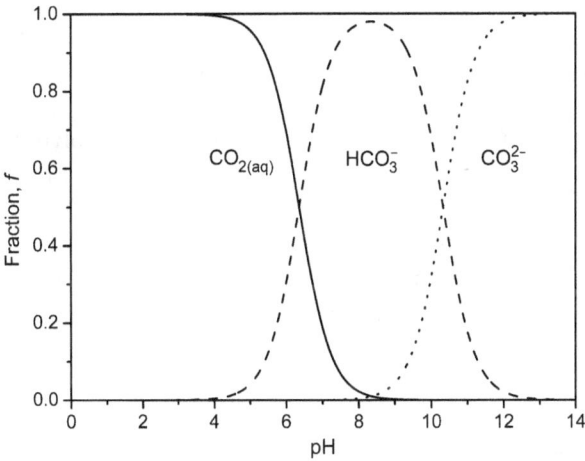

**Figure 7.4:** Speciation of carbonic acid (25 °C).

From the diagram in Figure 7.4, one could receive the impression that the distribution curves end at certain pH values where $f = 0$. Actually, the curves are asymptotic. This can be seen from a diagram that uses a logarithmic ordinate (Figure 7.5). Obviously, dissolved $CO_2$ also exists at relatively high pH values and carbonate also occurs at relatively low pH values. However, their fractions are very low under these conditions.

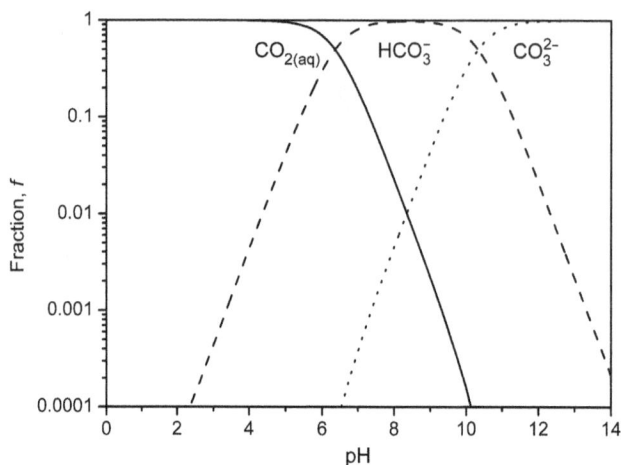

**Figure 7.5:** Speciation of carbonic acid in logarithmic representation (25 °C).

### 7.7.3 Determination of the Carbonic Acid Species by Acid/Base Titrations

Equations (7.128)–(7.130) can be used to calculate the fractions of the carbonic acid species but they give no information on the absolute concentrations. The absolute concentrations can be calculated from these equations only if the concentration of the total inorganic carbon is available from a separate measurement. In principle, a direct analytical determination of DIC is possible, but requires expensive analytical equipment. Therefore, in practice, an alternative method based on simple acid/base titrations is used to determine the concentrations of the carbonic acid species. In this analytical method, the acid or base consumption until a defined pH endpoint is determined and related to the concentrations of the carbonic acid species in the water sample. Solutions of hydrochloric acid and sodium hydroxide are typically used to carry out the acid/base titrations. It has to be noted that a direct relationship between the carbonic acid species concentrations and the results of the titration only exists if no other acid/base systems (e.g., phosphoric acid species or $NH_4^+/NH_3$) contribute significantly to the proton or hydroxide ion consumption. This condition is fulfilled in most cases because the concentrations of the potentially disturbing water constituents are usually significantly lower than the concentrations of the carbonic acid species. Such waters are also referred to as carbonate-dominated waters.

To illustrate the following discussion, the carbonic acid species distribution is shown once again in Figure 7.6, but in this case not for their fractions but for their absolute concentrations. The curves have been calculated for two total concentrations ($1 \times 10^{-3}$ mol/L and $1 \times 10^{-2}$ mol/L), which covers the practically relevant range. Furthermore, the respective $H^+$ and $OH^-$ concentrations are shown in the diagram.

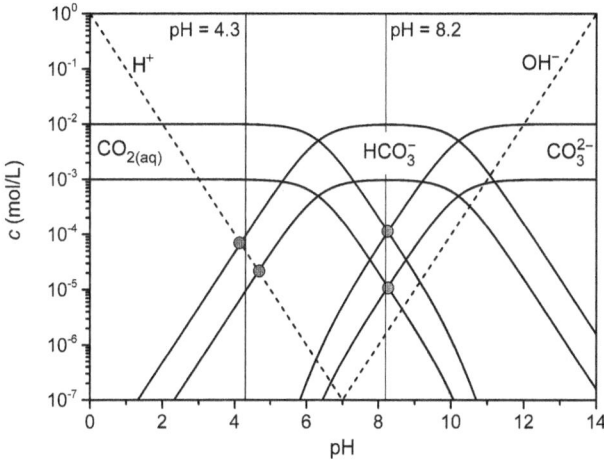

**Figure 7.6:** Speciation of carbonic acid (absolute concentrations) and equivalence points (gray circles) for the reference states $CO_2$ solution and sodium hydrogencarbonate solution (25 °C). The pH values 4.3 and 8.2 indicate the (defined) endpoints of the titrations that are applied to determine acid and base neutralizing capacities.

In order to relate the titration results to the concentrations, appropriate titration endpoints (equivalence points) have to be defined. Typical endpoints used in practice are pH = 4.3 and pH = 8.2. These endpoints are defined on the basis of the carbonic acid system under the assumption that other acid/base systems can be neglected. It follows from the speciation shown in Figure 7.6 that the titration endpoint pH = 4.3 represents the reference state of a carbon dioxide solution, whereas pH = 8.2 represents the reference state of a sodium hydrogencarbonate solution. At these points, the concentrations of the other carbonic acid species ($HCO_3^-$ and $CO_3^{2-}$ at pH = 4.3 or $CO_{2(aq)}$ and $CO_3^{2-}$ at pH = 8.2) are negligibly small. Accordingly, the titration results for a water with an initial pH that is different from the endpoints represents the amounts of $H^+$ or $OH^-$ that are necessary to transform other species nearly totally to $CO_2$ or to $HCO_3^-$. The concentrations of the transformed species can then be computed from the acid or base consumption by stoichiometric calculations.

Before starting the discussion on the relationships between the titration results and the carbonic acid species concentrations, we want to look at the proton balances at the titration endpoints. If we want to establish an exact proton balance for a $CO_2$ solution, we have to write:

$$c(H^+) = c(HCO_3^-) + 2c(CO_3^{2-}) + c(OH^-) \approx c(HCO_3^-) \tag{7.131}$$

which includes all species that results from the autoprotolysis of water ($H_2O \rightarrow H^+ + OH^-$) and the minor dissociation of $CO_2(CO_2 + H_2O \rightarrow H^+ + HCO_3^- \rightarrow 2H^+ + CO_3^{2-})$. As we can see from Figure 7.6, the contributions of $CO_3^{2-}$ and $OH^-$ to the proton balance are very small. Consequently, the theoretical equivalence point is located at the intersection of the $H^+$ and $HCO_3^-$ curves ($c(H^+) = c(HCO_3^-)$ in the speciation diagram). What we can also see from the diagram is that the equivalence point (given by the proton balance) is not really a fixed value but is determined by the total concentration of the carbonic acid species. It has further to be noted that the ionic strength and the temperature also influence the location of the equivalence point (not shown in the diagram).

The same effect, but not so strongly pronounced, can be found for the other equivalence point. For the hydrogencarbonate solution, we can write the following proton balance:

$$c(H^+) = c(CO_3^{2-}) + c(OH^-) - c(CO_2) \tag{7.132}$$

Equation (7.132) considers the ability of hydrogencarbonate to release protons (equivalent to the $CO_3^{2-}$ formation, $HCO_3^- \rightarrow H^+ + CO_3^{2-}$) and to accept protons (equivalent to the $CO_2$ formation, $HCO_3^- + H^+ \rightarrow H_2O + CO_2$). Since the $CO_2$ formation decreases the proton concentration in the solution, the term $c(CO_2)$ on the right-hand side of the proton balance equation is negative. Since the protons and the hydroxide ions do not significantly contribute to the balance, the equivalence point is approximately given by the condition $c(CO_2) = c(CO_3^{2-})$, which is shown in Figure 7.6 as the intersection of the respective curves.

Since the use of individual equivalence points for each water composition and temperature as titration endpoints would not be practicable, fixed titration endpoints are used in practice. These titration endpoints are defined in such a manner that they fall into the practically relevant range of the equivalence points. Here, we want to use pH = 4.3 and pH = 8.2 as proposed in German guidelines. Slightly different endpoints can be found in other guidelines and national standards (e.g., 4.5 instead of 4.3 or 8.3 instead of 8.2). However, these small differences have no relevant influence on the titration results. The minor sensitivity of the titration results with respect to the endpoints also allows using color indicators instead of a pH-meter to determine the endpoints (e.g., methyl orange for the lower endpoint or phenolphthalein for the higher endpoint) even though these indicators change their color in a pH interval and not at an exact pH value. The symbols $m$ and $p$, sometimes used for the titration results, have been derived from the names of the indicators.

It has to be noted that the given equations (7.131) and (7.132) describe reference states where all species come only from a pure solution of $CO_2$ or from a pure solution of $NaHCO_3$. In real waters, however, the considered species can also originate from other sources. This means that the water samples contain more bases or acids than in the reference state. Accordingly, the pH in a water sample can be lower or higher than the respective reference points. This means that in terms of the proton balance there is a proton excess or a proton deficiency when compared to the proton balances given in equations (7.131) and (7.132). A proton deficiency can be compensated by adding a strong acid, a proton excess by adding a strong base. This consideration leads to the definition of acid neutralizing capacities to pH = 4.3 and pH = 8.2 and base neutralizing capacities to pH = 4.3 and pH = 8.2. The acid neutralizing capacities are the amounts of a strong acid that are necessary to bring the water to pH = 4.3 or to pH = 8.2. The base neutralizing capacities are the amounts of a strong base that are necessary to bring the water to pH = 4.3 or to pH = 8.2. Acid neutralizing capacities are commonly referred to as alkalinities (with the qualifiers total or carbonate), whereas the base neutralizing capacities are referred to as acidities (with the qualifiers mineral or CO$_2$). Unfortunately, there are numerous different terms in use for the titration parameters, which can lead to confusion. An overview of commonly used terms is given in Table 7.4.

**Table 7.4:** Terms used to describe the titration parameters in carbonate-dominated aqueous systems.

| Titration with | Endpoint | | Terms in use | | |
|---|---|---|---|---|---|
| Acid | pH = 4.3 | Acid capacity (consumption) to pH = 4.3 | Acid neutralizing capacity to pH = 4.3 | (Total) Alkalinity | m alkalinity |
| Acid | pH = 8.2 | Acid capacity (consumption) to pH = 8.2 | Acid neutralizing capacity to pH = 8.2 | Phenolphthalein alkalinity, carbonate alkalinity | p alkalinity |
| Base | pH = 4.3 | Base capacity (consumption) to pH = 4.3 | Base neutralizing capacity to pH = 4.3 | Mineral acidity | −m alkalinity |
| Base | pH = 8.2 | Base capacity (consumption) to pH = 8.2 | Base neutralizing capacity to pH = 8.2 | Phenolphthalein acidity, CO$_2$ acidity | −p alkalinity |

Note: The term m (or m alkalinity) refers to the use of methyl orange as color indicator in the titration, the terms phenolphthalein alkalinity and p (or p alkalinity) refer to the use of phenolphthalein as color indicator in the titration. Depending on national regulations, also slightly different values of the pH endpoints are in use (e.g., pH = 4.5 and pH = 8.3).

In principle, four different titrations are possible. We will see later that one of these titrations is irrelevant for practical purposes. The possible titrations are shown in Figure 7.7. The arrows indicate the titration with acid (from right to left, decreasing pH) or base (from left to right, increasing pH). It can be seen from the scheme that only two of four titrations are possible for a given water sample. It depends on the initial pH value of the water sample which titrations can be carried out. For strongly acidic waters, both base neutralizing capacities can be determined, whereas for strongly basic waters, both acid neutralizing capacities are accessible. For the most relevant medium pH range, we can determine the acid neutralizing capacity to pH = 4.3 and the base neutralizing capacity to pH = 8.2.

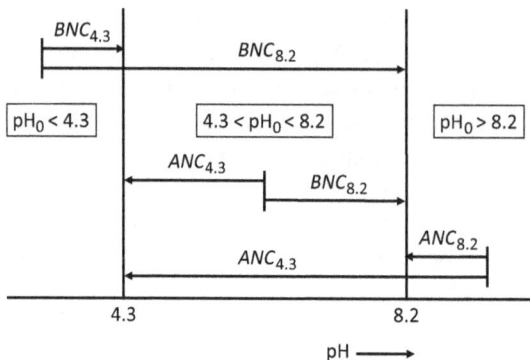

**Figure 7.7:** Determination of acid and base capacities by titration with an acid (typically hydrochloric acid) or a base (typically sodium hydroxide). The arrows indicate the consumption of acid or base and the related change of pH during the titration to the defined pH endpoints 4.3 and 8.2.

During the acid or base addition, the carbonic acid species concentrations given in equations (7.131) and (7.132) change their values due to the reaction with $H^+$ or $OH^-$. This gives the opportunity to relate the species concentrations to the acid or base consumption. In the following paragraphs, the titrations and the information that can be derived from their results are discussed in more detail. At first, we want to follow a practical approach that relates the titration results to the carbonic acid species on the basis of the relevant transformation reactions during the titration. Later, we want to consider more rigorous definitions of the alkalinities and acidities (Section 7.7.4).

### Acid (Neutralizing) Capacity to pH = 4.3 (Total Alkalinity, m Alkalinity)

The acid neutralizing capacity to pH = 4.3 (total alkalinity, $m$ alkalinity) can be determined for each water with an initial pH higher than 4.3. The proton consumption during the titration with a strong acid (addition of $H^+$) results from the following transformation reactions:

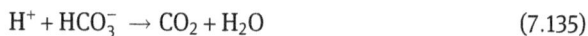

$$H^+ + OH^- \rightarrow H_2O \tag{7.133}$$

$$2H^+ + CO_3^{2-} \rightarrow CO_2 + H_2O \tag{7.134}$$

$$H^+ + HCO_3^- \rightarrow CO_2 + H_2O \tag{7.135}$$

Given that the residual concentrations of $OH^-$, $CO_3^{2-}$, and $HCO_3^-$ at the titration endpoint are negligibly small (Figure 7.6), we can set the proton consumption equal to the concentrations originally present in the solution. Consequently, the acid neutralizing capacity determined by titration is:

$$ANC_{4.3} = Alk_T = m = c(HCO_3^-) + 2c(CO_3^{2-}) + c(OH^-) \tag{7.136}$$

Depending on the initial pH of the water ($pH_0$), some simplifications are possible. At $pH_0 < 9.5$, the initial concentration of $OH^-$ can be neglected ($c < 10^{-4.5}$ mol/L) and at $pH_0 < 8.2$, the carbonate concentration can also be neglected (see the speciation in Figure 7.6):

$$ANC_{4.3} = Alk_T = m \approx c(HCO_3^-) + 2c(CO_3^{2-}) \qquad pH_0 < 9.5 \qquad (7.137)$$

$$ANC_{4.3} = Alk_T = m \approx c(HCO_3^-) \qquad pH_0 < 8.2 \qquad (7.138)$$

**Acid (Neutralizing) Capacity to pH = 8.2 (Phenolphthalein Alkalinity, $p$ Alkalinity)**
The proton consumption during titration of water with $pH_0 > 8.2$, where only the concentrations of $OH^-$ and $CO_3^{2-}$ are relevant (Figure 7.6), results from the (practically complete) transformation reactions:

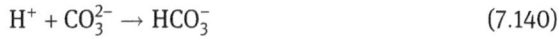

$$H^+ + OH^- \rightarrow H_2O \qquad (7.139)$$

$$H^+ + CO_3^{2-} \rightarrow HCO_3^- \qquad (7.140)$$

Accordingly, the experimentally determined acid neutralizing capacity to pH = 8.2 equals the sum of the concentrations of $OH^-$ and $CO_3^{2-}$:

$$ANC_{8.2} = Alk_P = p = c(CO_3^{2-}) + c(OH^-) \qquad (7.141)$$

Again, for $pH_0 < 9.5$ (negligible contribution of $OH^-$), equation (7.141) can be further simplified to:

$$ANC_{8.2} = Alk_P = p \approx c(CO_3^{2-}) \qquad (7.142)$$

This equation also explains the alternative name carbonate alkalinity.

**Base (Neutralizing) Capacity to pH = 4.3 (Mineral Acidity, $-m$ Alkalinity)**
The base neutralizing capacity to pH = 4.3 can only be determined for strong acidic solutions with $pH_0 < 4.3$. Under these pH conditions, $CO_2$ is the only relevant carbonic acid species, and up to pH = 4.3, no transformation occurs. The $OH^-$ consumption results from the reaction with $H^+$ only:

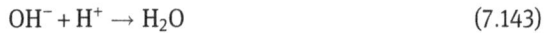

$$OH^- + H^+ \rightarrow H_2O \qquad (7.143)$$

and $BNC_{4.3}$ is simply a parameter that gives the concentration of protons released from strong (mineral) acids ("mineral acidity"):

$$BNC_{4.3} = Aci_M = -m = c(H^+) \qquad (7.144)$$

Since the pH can be measured directly and more easily with a pH-meter, $BNC_{4.3}$ has no practical relevance.

**Base (Neutralizing) Capacity to pH = 8.2 (Phenolphthalein Acidity, −*p* Alkalinity)**
Since the initial pH of the water must be lower than 8.2 to determine the base neutralizing capacity to pH = 8.2, the concentrations of $CO_3^{2-}$ and $OH^-$ can be neglected (Figure 7.6). Accordingly, the $OH^-$ consumption results only from the reactions:

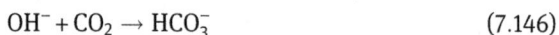

$$OH^- + H^+ \rightarrow H_2O \tag{7.145}$$

$$OH^- + CO_2 \rightarrow HCO_3^- \tag{7.146}$$

Accordingly, $BNC_{8.2}$ is related to the concentrations of $CO_2$ and $H^+$:

$$BNC_{8.2} = Aci_P = -p = c(CO_2) + c(H^+) \tag{7.147}$$

The proton concentration has only to be considered at initial pH values lower than about 4.5. Therefore, for the most practical cases, the simple relationship:

$$BNC_{8.2} = Aci_P = -p \approx c(CO_2) \tag{7.148}$$

can be applied. The base neutralizing capacity to pH = 8.2 is therefore also referred to as $CO_2$ acidity.

**Summary: Calculation of the Carbonic Acid Species Concentrations**
The general equations that have to be used to calculate the concentrations of the different carbonic acid species can be derived from equations (7.136), (7.141), and (7.147):

$$c(CO_2) = BNC_{8.2} - c(H^+) \tag{7.149}$$

$$c(HCO_3^-) = ANC_{4.3} - 2c(CO_3^{2-}) - c(OH^-) \tag{7.150}$$

$$c(CO_3^{2-}) = ANC_{8.2} - c(OH^-) \tag{7.151}$$

The concentrations of $H^+$ and $OH^-$ can be found by measuring the initial pH of the water sample before titration. Under certain conditions with respect to the initial pH of the considered water, some terms of the equations can be neglected. This is true for the proton concentration in equation (7.149) at $pH_0 > 4.5$, for the hydroxide ion concentrations in equations (7.150) and (7.151) at $pH_0 < 9.5$, and for the carbonate concentration in equation (7.150) at $pH_0 < 8.2$. Table 7.5 summarizes the equations that have to be used for the different pH conditions. If the concentration of hydrogencarbonate in a water sample with $pH_0 > 8.2$ (consideration of $CO_3^{2-}$) should be determined, equations (7.150) and (7.151) have to be combined.

**Table 7.5:** Relationships between the concentrations of the carbonic acid species and the titration parameters for different pH ranges.

| Species concentrations | $pH_0 < 4.5$ | $4.5 < pH_0 < 8.2$ | $8.2 < pH_0 < 9.5$ | $pH_0 > 9.5$ |
|---|---|---|---|---|
| $c(CO_2)$ | $BNC_{8.2} - c(H^+)$ | $BNC_{8.2}$ | $\approx 0$ | $\approx 0$ |
| $c(HCO_3^-)$ | $\approx 0$ | $ANC_{4.3}$ | $ANC_{4.3} - 2\,ANC_{8.2}$ | $ANC_{4.3} - 2\,ANC_{8.2} + c(OH^-)$ |
| $c(CO_3^{2-})$ | $\approx 0$ | $\approx 0$ | $ANC_{8.2}$ | $ANC_{8.2} - c(OH^-)$ |

### 7.7.4 General Definitions of the Alkalinities and Acidities on the Basis of Proton Balances

As shown in the previous section, both reference states ($CO_2$ solution and $NaHCO_3$ solution) defined for the acid/base titrations can be expressed by proton balances (equations (7.131) and (7.132)). These equations can also be written in the form:

$$0 = c(HCO_3^-) + 2\,c(CO_3^{2-}) + c(OH^-) - c(H^+) \quad (pH = 4.3) \tag{7.152}$$

$$0 = c(CO_3^{2-}) + c(OH^-) - c(CO_2) - c(H^+) \quad (pH = 8.2) \tag{7.153}$$

Waters with other compositions have a proton deficiency or excess in comparison to the reference states and therefore also respective acid or base neutralizing capacities. We can therefore define the alkalinities and acidities in a more general approach as deviations from the proton balances of the reference states. Accordingly, the general definition of the acid neutralizing capacity to pH = 4.3 (total alkalinity) reads:

$$ANC_{4.3} = Alk_T = m = c(HCO_3^-) + 2\,c(CO_3^{2-}) + c(OH^-) - c(H^+) \tag{7.154}$$

and the acid neutralizing capacity to pH = 8.2 (phenolphthalein alkalinity) is given by:

$$ANC_{8.2} = Alk_P = p = c(CO_3^{2-}) + c(OH^-) - c(CO_2) - c(H^+) \tag{7.155}$$

In the same manner, the base neutralizing capacity to pH = 4.3 (mineral acidity) is defined as:

$$BNC_{4.3} = Aci_M = -m = -c(HCO_3^-) - 2\,c(CO_3^{2-}) - c(OH^-) + c(H^+) \tag{7.156}$$

and the base neutralizing capacity to pH = 8.2 (phenolphthalein acidity) is given by:

$$BNC_{8.2} = Aci_P = -p = -c(CO_3^{2-}) - c(OH^-) + c(CO_2) + c(H^+) \tag{7.157}$$

The definitions given here include all involved species independent of their absolute concentrations. Therefore, they are generally valid. In contrast, in the definitions given in the previous section, species with negligibly small concentrations under the given pH conditions are omitted. As we can see from the general definitions, the equations of the acid and base neutralizing capacities for the same endpoint differ only in the signs.

### 7.7.5 The Conservative Character of Alkalinity

From the previous discussions, the question may arise: Why is it reasonable to define the alkalinities and acidities on the basis of the proton balances and not only on the basis of the simplified equations derived from the relevant $H^+$ or $OH^-$ consuming reactions during the titrations? In fact, the latter would be sufficient if we are only interested in the estimation of the carbonic acid species concentrations from the titration results. In contrast, the complete proton balances are necessary to understand the theoretical concept of alkalinity and, in particular, its conservative character. In the following explanations, we want to use the shorthand term $m$ for the total alkalinity.

To explain the conservative character, an alternative definition of the $m$ alkalinity can be used. As we have seen in Chapter 3 (Section 3.6), the condition of electroneutrality requires that the sum of the equivalent concentrations of the cations must equal the sum of the equivalent concentrations of the anions. Accordingly, the ion balance for the major ions and the ions $H^+$ and $OH^-$ reads:

$$c(Na^+) + c(K^+) + 2c(Ca^{2+}) + 2c(Mg^{2+}) + c(H^+)$$
$$= c(Cl^-) + 2c(SO_4^{2-}) + c(HCO_3^-) + 2c(CO_3^{2-}) + c(OH^-) \tag{7.158}$$

If we rearrange this equation in such a manner that the equivalent concentrations of all species, which behave conservative with respect to pH, are written on the left-hand side, we obtain:

$$c(Na^+) + c(K^+) + 2c(Ca^{2+}) + 2c(Mg^{2+}) - c(Cl^-) - 2c(SO_4^{2-})$$
$$= c(HCO_3^-) + 2c(CO_3^{2-}) + c(OH^-) - c(H^+) = m \tag{7.159}$$

where the right-hand side equals the $m$ alkalinity (see equation (7.154)). Thus, for a given water composition, the $m$ alkalinity is a characteristic parameter with a constant value. The $m$ value is independent of temperature, pressure, and ionic strength.

The addition of salts, such as NaCl or CaSO$_4$, does not change the $m$ alkalinity due to the opposite signs of the cation and anion concentrations on the left-hand side of equation (7.159). Moreover, addition or release of $CO_2$ has no influence on $m$. If $CO_2$ is introduced into the water, a part of this $CO_2$ dissociates according to:

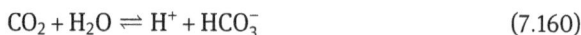

$$CO_2 + H_2O \rightleftharpoons H^+ + HCO_3^- \tag{7.160}$$

As a result of the $CO_2$ dissolution, there is an increase in the $CO_2$ concentration (without any impact on $m$), an increase in the proton concentration and an equivalent increase in the hydrogencarbonate concentration. Due to the opposite signs of $c(HCO_3^-)$ and $c(H^+)$ in equation (7.159), there is no change in $m$. If $CO_2$ is removed from the water, the equilibrium given in equation (7.160) is shifted to the left. Again, the concentrations of $H^+$ and $HCO_3^-$ change to the same extent and the effects on $m$ compensate each other due to the opposite signs. We come to the same conclusion if we look at the left-hand side of equation (7.159). The ion concentrations on the left-hand side will not be influenced by $CO_2$ dissolution or removal and therefore no change of $m$ will occur.

If two waters are mixed, the $m$ alkalinity of the mixed water is the weighted average of the alkalinities of the original waters. This is also true for all single concentrations on the left-hand side of equation (7.156) (conservative ions) but not for the single concentrations of the weak electrolytes on the right-hand side. Here, mixing of waters typically causes a change in the concentration distribution.

### 7.7.6 Determination of Dissolved Inorganic Carbon

Based on the general definition equations given in Section 7.7.4, we can derive a relationship between $m$ alkalinity, $p$ alkalinity, and the total concentration of the carbonic acid species, $c(DIC)$:

$$m - p = c(HCO_3^-) + 2c(CO_3^{2-}) + c(OH^-) - c(H^+)$$

$$- c(CO_3^{2-}) - c(OH^-) + c(CO_2) + c(H^+) \tag{7.161}$$

$$m - p = c(CO_2) + c(HCO_3^-) + c(CO_3^{2-}) = c(DIC) \tag{7.162}$$

Since the values of $m$ and $p$ can be determined experimentally by titrations as shown in Section 7.7.3, also $c(DIC)$ is easily accessible. The respective relationships are $m = ANC_{4.3}$, $p = ANC_{8.2}$ (for $pH_0 > 8.2$), and $-p = BNC_{8.2}$ (for $pH_0 < 8.2$).

Alternatively, it is also possible to determine $c(DIC)$ only by means of the $m$ alkalinity. For that, it is necessary to substitute the concentrations of $HCO_3^-$ and $CO_3^{2-}$ in the definition of $m$ by the products $f(HCO_3^-) \cdot c(DIC)$ and $f(CO_3^{2-}) \cdot c(DIC)$, respectively. The fractions, $f$, of the hydrogencarbonate and carbonate ions can be computed by using the equations (7.129) and (7.130). For the sake of simplicity, we want to set $f(HCO_3^-) = f_1$ and $f(CO_3^{2-}) = f_2$:

$$m = c(HCO_3^-) + 2c(CO_3^{2-}) + c(OH^-) - c(H^+) \tag{7.163}$$

$$m = f_1 c(DIC) + 2f_2 c(DIC) + c(OH^-) - c(H^+) \tag{7.164}$$

$$m = c(DIC)(f_1 + 2f_2) + c(OH^-) - c(H^+) \tag{7.165}$$

$$c(DIC) = \frac{m - c(OH^-) + c(H^+)}{f_1 + 2f_2} \tag{7.166}$$

In the medium pH range, the concentrations of $H^+$ and $OH^-$ are much lower than $m$ and can therefore be neglected. The simplified equation reads:

$$c(DIC) = \frac{m}{f_1 + 2f_2} \tag{7.167}$$

To apply equations (7.166) or (7.167), the acidity constants of the carbonic acid system at the given temperature have to be known (see also Table 9.2 in Chapter 9).

---

**Example 7.6**

What is the concentration of DIC in a water that is characterized by $m$ = 5.3 mmol/L, pH = 7, and $\vartheta$ = 10 °C? The acidity constants of the carbonic acid system at 10 °C are $K_{a1}$ = 3.43 × 10$^{-7}$ mol/L and $K_{a2}$ = 3.25 × 10$^{-11}$ mol/L; the dissociation constant for water at the same temperature is 2.9 × 10$^{-15}$ mol$^2$/L$^2$.

**Solution:**

At first, the fractions of hydrogencarbonate and carbonate have to be calculated by using equations (7.129) and (7.130). With the concentration of protons related to pH = 7 ($c(H^+)$ = 1 × 10$^{-7}$ mol/L) and the product of the constants $K_{a1} K_{a2}$ = (3.43 × 10$^{-7}$ mol/L) (3.25 × 10$^{-11}$ mol/L) = 1.11 × 10$^{-17}$ mol$^2$/L$^2$, we receive:

$$f(HCO_3^-) = f_1 = \cfrac{1}{\cfrac{1 \times 10^{-7}\,\text{mol/L}}{3.43 \times 10^{-7}\,\text{mol/L}} + 1 + \cfrac{3.25 \times 10^{-11}\,\text{mol/L}}{1 \times 10^{-7}\,\text{mol/L}}}$$

$$f_1 = \frac{1}{0.29 + 1 + 3.25 \times 10^{-4}} = 0.775$$

$$f(CO_3^{2-}) = f_2 = \cfrac{1}{\cfrac{1 \times 10^{-14}\,\text{mol}^2/\text{L}^2}{1.11 \times 10^{-17}\,\text{mol}^2/\text{L}^2} + \cfrac{1 \times 10^{-7}\,\text{mol/L}}{3.25 \times 10^{-11}\,\text{mol/L}} + 1}$$

$$f_2 = \frac{1}{900.9 + 3\,076.9 + 1} = 2.51 \times 10^{-4}$$

As expected, the fraction of carbonate at pH = 7 is very low and the system is dominated by hydrogencarbonate. The $OH^-$ concentration can be calculated according to:

$$c(OH^-) = \frac{K_w}{c(H^+)} = \frac{2.9 \times 10^{-15}\,\text{mol}^2/\text{L}^2}{1 \times 10^{-7}\,\text{mol/L}} = 2.9 \times 10^{-8}\,\text{mol/L}$$

Now, we can calculate the DIC concentration:

$$c(DIC) = \frac{m - c(OH^-) + c(H^+)}{f_1 + 2f_2} = \frac{5.3 \times 10^{-3}\,\text{mol/L} - 2.9 \times 10^{-8}\,\text{mol/L} + 1 \times 10^{-7}\,\text{mol/L}}{0.775 + 5.02 \times 10^{-4}}$$

$$c(DIC) = \frac{5.3 \times 10^{-3}\,\text{mol/L}}{0.7755} = 6.83 \times 10^{-3}\,\text{mol/L}$$

As we can see, the concentrations of the protons and the hydroxide ions have no impact on the result due to their low concentrations in comparison to the $m$ alkalinity. This is generally true in the medium pH range.

### 7.7.7 pH of Pristine Rain Water

In the absence of other acidic or basic gases, the pH of rain water is determined by atmospheric carbon dioxide due to its dissolution and subsequent dissociation. Due to the high value of $pK_{a2}$, the formation of carbonate can be neglected and $CO_2$ can be treated here in a simplified manner as a monoprotic acid. Accordingly, the relevant reactions are:

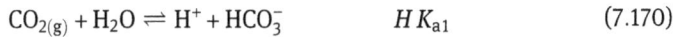

$$CO_{2(g)} \rightleftharpoons CO_{2(aq)} \qquad\qquad H \qquad\qquad (7.168)$$

$$CO_{2(aq)} + H_2O \rightleftharpoons H^+ + HCO_3^- \qquad\qquad K_{a1} \qquad\qquad (7.169)$$

$$\overline{CO_{2(g)} + H_2O \rightleftharpoons H^+ + HCO_3^-} \qquad\qquad H K_{a1} \qquad\qquad (7.170)$$

The dissolution of $CO_2$ is described by the Henry constant, $H$ (Chapter 6, Section 6.2). Given that the same amounts of $H^+$ and $HCO_3^-$ are formed ($c(H^+) = c(HCO_3^-)$), the law of mass action for the overall reaction can be written as:

$$H K_{a1} = \frac{c(H^+)\,c(HCO_3^-)}{p(CO_2)} = \frac{c^2(H^+)}{p(CO_2)} \qquad\qquad (7.171)$$

Consequently, the proton concentration is given by:

$$c(H^+) = \sqrt{H K_{a1}\, p(CO_2)} \qquad\qquad (7.172)$$

---

**Example 7.7**

What is the pH of pristine rain water if we assume a temperature of 25 °C, a total atmospheric pressure of 1 bar, and a $CO_2$ content in the atmosphere of 415 ppm$_v$? The equilibrium constants for 25 °C are $H(CO_2) = 0.033$ mol/(L·bar) and $K_{a1} = 4.4 \times 10^{-7}$ mol/L.

**Solution:**

415 ppm$_v$ equals 0.0415 vol.% and $y = 0.000415$. Note that the gas-phase mole fraction, $y$, can be found by dividing the volume percentage by 100% (see Section 3.5 in Chapter 3). The partial pressure is given as the product of the gas-phase mole fraction and the total pressure:

$$p(CO_2) = y(CO_2)\, p_{total} = 0.000415\,(1\,\text{bar}) = 0.000415\,\text{bar}$$

With equation (7.172), we find:

$$c(H^+) = \sqrt{\left(0.033\,\frac{\text{mol}}{\text{L·bar}}\right)\left(4.4 \times 10^{-7}\,\frac{\text{mol}}{\text{L}}\right)(0.000415\,\text{bar})} = 2.45 \times 10^{-6}\,\text{mol/L}$$

$$pH = -\log(2.45 \times 10^{-6}) = 5.61$$

---

### 7.7.8 Photosynthesis and Carbonic Acid System

The production of biomass from inorganic compounds with light as energy source is referred to as photosynthesis. In water bodies, photosynthesis is carried out by algae and aquatic plants. The process of photosynthesis in water bodies can be described in a simplified manner by the reaction equation:

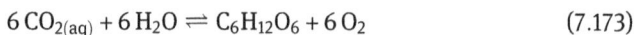

$$6\,CO_{2(aq)} + 6\,H_2O \rightleftharpoons C_6H_{12}O_6 + 6\,O_2 \tag{7.173}$$

where $C_6H_{12}O_6$ is the organic compound glucose that is produced from the inorganic compounds carbon dioxide and water. The reverse process describes the degradation of organic material by aerobic respiration (consumption of oxygen and production of $CO_2$). Both back and forth reactions are redox reactions in the natural carbon cycle (see Chapter 10, Section 10.6.2). Here, we are interested in the impact of photosynthesis on the carbonic acid equilibria in water bodies.

Due to the need for light, the reaction shows a pronounced daily cycle. In particular, on hot summer days with strong sunlight, the equilibrium is strongly shifted to the right, whereas in the night, the reverse reaction dominates. The shift in the reaction equilibrium, coupled with a decrease or increase of dissolved $CO_2$, has an influence on the total carbonic acid system and therefore also on the pH of the water body. The possible uptake of $CO_2$ from the atmosphere or the release into the atmosphere is not considered here because the mass transfer between surface water and atmosphere is very slow and we want to discuss only short-term effects.

From the previous discussion on the conservative character of the alkalinity (Section 7.7.5) and the relationship between alkalinity and DIC (Section 7.7.6), we can draw the following conclusions:

Since the alkalinity ($m$ value) in a carbonate-dominated system is given by:

$$m = c(HCO_3^-) + 2\,c(CO_3^{2-}) + c(OH^-) - c(H^+) \tag{7.174}$$

a change in the $CO_2$ concentration has no influence on the alkalinity. This is also true when $CO_{2(aq)}$ is formed from $H^+$ and $HCO_3^-$ to compensate for a $CO_{2(aq)}$ consumption during photosynthesis or, in reverse direction, if produced $CO_{2(aq)}$ dissociates into $H^+$ and $HCO_3^-$ because always equivalent amounts of oppositely charged species are consumed or produced.

However, the consumption of $CO_{2(aq)}$ during photosynthesis or the production of $CO_{2(aq)}$ during respiration leads to a shift in the carbonic acid equilibria and the distribution of the carbonic acid species. Therefore, the consumption or production cannot be expressed by the $CO_2$ concentration alone but has to be expressed in a more general form by the total concentration of dissolved carbonic acid species, $c(DIC)$:

$$c(DIC) = c(CO_2) + c(HCO_3^-) + c(CO_3^{2-}) \tag{7.175}$$

Combining the definitions of alkalinity and DIC gives:

$$m = c(\text{DIC})(f_1 + 2f_2) + c(\text{OH}^-) - c(\text{H}^+) \qquad (7.176)$$

with

$$f_1 = f(\text{HCO}_3^-) = \frac{c(\text{HCO}_3^-)}{c(\text{DIC})} = \frac{1}{\dfrac{c(\text{H}^+)}{K_{a1}} + 1 + \dfrac{K_{a2}}{c(\text{H}^+)}} \qquad (7.177)$$

$$f_2 = f(\text{CO}_3^{2-}) = \frac{c(\text{CO}_3^{2-})}{c(\text{DIC})} = \frac{1}{\dfrac{c^2(\text{H}^+)}{K_{a1}K_{a2}} + \dfrac{c(\text{H}^+)}{K_{a2}} + 1} \qquad (7.178)$$

In principle, two types of problems can be solved by means of equation (7.176). Either the decrease or increase of DIC (photosynthesis or respiration) is known and the change of pH should be calculated or the change of pH has been measured and the impact of the pH shift on the carbonic acid speciation is of interest. In both cases, the (constant) alkalinity has to be known. The first type of calculation is more complicated because the equation (7.176) cannot be simply rearranged for $c(\text{H}^+)$. Therefore, an iterative procedure has to be applied. Example 7.8 shows such an iteration method. The second type of calculation is considered in Problem 7.12.

---

**Example 7.8**
In a lake, the following data were found in the morning: $c(\text{DIC}) = 1.15 \times 10^{-3}$ mol/L, $m = 1.14 \times 10^{-3}$ mol/L, and pH = 8.2. What is the pH in the evening if during the day the DIC concentration is reduced to 50% of its initial value as a result of photosynthetic activity in the water body? The temperature is 15 °C, and the corresponding acidity constants are $K_{a1} = 3.78 \times 10^{-7}$ mol/L and $K_{a2} = 3.72 \times 10^{-11}$ mol/L. The dissociation constant of water at 15 °C is $K_w = 4.51 \times 10^{-15}$ mol$^2$/L$^2$.

**Solution:**
Introducing the given data into equations (7.176), we get:

$$1.14 \times 10^{-3}\ \text{mol/L} = (5.75 \times 10^{-4}\ \text{mol/L})(f_1 + 2f_2) + c(\text{OH}^-) - c(\text{H}^+)$$

Note that the $m$ value was kept constant ($m = 1.14 \times 10^{-3}$ mol/L) and $c(\text{DIC})$ was reduced to 50% of its initial value ($c(\text{DIC}) = (0.5)(1.15 \times 10^{-3}$ mol/L$) = 5.75 \times 10^{-4}$ mol/L).

The fractions of hydrogencarbonate and carbonate, $f_1$ and $f_2$, can be found from equations (7.177) and (7.178):

$$f_1 = \frac{1}{\dfrac{c(\text{H}^+)}{3.78 \times 10^{-7}\ \text{mol/L}} + 1 + \dfrac{3.72 \times 10^{-11}\ \text{mol/L}}{c(\text{H}^+)}}$$

$$f_2 = \frac{1}{\dfrac{c^2(\text{H}^+)}{1.41 \times 10^{-17}\ \text{mol}^2/\text{L}^2} + \dfrac{c(\text{H}^+)}{3.72 \times 10^{-11}\ \text{mol/L}} + 1}$$

Now, the pH has to be varied until the right-hand side (RHS) of the first equation equals the value of the left-hand side ($1.14 \times 10^{-3}$ mol/L). Since $CO_2$ is consumed, we expect an increase of pH. Therefore, we start our iteration with pH values higher than 8.2.

The corresponding $OH^-$ concentration is found from:

$$c(OH^-) = = \frac{K_w}{c(H^+)} = \frac{4.51 \times 10^{-15} \text{ mol}^2/L^2}{10^{-pH} \text{ mol}/L}$$

| pH | $c(H^+)$ (mol/L) | $c(OH^-)$ (mol/L) | $f_1$ | $f_2$ | RHS (mol/L) |
|---|---|---|---|---|---|
| 9 | $1 \times 10^{-9}$ | $4.510 \times 10^{-6}$ | 0.9617 | 0.0358 | $5.986 \times 10^{-4}$ |
| 10 | $1 \times 10^{-10}$ | $4.510 \times 10^{-5}$ | 0.7287 | 0.2711 | $7.759 \times 10^{-4}$ |
| 11 | $1 \times 10^{-11}$ | $4.510 \times 10^{-4}$ | 0.2119 | 0.7881 | $1.479 \times 10^{-3}$ |
| 10.5 | $3.162 \times 10^{-11}$ | $1.426 \times 10^{-4}$ | 0.4595 | 0.5405 | $1.028 \times 10^{-3}$ |
| 10.7 | $1.995 \times 10^{-11}$ | $2.260 \times 10^{-4}$ | 0.3491 | 0.6509 | $1.175 \times 10^{-3}$ |
| 10.65 | $2.239 \times 10^{-11}$ | $2.015 \times 10^{-4}$ | 0.3757 | 0.6243 | $1.135 \times 10^{-3}$ |

During the course of the day, the pH increases from 8.2 to about 10.7 due to the consumption of $CO_2$ by photosynthesis.

## 7.8 Problems

Note: It is assumed that ideal conditions exist in all cases ($a = c$, $K^* = K$). If not otherwise stated, the constants are valid for 25 °C.

**7.1.** What are the concentrations of protons and hydroxide ions at pH = 8.5 and $\vartheta = 25$ °C ($pK_w = 14$)?

**7.2.** The $pK_a$ of 4-chlorophenol ($C_6H_4OHCl$, abbreviation: 4-CP) is 9.4. Calculate the pH of 4-CP solutions with initial concentrations of 5 and 0.5 g/L. $M$(4-CP) = 128.6 g/mol.

**7.3.** What is the pH of a dilute phenol solution with $c_0 = 0.01$ mmol/L? Calculate the pH with and without consideration of the autoprotolysis of water. Check the plausibility of the results. $pK_a$ (phenol) = 9.9, $pK_w = 14$.

**7.4.** The $pK_a$ of the protonated aniline ($C_6H_5NH_3^+$) is 4.58, and the ion product of water is $K_w = 1 \times 10^{-14}$ mol²/L².
a.  What is the basicity constant, $K_b$, of the conjugate base aniline ($C_6H_5NH_2$)?
b.  What is the ratio of protonated and neutral aniline at pH = 5.5?
c.  What is the equilibrium constant of the protonation reaction written in the form

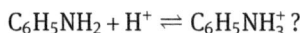

$$C_6H_5NH_2 + H^+ \rightleftharpoons C_6H_5NH_3^+ ?$$

**7.5.** What is the pH of a sodium carbonate solution with $c_0 = 2$ mol/L? It is assumed that only the transformation of carbonate to hydrogencarbonate has to be considered as pH-determining reaction? $pK_a$ ($HCO_3^-$) = 10.33, $pK_w = 14$.

**7.6.** What is the pH of an iron(III) chloride solution with $c_0 = 0.05$ mol/L? It is assumed that only the first protolysis step determines the pH of the solution. The respective $pK_a$ is 2.2.

**7.7.** What is the degree of protolysis of an arbitrary monoprotic acid at pH = $pK_a + 1$?

**7.8.** In the following example, we want to consider the buffer system $H_2PO_4^-/HPO_4^{2-}$. The concentrations of both components are assumed to be 0.1 mol/L and the $pK_a$ of $H_2PO_4^-$ is 7.12.
a. What is the pH of the buffer solution?
b. Will the buffer pH change if the buffer solution is diluted by a factor of 4?
c. How does the pH change if 0.005 mol/L of the very strong hydrochloric acid, HCl, is added to the original and to the diluted buffer solution?
d. What would be the pH of the 0.005 M HCl solution without any buffer?

**7.9.** Calculate the fractions of $CO_{2(aq)}$, $HCO_3^-$, and $CO_3^{2-}$ related to the dissolved inorganic carbon, DIC, at pH = 7. $pK_a$ ($CO_2$) = 6.36, $pK_a$ ($HCO_3^-$) = 10.33.

**7.10.** In a carbonate-dominated water with pH = 7.7, the acid neutralizing capacity to pH = 4.3 ($ANC_{4.3}$) and the base neutralizing capacity to pH = 8.2 ($BNC_{8.2}$) were determined to be 2.1 and 0.08 mmol/L, respectively. What are the concentrations of $CO_2$, $HCO_3^-$, $CO_3^{2-}$, and DIC?

**7.11.** The pH of pristine rain water at 25 °C calculated from the Henry constant ($H = 0.033$ mol/(L·bar)) and the dissociation constant ($K_{a1} = 4.4 \times 10^{-7}$ mol/L) of $CO_2$ amounts to 5.61 (see Example 7.7). What would be the pH if the $CO_2$ concentration in the atmosphere would increase from the current value 415 to 600 $ppm_v$? The total pressure of the atmosphere is assumed to be $p_{total} = 1$ bar.

**7.12.** The pH in a lake increases from 8.2 in the morning to 10.5 in the evening due to the process of photosynthesis. Calculate the carbonic acid species distribution for both cases. The temperature is assumed to be constant (20 °C), and the respective equilibrium constants at this temperature are $K_{a1} = 4.11 \times 10^{-7}$ mol/L, $K_{a2} = 4.21 \times 10^{-11}$ mol/L, and $K_w = 6.82 \times 10^{-15}$ mol²/L². The alkalinity ($m$ value) was found to be $1.05 \times 10^{-3}$ mol/L.

# 8 Precipitation/Dissolution Equilibria

## 8.1 Introduction

Natural water bodies are often in contact with solid phases that contain inorganic minerals as constituents. In contact with water, such minerals can be dissolved which leads, as a consequence of dissociation during dissolution, to an input of ionic species into the water body. On the other hand, if the concentrations of specific cations and anions in an aqueous system exceed a certain level, these ions can combine to form a solid that precipitates. In summary, it can be stated that dissolution and precipitation are important input and output pathways for ionic inorganic water constituents. Furthermore, precipitation processes can be utilized in drinking water treatment and in wastewater treatment to remove undesired ionic water constituents such as hardness-causing ions ($Ca^{2+}$ and $Mg^{2+}$), phosphate ions, or heavy metal ions.

In this context, it is of practical interest to define the conditions under which precipitation or dissolution will occur. As for other reaction equilibria, the law of mass action can be used to describe the equilibrium conditions. In the case of precipitation/dissolution, the characteristic equilibrium constant is referred to as the solubility product or solubility product constant (Section 8.2). The solubility product as an equilibrium constant should not be confused with the solubility that describes the amount of a solid that can be dissolved. A relationship between both parameters can be derived from the law of mass action and material balances (Section 8.3). Besides the different numerical values, the main difference between the parameters solubility product and solubility consists in the fact that the solubility product as a thermodynamic constant is only influenced by the temperature (pressure dependence can be neglected here), whereas the solubility is further influenced by the ionic strength and possible side reactions with other water constituents.

Frequently, it is also of interest to know if in a considered aqueous system precipitation or dissolution of a mineral compound can be expected or if the system is just in the state of equilibrium. This assessment can be carried out on the basis of a comparison of equilibrium and actually measured concentrations or activities (Section 8.4).

## 8.2 The Solubility Product

Let us consider an inorganic salt $C_mA_{n(s)}$ consisting of the cation $C^{i \cdot n+}$ and the anion $A^{i \cdot m-}$. The reaction equation for the dissolution/precipitation equilibrium is:

$$C_mA_{n(s)} \rightleftharpoons m\,C^{i \cdot n+} + n\,A^{i \cdot m-} \tag{8.1}$$

where $m$ and $n$ are the stoichiometric factors. Due to the electroneutrality condition, the charges of the ions must be reflected in the opposite stoichiometric factors. The

https://doi.org/10.1515/9783110758788-008

factor $i$ is a multiplier that takes into account the fact that in cases where $m = n$, the charges do not necessarily have to be $n+$ and $m-$, but can also be a multiple of them (e.g., $2n+$ and $2m-$). In cases where $m \neq n$, the factor $i$ is typically 1. Some examples for solids with different composition are given in Table 8.1.

**Table 8.1:** Examples of the stoichiometric factors, $m$ and $n$, and the related multiplier of the charge value, $i$.

| Solid | $m$ | $n$ | $i$ |
|---|---|---|---|
| $NaCl \rightleftharpoons Na^+ + Cl^-$ | 1 | 1 | 1 |
| $CaF_2 \rightleftharpoons Ca^{2+} + 2\,F^-$ | 1 | 2 | 1 |
| $Al(OH)_3 \rightleftharpoons Al^{3+} + 3\,OH^-$ | 1 | 3 | 1 |
| $CaCO_3 \rightleftharpoons Ca^{2+} + CO_3^{2-}$ | 1 | 1 | 2 |
| $Ca_3(PO_4)_2 \rightleftharpoons 3\,Ca^{2+} + 2\,PO_4^{3-}$ | 3 | 2 | 1 |
| $AlPO_4 \rightleftharpoons Al^{3+} + PO_4^{3-}$ | 1 | 1 | 3 |

The reaction from left to right describes the dissolution, whereas the reaction from right to left describes the precipitation. The law of mass action related to equation (8.1) reads:

$$K_{sp}^* = a^m(C^{i \cdot n+})\, a^n(A^{i \cdot m-}) \tag{8.2}$$

where $K_{sp}^*$ is the (thermodynamic) solubility product constant. As discussed in Chapter 5, the solid does not occur in the law of mass action. After introducing the conditional solubility product constant, the law of mass action can be rewritten as:

$$K_{sp} = \frac{K_{sp}^*}{\gamma^m(C^{i \cdot n+})\,\gamma^n(A^{i \cdot m-})} = c^m(C^{i \cdot n+})\,c^n(A^{i \cdot m-}) \tag{8.3}$$

In ideal dilute solutions, $K_{sp}$ equals $K_{sp}^*$; otherwise, $K_{sp}^*$ has to be corrected by the activity coefficients to find $K_{sp}$.

Instead of $K_{sp}^*$ (or $K_{sp}$), often the solubility exponent is used, which is defined by:

$$pK_{sp}^* = -\log K_{sp}^* \qquad pK_{sp} = -\log K_{sp} \tag{8.4}$$

Table 8.2 lists solubility exponents for selected minerals. Following the definition of the solubility product, high values of $K_{sp}^*$ or low values of $pK_{sp}^*$ stand for high solubility (high equilibrium activities or concentrations of the ions), whereas hardly soluble compounds are characterized by low values of $K_{sp}^*$ and high values of $pK_{sp}^*$ (low equilibrium activities or concentrations of the ions). The exact relationship between solubility product and solubility is discussed in the next section.

**Table 8.2:** Selected solubility exponents at 25 °C.

| Solid | $pK_{sp}^*$ | Solid | $pK_{sp}^*$ |
|---|---|---|---|
| $CaSO_4$ | 4.3 | $Mn(OH)_2$ | 12.7 |
| $CaSO_4 \cdot 2\,H_2O$ | 4.5 | $Cd(OH)_2$ | 13.8 |
| $CaHPO_4$ | 6.7 | $Ni(OH)_2$ | 13.8 |
| $CaCO_3$ | 8.5 | $Fe(OH)_2$ | 16.3 |
| $MgCO_3$ | 5.2 | $Zn(OH)_2$ | 16.4 |
| $Ca(OH)_2$ | 5.3 | $FeS$ | 18.1 |
| $MgCO_3 \cdot 3\,H_2O$ | 5.6 | $Cu(OH)_2$ | 19.3 |
| $NiCO_3$ | 6.8 | $AlPO_4$ | 20.0 |
| $BaSO_4$ | 9.9 | $FePO_4$ | 21.9 |
| $FeCO_3$ | 10.5 | $Ca_3(PO_4)_2$ | 32.5 |
| $MnCO_3$ | 10.6 | $Al(OH)_3$ | 33.7 |
| $Mg(OH)_2$ | 11.3 | $Fe(OH)_3$ | 38.7 |

## 8.3 Solubility Product and Solubility

### 8.3.1 Relationship Between Solubility Product and Solubility

The relationship between solubility and solubility product depends on the stoichiometric composition of the solid. Let us first consider the example barium sulfate, $BaSO_4$. The reaction equation for dissolution is:

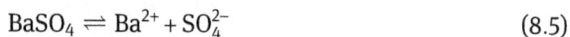

$$BaSO_4 \rightleftharpoons Ba^{2+} + SO_4^{2-} \tag{8.5}$$

It follows from the equation that per mol $BaSO_4$ that is dissolved, 1 mol $Ba^{2+}$ and 1 mol $SO_4^{2-}$ are introduced into the water. Given that the solubility describes the amount of the solid $BaSO_4$ that can be dissolved in water, we can therefore write:

$$c_s = c(Ba^{2+}) = c(SO_4^{2-}) \tag{8.6}$$

where $c_s$ is the solubility that, in this case, equals the saturation concentration of the cation and the saturation concentration of the anion in the state of equilibrium.

If we assume ideal conditions ($a = c$, $K_{sp}^* = K_{sp}$), the law of mass action for the considered reaction is:

$$K_{sp} = c(Ba^{2+})\,c(SO_4^{2-}) \tag{8.7}$$

Introducing equation (8.6) into equation (8.7) gives:

$$K_{sp} = c_s^2 \tag{8.8}$$

or

$$c_s = \sqrt{K_{sp}} \qquad (8.9)$$

Equation (8.9) gives the relationship between solubility and solubility product for solids with a stoichiometric composition cation : anion $= 1 : 1$.

---

**Example 8.1**

What is the solubility of $BaSO_4$ ($M = 233.4$ g/mol) at 25 °C in mmol/L and mg/L under the assumption of an ideal dilute solution? The solubility exponent for the given temperature is $pK_{sp} = 9.9$.

**Solution:**

The solubility product is:

$$K_{sp} = c(Ba^{2+})\, c(SO_4^{2-}) = 10^{-pK_{sp}}\ mol^2/L^2 = 1.26 \times 10^{-10}\ mol^2/L^2$$

The molar solubility can be calculated according to equation (8.9):

$$c_s = \sqrt{K_{sp}} = \sqrt{1.26 \times 10^{-10}\ mol^2/L^2} = 1.12 \times 10^{-5}\ mol/L = 1.12 \times 10^{-2}\ mmol/L$$

and the related mass concentration, $\rho_s^*$, can be found as follows:

$$\rho_s^* = M\, c_s = (233.4\ g/mol)\,(1.12 \times 10^{-5}\ mol/L) = 2.61 \times 10^{-3}\ g/L = 2.61\ mg/L$$

---

For solids with other compositions, different relationships result. This will be demonstrated for calcium fluoride, $CaF_2$, as an example. The reaction equation for the dissolution is:

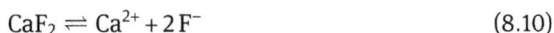

$$CaF_2 \rightleftharpoons Ca^{2+} + 2\,F^- \qquad (8.10)$$

Here, per mol $CaF_2$ that is dissolved, 1 mol $Ca^{2+}$ and 2 mol $F^-$ are introduced into the water. Therefore, the solubility can be expressed as:

$$c_s = c(Ca^{2+}) = 0.5\, c(F^-) \qquad (8.11)$$

and the law of mass action reads:

$$K_{sp} = c(Ca^{2+})\, c^2(F^-) = c_s\,(2\,c_s)^2 = 4\,c_s^3 \qquad (8.12)$$

Consequently, the relationship between $c_s$ and $K_{sp}$ is:

$$c_s = \sqrt[3]{\frac{K_{sp}}{4}} \qquad (8.13)$$

A general relationship between solubility and solubility product for a solid with the composition $C_mA_n$ can be formulated as:

$$c_S = \sqrt[m+n]{\frac{K_{sp}}{m^m\,n^n}} \tag{8.14}$$

where $m$ and $n$ are the formula indices of the compound $C_mA_n$. As can be easily proved, equations (8.9) and (8.13) are special cases of equation (8.14) with $m = 1$, $n = 1$ and $m = 1$, $n = 2$, respectively.

The general relationship between the solubility of a compound $C_mA_n$ and the equilibrium concentrations of the cation ($C^{i\cdot n+}$) and the anion ($A^{i\cdot m-}$) is given by:

$$c_S = \frac{c(C^{i\cdot n+})}{m} = \frac{c(A^{i\cdot m-})}{n} \tag{8.15}$$

Obviously, equation (8.15) includes equations (8.6) and (8.11) as special cases.

Equation (8.14) can directly be used with the thermodynamic constants if the solution is considered an ideal dilute solution where $K_{sp}$ equals $K_{sp}^*$. Otherwise, the activity coefficients have to be introduced. This case will be discussed in more detail in the next section.

### 8.3.2 Influence of the Ionic Strength on the Solubility

For the sake of simplicity, we want to consider a salt CA consisting of the cation $C^+$ and the anion $A^-$ ($m = n = 1$) to demonstrate the influence of the ionic strength on the solubility. The thermodynamic solubility product for this salt reads:

$$K_{sp}^* = a(C^+)\,a(A^-) = \gamma(C^+)\,c(C^+)\,\gamma(A^-)\,c(A^-) \tag{8.16}$$

or

$$\frac{K_{sp}^*}{\gamma(C^+)\,\gamma(A^-)} = K_{sp} = c(C^+)\,c(A^-) \tag{8.17}$$

The relationship between the solubility and the solubility product can be found in the same manner as shown in Section 8.3.1:

$$c_S = c(C^+) = c(A^-) = \sqrt{K_{sp}} = \sqrt{\frac{K_{sp}^*}{\gamma(C^+)\,\gamma(A^-)}} \tag{8.18}$$

From equation (8.18), we can draw a conclusion about the influence of the ionic strength on the solubility. If the ionic strength increases, the activity coefficients decrease from 1 (in an ideal solution) to values lower than 1 and therefore the value of the expression under the root sign increases. Accordingly, the solubility also increases. We can conclude from this example that the solubility of a given salt increases with increasing ionic strength of the aqueous solution. This means that other ionic water constituents can also influence the solubility of a considered solid even though they do not react directly with the considered solid or with its ions produced by dissolution.

---

**Example 8.2**
What is the solubility of $CaSO_4$
a) under ideal conditions (assumption of an ideal dilute solution) and
b) in a water with the ionic strength $I = 0.04$ mol/L?
The solubility product of $CaSO_4$ is $K_{sp}^* = 5 \times 10^{-5}$ mol$^2$/L$^2$.

**Solution:**
a) The solubility for a salt with the composition cation : anion = 1 : 1 can be found from:

$$c_S = \sqrt{K_{sp}}$$

(see Example 8.1)
For an ideal dilute solution $K_{sp}^*$ equals $K_{sp}$ and we get:

$$c_S = \sqrt{K_{sp}} = \sqrt{5 \times 10^{-5} \text{ mol}^2/\text{L}^2} = 7.07 \times 10^{-3} \text{ mol/L}$$

b) If we want to consider the ionic strength, we have to calculate the conditional solubility product by:

$$K_{sp} = \frac{K_{sp}^*}{\gamma(Ca^{2+})\,\gamma(SO_4^{2-})}$$

The activity coefficients can be calculated by means of the Güntelberg equation (Chapter 3, Section 3.8):

$$\log \gamma_z = -0.5\,z^2\,\frac{\sqrt{\dfrac{I}{\text{mol/L}}}}{1 + 1.4\sqrt{\dfrac{I}{\text{mol/L}}}}$$

For a bivalent ion and the given ionic strength, we get:

$$\log \gamma_2 = (-0.5)\,4\,\frac{0.2}{1 + 1.4(0.2)} = -0.313$$

$$\gamma_2 = 0.487$$

With the activity coefficients for the cation and the anion, we can calculate the conditional solubility product:

$$K_{sp} = \frac{5 \times 10^{-5} \text{ mol}^2/\text{L}^2}{(0.487)\,(0.487)} = 2.11 \times 10^{-4} \text{ mol}^2/\text{L}^2$$

The solubility is then:

$$c_S = \sqrt{K_{sp}} = \sqrt{2.11 \times 10^{-4} \text{ mol}^2/\text{L}^2} = 1.45 \times 10^{-2} \text{ mol/L}$$

The solubility for the given ionic strength is about two times higher than the solubility calculated for ideal conditions.

---

### 8.3.3 Influence of Side Reactions on the Solubility

An influence on the solubility can also be expected if the cation or the anion or both react with other water constituents after the dissolution of the solid. Acid/base reactions or complex formations are typical side reactions that may influence the solubility of a considered solid. The effect of a side reaction will be demonstrated below by using calcium carbonate as an example. If calcium carbonate is dissolved in water, at first calcium and carbonate ions are formed as dissociation products. Carbonate, as a component of the carbonic acid system, can subsequently react in a side reaction with the protons from the water dissociation with the result that a part of carbonate is transformed to hydrogencarbonate and possibly also further to $CO_2$ (Section 7.7). The extent of the transformation depends on the pH. Thus, the relevant reactions are:

$$CaCO_3 \rightleftharpoons Ca^{2+} + CO_3^{2-} \tag{8.19}$$

$$CO_3^{2-} + H^+ \rightleftharpoons HCO_3^- \tag{8.20}$$

$$HCO_3^- + H^+ \rightleftharpoons CO_{2(aq)} + H_2O \tag{8.21}$$

The reactions described by equations (8.20) and (8.21) are the reverse reactions to the respective acid dissociations. The laws of mass action for the dissolution and the acid/base reactions under ideal conditions are:

$$K_{sp} = c(Ca^{2+}) \, c(CO_3^{2-}) \tag{8.22}$$

$$K_{a2} = \frac{c(H^+) \, c(CO_3^{2-})}{c(HCO_3^-)} \tag{8.23}$$

$$K_{a1} = \frac{c(H^+) \, c(HCO_3^-)}{c(CO_2)} \tag{8.24}$$

According to equation (8.19), the solubility equals the saturation concentration of $Ca^{2+}$ and the primary concentration of $CO_3^{2-}$. Given that after the dissolution a part of $CO_3^{2-}$ is transformed to $HCO_3^-$ and possibly (at low pH) also to $CO_2$, the solubility must be the sum of the residual concentration of $CO_3^{2-}$ and the concentrations of the formed $HCO_3^-$ and $CO_2$:

$$c_s = c(Ca^{2+}) = c(CO_3^{2-}) + c(HCO_3^-) + c(CO_2) \tag{8.25}$$

To link the solubility to the solubility product, we have to substitute all concentrations that do not occur in the solubility product by means of the equilibrium relationships:

$$C_s = c(\text{Ca}^{2+}) = c(\text{CO}_3^{2-}) + \frac{c(\text{CO}_3^{2-})\, c(\text{H}^+)}{K_{a2}} + \frac{c(\text{CO}_3^{2-})\, c^2(\text{H}^+)}{K_{a1}\, K_{a2}} \quad (8.26)$$

$$C_s = c(\text{Ca}^{2+}) = c(\text{CO}_3^{2-})\left(1 + \frac{c(\text{H}^+)}{K_{a2}} + \frac{c^2(\text{H}^+)}{K_{a1}\, K_{a2}}\right) \quad (8.27)$$

Thus, the solubility product can be written as:

$$K_{sp} = c(\text{Ca}^{2+})\, c(\text{CO}_3^{2-}) = C_s\, \frac{C_s}{1 + \dfrac{c(\text{H}^+)}{K_{a2}} + \dfrac{c^2(\text{H}^+)}{K_{a1}\, K_{a2}}} \quad (8.28)$$

and for the relationship between solubility and solubility product, we get:

$$C_s = \sqrt{K_{sp}\left(1 + \frac{c(\text{H}^+)}{K_{a2}} + \frac{c^2(\text{H}^+)}{K_{a1}\, K_{a2}}\right)} \quad (8.29)$$

Given that $K_{a2}$ and $K_{a1}\, K_{a2}$ are constants, the solubility increases with increasing $c(\text{H}^+)$ or decreasing pH. On the other hand, with increasing pH (decreasing $c(\text{H}^+)$) at first the third term within the brackets can be neglected (if pH $\gg$ 0.5 $(pK_{a1} + pK_{a2}) = 8.35$) which means that the transformation of $CO_3^{2-}$ to $CO_2$ can be neglected and only the transformation to $HCO_3^-$ has to be taken into account. With further increase of pH, the second term can also be neglected (if pH $\gg pK_{a2} = 10.33$). In this case, there is also no substantial transformation of $CO_3^{2-}$ to $HCO_3^-$ and the relationship between solubility and solubility product is the same as for a system without side reactions (see equation (8.9)).

---

**Example 8.3**

What is the solubility of calcium carbonate at pH = 8.5 and pH = 11? The solubility exponent of calcium carbonate is $pK_{sp} = 8.48$. To consider the side reactions of carbonate to hydrogencarbonate and carbon dioxide, the acidity constants of the carbonic acid system are necessary, which are given as $pK_{a1} = 6.36$ and $pK_{a2} = 10.33$.

**Solution:**

To solve the problem, we have to apply equation (8.29).
With $c(\text{H}^+) = 10^{-pH}$ mol/L $= 3.16 \times 10^{-9}$ mol/L, $K_{a1} = 10^{-pK_{a1}}$ mol/L $= 4.37 \times 10^{-7}$ mol/L, $K_{a2} = 10^{-pK_{a2}}$ mol/L $= 4.68 \times 10^{-11}$ mol/L, $K_{a1}\, K_{a2} = 2.05 \times 10^{-17}$ mol$^2$/L$^2$, and $K_{sp} = 10^{-pK_{sp}}$ mol$^2$/L$^2 = 3.31 \times 10^{-9}$ mol$^2$/L$^2$, we find for pH = 8.5:

$$C_s = \sqrt{K_{sp}\left(1 + \frac{c(\text{H}^+)}{K_{a2}} + \frac{c^2(\text{H}^+)}{K_{a1}\, K_{a2}}\right)}$$

$$C_s = \sqrt{3.31 \times 10^{-9}\ \text{mol}^2/\text{L}^2\left(1 + \frac{3.16 \times 10^{-9}\ \text{mol/L}}{4.68 \times 10^{-11}\ \text{mol/L}} + \frac{9.99 \times 10^{-18}\ \text{mol}^2/\text{L}^2}{2.05 \times 10^{-17}\ \text{mol}^2/\text{L}^2}\right)}$$

$$C_s = \sqrt{3.31 \times 10^{-9}\ \text{mol}^2/\text{L}^2\, (1 + 67.52 + 0.49)} = 4.78 \times 10^{-4}\ \text{mol/L}$$

As we can derive from the values of the three terms within the brackets, the solubility at pH = 8.5 is strongly influenced by the transformation of carbonate to hydrogencarbonate (great value of the second summand), whereas the transformation to carbon dioxide plays only a minor role (third summand) and could be neglected.

For pH = 11, the solubility is given by:

$$c_s = \sqrt{3.31 \times 10^{-9}\ mol^2/L^2 \left(1 + \frac{1 \times 10^{-11}\ mol/L}{4.68 \times 10^{-11}\ mol/L} + \frac{1 \times 10^{-22}\ mol^2/L^2}{2.05 \times 10^{-17}\ mol^2/L^2}\right)}$$

$$c_s = \sqrt{3.31 \times 10^{-9}\ mol^2/L^2\ (1 + 0.21 + 4.88 \times 10^{-6})} = 6.33 \times 10^{-5}\ mol/L$$

which is about tenfold lower than the solubility at pH = 8.5. Without consideration of the side reactions, we would find:

$$c_s = \sqrt{K_{sp}} = \sqrt{3.31 \times 10^{-9}\ mol^2/L^2} = 5.75 \times 10^{-5}\ mol/L$$

which demonstrates that the side reactions play only a minor role at pH = 11, because the pH is greater than $pK_{a2}$.

## 8.4 Assessment of the Saturation State of a Solution

To assess the saturation state of a solution with respect to a considered ionic solid (e.g., salt or hydroxide), we can follow the general method described in Chapter 5 (Section 5.4, Table 5.1). This general method is based on the comparison of the reaction quotient, $Q$, with the equilibrium constant, $K^*$, where $Q$ is calculated from an expression equal to the law of mass action but with measured activities (or concentrations) instead of the equilibrium activities (concentrations). Although in the case of dissolution/precipitation, the law of mass action and therefore also $Q$ is not a quotient but a product, we want to keep the letter $Q$ also for these specific types of reaction. Note that in literature sometimes the term ion activity product ($IAP$) is used instead of $Q$.

For the dissolution/precipitation of a salt CA (stoichiometric factors $m = n = 1$) under ideal conditions ($K_{sp} = K_{sp}^*$, $c = a$), we can calculate $Q$ by:

$$Q = c_{meas}(C^{i \cdot n+})\, c_{meas}(A^{i \cdot m-}) \tag{8.30}$$

After that, $Q$ has to be compared with $K_{sp}$:

$$K_{sp} = c_{eq}(C^{i \cdot n+})\, c_{eq}(A^{i \cdot m-}) \tag{8.31}$$

to find the saturation state. The subscript "eq" is here introduced to indicate the difference to the measured concentrations.

If $Q > K_{sp}$, the product of the measured concentrations is greater than the product of the equilibrium concentrations. This means that the solution is supersaturated and therefore precipitation of CA should occur. It has to be noted that in practice sometimes the state of supersaturation can exist over longer periods if the precipitation is kinetically hindered. In engineered processes, often seed crystals are introduced into the system to break the metastable state of supersaturation and to allow the solid to precipitate. Generally, after a shorter or longer time, precipitation from a supersaturated solution will occur until the solution reaches the state of equilibrium where $Q = K_{sp}$.

If $Q < K_{sp}$, the product of the measured concentrations is lower than the product of the equilibrium concentrations, which means that the solution is undersaturated. If the solid material (CA in our case) is present in the system (e.g., as sediment in a water body), dissolution can be expected as long as the state of equilibrium ($Q = K_{sp}$) is reached or all solid material is consumed.

If in the considered water $Q$ equals $K_{sp}$, the solution is just in the state of equilibrium and neither precipitation nor dissolution will occur.

---

**Example 8.4**

What is the saturation state of a water with respect to $Cd(OH)_2$ if the $Cd^{2+}$ concentration in the water is $c(Cd^{2+}) = 1 \times 10^{-8}$ mol/L and the pH is measured to be pH = 8? $pK_{sp}$ ($Cd(OH)_2$) = 13.8, $pK_w$ = 14.

**Solution:**

To solve the problem, first we have to find $c(OH^-)$ and $K_{sp}$ by:

$$pOH = pK_w - pH = 14 - 8 = 6$$

$$c(OH^-) = 10^{-pOH} \text{ mol/L} = 1 \times 10^{-6} \text{ mol/L}$$

$$K_{sp} = c(Cd^{2+})\, c^2(OH^-) = 10^{-pK_{sp}} \text{ mol}^3/L^3 = 1.58 \times 10^{-14} \text{ mol}^3/L^3$$

Then, we can calculate $Q$:

$$Q = c_{meas}(Cd^{2+})\, c^2_{meas}(OH^-) = (1 \times 10^{-8} \text{ mol/L}) (1 \times 10^{-12} \text{ mol}^2/L^2) = 1 \times 10^{-20} \text{ mol}^3/L^3$$

Obviously, $Q$ is lower than $K_{sp}$ and, accordingly, the water is undersaturated with respect to $Cd(OH)_2$.

---

# 8.5 Problems

Note: We assume that ideal conditions exist in all cases ($a = c$, $K^* = K$). If not stated otherwise, the constants are valid for 25 °C.

**8.1.** In water treatment and other technical processes, alkaline solutions of calcium hydroxide are frequently used for precipitation or neutralization reactions. What is the pH of a saturated $Ca(OH)_2$ solution? The solubility product of calcium hydroxide is $K_{sp} = 5 \times 10^{-6}$ mol$^3$/L$^3$ and the ion product of water is $K_w = 1 \times 10^{-14}$ mol$^2$/L$^2$.

**8.2.** We want to assume that nickel has to be removed from wastewater by precipitation as hydroxide. The mass concentration of $Ni^+$ after the treatment should not be higher than 0.5 mg/L. What is the minimum value the pH of the water has to be raised to in order to meet this limiting concentration? The solubility exponent of nickel hydroxide is $pK_{sp}(Ni(OH)_2) = 13.8$, the molecular weight of nickel is $M(Ni) = 58.7$ g/mol, and the ion product of water is $K_w = 1 \times 10^{-14}$ mol$^2$/L$^2$.

**8.3.** Is a pH value of 9 high enough to precipitate Cd$^{2+}$ as Cd(OH)$_2$ from a wastewater to such an extent that the residual concentration is lower than 0.1 mg/L? The solubility product of cadmium hydroxide is $K_{sp} = 1.6 \times 10^{-14}$ mol$^3$/L$^3$, the molecular weight of cadmium is $M(Cd) = 112.4$ g/mol, and the ion product of water is $K_w = 1 \times 10^{-14}$ mol$^2$/L$^2$.

**8.4.** An alkaline water with pH = 11 contains 2 mmol/L $CO_3^{2-}$ and 0.5 mmol/L Mg$^{2+}$. Can we expect that precipitation of MgCO$_3$ and/or Mg(OH)$_2$ occurs? How is the situation for Ca$^{2+}$ compared to Mg$^{2+}$ under the same conditions (same concentrations and pH)? $pK_{sp}(MgCO_3) = 5.2$, $pK_{sp}(CaCO_3) = 8.5$, $pK_{sp}(Mg(OH)_2) = 11.3$, $pK_{sp}(Ca(OH)_2) = 5.3$.

# 9 Calco-Carbonic Equilibrium

## 9.1 Introduction

The calco-carbonic equilibrium (also referred to as the lime/carbonic acid equilibrium) is one of the fundamental reaction equilibria in the hydrosphere and a descriptive example for combined reaction equilibria as are typical for aquatic systems. It links the carbonic acid system with the precipitation/dissolution of calcium carbonate (mineral name: calcite). The major ions calcium and hydrogencarbonate, present in all natural waters in relatively high concentrations, as well as the dissolved carbon dioxide are involved in this system. Through the dissociation of the carbonic acid, a direct link exists to the pH of the water.

Dissolution of calcite is one of the main sources of the calcium and hydrogencarbonate input into natural waters. On the other hand, under certain conditions, calcium and hydrogencarbonate ions can combine to form calcite which precipitates. In particular, precipitation may occur when $CO_2$ is removed from the water by consumption in chemical reactions or by heating. The latter process is also known as scaling (see also carbonate hardness, Section 3.7 in Chapter 3).

Since dissolved $CO_2$ behaves as a diprotic acid ("carbonic acid"), the ratio of $CO_2$ and hydrogencarbonate as well as the ratio of hydrogencarbonate and carbonate is related to the pH (Chapter 7). As we have already seen in Chapter 8, the pH also influences the dissolution/precipitation equilibrium of calcium carbonate.

When going into more detail, we can see that this complex system is composed of three elementary reactions, which are the first and the second dissociation step of carbonic acid and the precipitation/dissolution equilibrium of calcite. From the addition of these elementary reactions, an overall reaction equation for the calco-carbonic equilibrium can be derived in the following manner:

$$CO_{2(aq)} + H_2O \rightleftharpoons H^+ + HCO_3^- \qquad\qquad K_{a1}^* \qquad\qquad (9.1)$$

$$H^+ + CO_3^{2-} \rightleftharpoons HCO_3^- \qquad\qquad 1/K_{a2}^* \qquad\qquad (9.2)$$

$$CaCO_{3(s)} \rightleftharpoons Ca^{2+} + CO_3^{2-} \qquad\qquad K_{sp}^* \qquad\qquad (9.3)$$

$$\overline{CaCO_{3(s)} + CO_{2(aq)} + H_2O \rightleftharpoons Ca^{2+} + 2\,HCO_3^- \qquad\qquad K_{a1}^* K_{sp}^*/K_{a2}^* \qquad (9.4)}$$

Note that the second dissociation equation has to be written in the reverse direction to find the overall equation. Accordingly, the reciprocal acidity constant is valid for this reaction.

When a water is in the state of the calco-carbonic equilibrium, calcium carbonate will neither be dissolved nor precipitated. Due to the complexity of the system, there is no single equilibrium state but an infinite number of possible equilibrium states, each of them characterized by a specific combination of the following

https://doi.org/10.1515/9783110758788-009

parameters: $c(CO_2)$, $c(HCO_3^-)$, $c(CO_3^{2-})$, $c(Ca^{2+})$, and pH (Section 9.3). The different situations that may occur in aqueous systems can be depicted by means of the given overall equation.

If a water contains a higher concentration of $CO_2$ than in the state of equilibrium, calcium carbonate will be dissolved under consumption of $CO_2$ and formation of $Ca^{2+}$ and $HCO_3^-$ (reaction from left to right) until the equilibrium state is reached. Higher concentrations of $CO_2$ are often found in seepage water and groundwater as a result of biological degradation processes in the soil layer. The formed $CO_2$ is dissolved and transported with the seepage water to the groundwater. The establishment of a new equilibrium state (related to the higher $CO_2$ concentration) by dissolution of carbonate is only possible if solid calcium carbonate is available in the subsurface and the contact time of the water with the carbonate is long enough to allow the dissolution of the required amount of carbonate. If these conditions are not fulfilled, the water remains in a non-equilibrium state with an excess concentration of $CO_2$. Since $CO_2$ is corrosive, such water cannot be distributed as drinking water without a specific treatment that removes the excess concentration of $CO_2$. This treatment is referred to as deacidification and is often applied in waterworks that use groundwater as raw water source.

If the water contains a lower concentration of $CO_2$ in comparison to the equilibrium state, calcium and hydrogencarbonate ions form solid calcium carbonate that precipitates until the equilibrium state is reached (reaction from right to left). Such a situation can be found, for instance, in lakes with a high photosynthetic activity of submerged aquatic plants and phytoplankton because $CO_2$ is consumed during photosynthesis. The establishment of a new equilibrium state by formation of $CO_2$ and simultaneous precipitation of $CaCO_3$ can be observed in calcium-rich lakes as turbidity caused by fine white carbonate particles. This effect is also known as biogenic decalcification.

From the equilibrium relationships of the single reactions (equations (9.1)–(9.3)), specific equations can be derived that can be used as the basis for a graphical representation of the calco-carbonic equilibrium (Section 9.3) and for the assessment of the calcite saturation state of the water (Section 9.4).

## 9.2 Basic Equations

The laws of mass action for the single reactions of the calco-carbonic equilibrium are:

$$K_{a1}^* = \frac{a(H^+)\,a(HCO_3^-)}{a(CO_2)} \tag{9.5}$$

$$K_{a2}^* = \frac{a(H^+)\,a(CO_3^{2-})}{a(HCO_3^-)} \tag{9.6}$$

$$K_{sp}^* = a(\text{Ca}^{2+})\, a(\text{CO}_3^{2-}) \tag{9.7}$$

For practical applications, the laws of mass action have to be rewritten with measurable concentrations and activity coefficients instead of activities ($a = y\,c$). There are two exceptions: i) the proton activity is directly measurable by a pH-meter and must therefore not be substituted, and ii) in contrast to the ions, the behavior of neutral $CO_2$ can be considered as ideal with the consequence that $a(CO_2)$ equals $c(CO_2)$. Furthermore, in deviation from the conventions made in Section 5.3 (Chapter 5), the liquid-phase concentration instead of the partial pressure is used here for the dissolved gas $CO_2$. Under these conditions, the set of equations reads:

$$\frac{K_{a1}^*}{y(\text{HCO}_3^-)} = \frac{K_{a1}^*}{f_{a1}} = \frac{a(\text{H}^+)\, c(\text{HCO}_3^-)}{c(\text{CO}_2)} \tag{9.8}$$

$$\frac{K_{a2}^*\, y(\text{HCO}_3^-)}{y(\text{CO}_3^{2-})} = \frac{K_{a2}^*}{f_{a2}} = \frac{a(\text{H}^+)\, c(\text{CO}_3^{2-})}{c(\text{HCO}_3^-)} \tag{9.9}$$

$$\frac{K_{sp}^*}{y(\text{Ca}^{2+})\, y(\text{CO}_3^{2-})} = \frac{K_{sp}^*}{f_{sp}} = c(\text{Ca}^{2+})\, c(\text{CO}_3^{2-}) \tag{9.10}$$

The activity coefficients occurring in the equations are summarized to $f_{a1}$, $f_{a2}$, and $f_{sp}$. As shown in Chapter 3, the activity coefficient of an ion with the charge $z$ is related to the activity coefficient of a univalent ion, $y_1$, by:

$$y_z = y_1^{z^2} \qquad \log y_z = z^2 \log y_1 \tag{9.11}$$

Accordingly, the following relationships for $f_{a1}$, $f_{a2}$, and $f_{sp}$ can be derived:

$$f_{a1} = y(\text{HCO}_3^-) = y_1 \qquad\qquad \log f_{a1} = \log y_1 \tag{9.12}$$

$$f_{a2} = \frac{y(\text{CO}_3^{2-})}{y(\text{HCO}_3^-)} = \frac{y_1^4}{y_1} = y_1^3 \qquad\qquad \log f_{a2} = 3\log y_1 \tag{9.13}$$

$$f_{sp} = y(\text{Ca}^{2+})\, y(\text{CO}_3^{2-}) = y_1^4\, y_1^4 = y_1^8 \qquad \log f_{sp} = 8\log y_1 \tag{9.14}$$

The activity coefficient of a univalent ion, $y_1$, can be calculated on the basis of the ionic strength by means of the Güntelberg equation, where the ionic strength can be calculated from the concentrations of the major ions or can be estimated from the electrical conductivity (details are given in Chapter 3, Section 3.8).

For an exact description of the calco-carbonic equilibrium, it is not only necessary to consider the activity coefficients but also the temperature dependence of the equilibrium constants. The latter can be expressed by a polynomial with $A$ and $B$ as characteristic parameters for the different equilibrium constants (Table 9.1):

$$\log K^*(T) = \log K^*(T_0) + A\left(\frac{1}{T_0} - \frac{1}{T}\right) + B\left[\ln\left(\frac{T}{T_0}\right) + \left(\frac{T_0}{T}\right) - 1\right] \qquad (9.15)$$

In equation (9.15), the temperature, $T$, has to be expressed as absolute temperature in Kelvin, and $T_0$ is 298.15 K (25 °C). Equation (9.15) is equivalent to equation (5.14) (Chapter 5, Section 5.6) with $A = \Delta_R H^0/(2.303\ R)$ and $B = c_p/(2.303\ R)$, where $\Delta_R H^0$ is the molar standard enthalpy of reaction, $c_p$ is the molar heat capacity of reaction, and $R$ is the gas constant. The factor 2.303 results from the conversion of the natural into the decimal logarithm.

Table 9.2 lists equilibrium constants calculated for selected temperatures. Equations (9.8)–(9.10) and (9.12)–(9.15) provide the basis for performing calco-carbonic equilibrium calculations.

**Table 9.1:** Equilibrium constants at 25 °C for the main reactions of the calco-carbonic equilibrium and parameters needed in equation (9.15) for estimating constants at other temperatures (taken from German standard DIN 38404-10).

| Reaction | Constant | log $K_0$ (25 °C) | A | B |
|---|---|---|---|---|
| $CaCO_{3(s)} \rightleftharpoons Ca^{2+} + CO_3^{2-}$ | $K_{sp}^*$ | −8.481 | −522.3 | −13.06 |
| $CO_{2(aq)} + H_2O \rightleftharpoons H^+ + HCO_3^-$ | $K_{a1}^*$ | −6.356 | 483.1 | −17.23 |
| $HCO_3^- \rightleftharpoons H^+ + CO_3^{2-}$ | $K_{a2}^*$ | −10.329 | 780.8 | −15.15 |
| $H_2O \rightleftharpoons H^+ + OH^-$ | $K_w^*$ | −13.996 | 2 952.5 | −10.29 |

**Table 9.2:** Constants of the calco-carbonic equilibrium for selected temperatures, including Tillmans constant and Langelier constant defined in Sections 9.3 and 9.4, respectively (taken from German standard DIN 38404-10).

| Temperature (°C) | log $K_{sp}^*$ | log $K_{a1}^*$ | log $K_{a2}^*$ | log $K_w^*$ | log $K_T$ | log $K_{La}$ |
|---|---|---|---|---|---|---|
| 5 | −8.387 | −6.515 | −10.555 | −14.733 | 4.347 | −2.168 |
| 10 | −8.406 | −6.465 | −10.488 | −14.535 | 4.383 | −2.082 |
| 15 | −8.428 | −6.422 | −10.429 | −14.346 | 4.421 | −2.001 |
| 20 | −8.453 | −6.386 | −10.376 | −14.166 | 4.463 | −1.923 |
| 25 | −8.481 | −6.356 | −10.329 | −13.996 | 4.508 | −1.848 |

## 9.3 Graphical Representation of the Calco-Carbonic Equilibrium: Tillmans Curve

As mentioned in Section 9.1, each equilibrium state in the calco-carbonic system is characterized by the following set of variables: $c(CO_2)$, $c(HCO_3^-)$, $c(CO_3^{2-})$, $c(Ca^{2+})$, pH. However, according to the overall reaction equation (equation (9.4)), only three variables are necessary to describe the equilibrium uniquely. These variables are the carbonic acid concentration, the hydrogencarbonate concentration, and the calcium concentration. If these concentrations are known, the pH is given by the ratio of $c(CO_2)/c(HCO_3^-)$ and the related acidity constant (equation (9.8)), whereas the carbonate concentration is given by the solubility product and the calcium concentration (equation (9.10)). If we further set the calcium concentration to a fixed value, then the calco-carbonic equilibrium can be represented in a two-dimensional diagram as a curve $c(CO_2) = f(c(HCO_3^-))$ with $c(Ca^{2+})$ as the curve parameter. In practice, the curve parameter is typically given as deviation from the stoichiometric ratio between hydrogencarbonate and calcium ions in the form $c(HCO_3^-) - 2\,c(Ca^{2+})$, or, more general, $m - 2\,c(Ca^{2+})$ where $m$ is the $m$ alkalinity. This difference has the value 0 if both species occur in the exact stoichiometric ratio given by equation (9.4): $c(HCO_3^-) = 2\,c(Ca^{2+})$. For the definition of $m$ and its relationship to $c(HCO_3^-)$, see Section 7.7.3 in Chapter 7.

The equation for the equilibrium curve can be derived after rearranging equation (9.8) and substituting $a(H^+)$ by equation (9.9) and $c(CO_3^{2-})$ by equation (9.10):

$$c(CO_2) = \frac{a(H^+)\,c(HCO_3^-)\,f_{a1}}{K_{a1}^*} = \frac{K_{a2}^*\,c^2(HCO_3^-)\,f_{a1}}{f_{a2}\,c(CO_3^{2-})\,K_{a1}^*} = \frac{K_{a2}^*\,c^2(HCO_3^-)\,f_{a1}\,f_{sp}\,c(Ca^{2+})}{f_{a2}\,K_{sp}^*\,K_{a1}^*} \qquad (9.16)$$

$$c(CO_2) = \frac{K_{a2}^*\,f_{a1}\,f_{sp}}{K_{a1}^*\,K_{sp}^*\,f_{a2}}\,c^2(HCO_3^-)\,c(Ca^{2+}) \qquad (9.17)$$

This equation was first described by Josef Tillmans and is therefore referred to as the Tillmans equation and the corresponding graph is called the Tillmans curve. A simpler mathematical form can be derived by summarizing the constants and the activity coefficients:

$$c(CO_2) = \frac{K_T}{f_T}\,c^2(HCO_3^-)\,c(Ca^{2+}) \qquad (9.18)$$

where $K_T$ is the Tillmans constant $(= K_{a2}^*/(K_{a1}^*\,K_{sp}^*)$, see Table 9.2) and $f_T$ is the summarized activity coefficient of the Tillmans equation. The activity coefficient $f_T$ can be calculated from the activity coefficient of a univalent ion under consideration of equations (9.12)–(9.14):

$$f_T = \frac{f_{a2}}{f_{a1}\,f_{sp}} = \frac{\gamma_1^3}{\gamma_1\,\gamma_1^8} = \frac{1}{\gamma_1^6} \qquad (9.19)$$

To determine the activity coefficient $\gamma_1$, the ionic strength has to be known. The ionic strength can be calculated exactly from the concentrations of all (major) ions or can be approximated by an empirical equation based on the electrical conductivity (Chapter 3, Section 3.8).

**Example 9.1**

What is the equilibrium concentration of $CO_2$ in a groundwater for which the following data are known: $\vartheta = 15$ °C, $c(Ca^{2+}) = 1.2$ mmol/L, $c(HCO_3^-) = 2.5$ mmol/L, $I = 5.2$ mmol/L? The Tillmans constant at 15 °C is given as $\log K_T = 4.421$.

**Solution:**

The equilibrium concentration of $CO_2$ can be calculated by means of the Tillmans equation:

$$c(CO_2) = \frac{K_T}{f_T} \, c^2(HCO_3^-) \, c(Ca^{2+})$$

$K_T$ can be derived from the decimal logarithm ($\log K_T = 4.421$) by:

$$K_T = 10^{4.421} \, L^2/mol^2 = 2.636 \times 10^4 \, L^2/mol^2$$

To calculate $f_T$, we need the activity coefficient of a univalent ion, which can be found from the ionic strength (in mol/L!) by means of the Güntelberg equation:

$$\log \gamma_1 = -0.5 \frac{\sqrt{\dfrac{I}{mol/L}}}{1+1.4\sqrt{\dfrac{I}{mol/L}}} = -0.5 \frac{\sqrt{5.2 \times 10^{-3}}}{1+1.4\sqrt{5.2 \times 10^{-3}}} = -0.5 \frac{0.072}{1+(1.4)\,(0.072)} = -0.0327$$

$$\gamma_1 = 0.927$$

With $\gamma_1$, we can calculate the activity coefficient of the Tillmans equation:

$$f_T = \frac{1}{\gamma_1^6} = 1.576$$

Now, we can insert all data into the Tillmans equation to find $c(CO_2)$:

$$c(CO_2) = \frac{2.636 \times 10^4 \, L^2/mol^2}{1.576} (2.5 \times 10^{-3} \, mol/L)^2 \, (1.2 \times 10^{-3} \, mol/L)$$

$$c(CO_2) = 1.254 \times 10^{-4} \, mol/L = 0.1254 \, mmol/L$$

A graphical representation of the Tillmans equation is shown in Figure 9.1. The curve represents the equilibrium states, here calculated for the conditions $c(HCO_3^-) -$ $2\, c(Ca^{2+}) = 0$ and $\vartheta = 25$ °C and under the simplifying assumption $f_T = 1$. In the area above the curve, we find all water compositions (with respect to $CO_2$ and $HCO_3^-$) where the $CO_2$ concentration is higher than in the state of equilibrium. Such waters can reach the equilibrium state by dissolving calcium carbonate (calcite) as can be seen from equation (9.4) (reaction from left to right). They are therefore referred to as carbonate-dissolving or calcite-dissolving waters. Waters with compositions that fall into the area below the equilibrium curve contain lower $CO_2$ concentrations in comparison to the equilibrium. They can reach the equilibrium state by precipitating calcium carbonate (calcite) (equation (9.4), reaction from right to left). Such waters are referred to as carbonate-precipitating or calcite-precipitating waters.

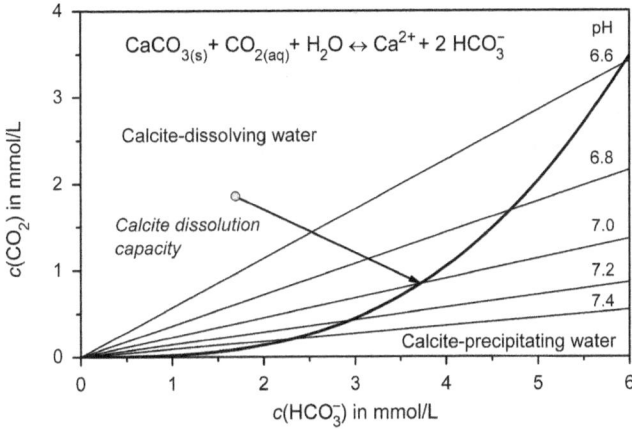

**Figure 9.1:** Tillmans curve, calculated for $c(HCO_3^-) - 2\, c(Ca^{2+}) = 0$, $f_T = 1$, and $\vartheta = 25\,°C$. The gray point indicates the composition of a calcite-dissolving water and the arrow shows the change of the water composition when the equilibrium is reached by dissolution of calcite according to the given reaction equation. The slope of the arrow (−0.5) results from the stoichiometry of the reaction. The length of the arrow is a measure of the calcite dissolution capacity.

The lines in the sketch represent water compositions with equal pH. The respective linear equations can be derived from equation (9.8):

$$c(CO_2) = \frac{a(H^+)\, f_{a1}}{K_{a1}^*}\, c(HCO_3^-) \tag{9.20}$$

As can be seen from equation (9.20), the slope of the line is determined by the proton activity. The higher the proton activity or the lower the pH is, the steeper the line is. The lines of constant pH intersect the equilibrium curve at different points. Consequently, each equilibrium state is characterized by its own specific pH, referred to as equilibrium pH ($pH_{eq}$) or pH of calcite saturation ($pH_c$). Therefore, the difference between the actually measured pH and the equilibrium pH can be used to assess the saturation state (Section 9.4).

To quantify the distance from the equilibrium state, the parameter "calcite dissolution capacity," which is directly linked to the $CO_2$ excess, is often used (Section 9.4). The higher the $CO_2$ excess is, the more calcite can be dissolved (see equation (9.4)) or, in other words, the higher is the calcite dissolution capacity. In an analogous manner, for waters with $CO_2$ deficit the parameter "calcite precipitation capacity" is defined.

Figure 9.2 shows the influence of the deviation from the stoichiometric ratio of $Ca^{2+}$ and $HCO_3^-$ on the position of the Tillmans curve. The deviations are here expressed by the parameter $c(HCO_3^-) - 2\, c(Ca^{2+})$. The reason for the deviations from the stoichiometric ratio in natural waters is that hydrogencarbonate and calcium ions can be introduced into water not only by dissolution of calcite but also partly from other sources. Consequently, the exact stoichiometric ratio is not necessarily found in each type of water. If $c(HCO_3^-) > 2\ c(Ca^{2+})$, the curve is shifted to the right in comparison to the curve for the exact stoichiometric ratio. This means that for the same hydrogencarbonate concentration, the equilibrium $CO_2$ concentration is lower and the pH is higher under these conditions. In contrast, if $c(HCO_3^-) < 2\ c(Ca^{2+})$, the curve is shifted to the left and the equilibrium $CO_2$ concentration is higher and the pH is lower than in the reference state.

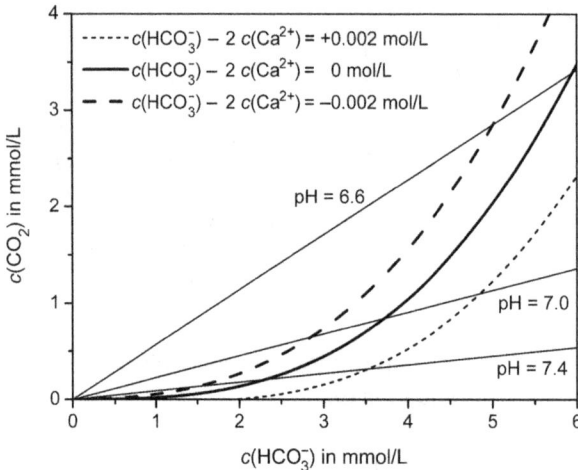

**Figure 9.2:** Tillmans curves ($f_T = 1$, $\vartheta = 25\ °C$) for varying equivalent concentration differences, $c(HCO_3^-) - 2\ c(Ca^{2+})$.

It has to be noted that the exact position of the Tillmans curve also depends on the temperature (due to the temperature dependence of the equilibrium constant, $K_T$) and on the ionic strength (due to the impact on the activity coefficient, $f_T$).

## 9.4 Assessment of the Calcite Saturation State

Since calcite-dissolving waters (waters with $CO_2$ excess) cause corrosion problems in the water distribution system, drinking water should be in the state of calco-carbonic equilibrium when it is delivered by the waterworks. If necessary, a specific treatment process (mechanical or chemical deacidification) has to be applied to remove the excess concentration of $CO_2$. To control the raw water quality and the quality of the treated water, an assessment method is required that allows answering of the question of whether the considered water is in the state of calco-carbonic equilibrium or if the water is calcite-dissolving or, in contrast, calcite-precipitating.

In principle, an assessment of the calcite saturation could be done on the basis of the Tillmans equation. For given calcium and hydrogencarbonate concentrations, the equilibrium concentration of $CO_2$ can be calculated by equation (9.18). This $CO_2$ concentration has to then be compared with the measured concentration. However, the determination of dissolved $CO_2$ by titration (see Section 7.7 in Chapter 7) is laborious and prone to error. Against this background, it is a better alternative to use the pH as an assessment parameter, because it can be easily and precisely measured with a pH-meter. Accordingly, the respective assessment principle is to calculate the equilibrium pH and to compare this value with the measured pH in the given water sample.

An equation that allows the calculation of the equilibrium pH can be derived by combining equations (9.9) and (9.10):

$$a(H^+) = \frac{K_{a2}^* f_{sp}}{K_{sp}^* f_{a2}} c(HCO_3^-) c(Ca^{2+}) \tag{9.21}$$

This equation is known as the Langelier equation, sometimes also referred to as the Langelier–Strohecker equation. After summarizing the constants and the activity coefficients, we receive:

$$a(H^+) = K_{La} f_{La} c(HCO_3^-) c(Ca^{2+}) \tag{9.22}$$

where $K_{La}$ $(= K_{a2}^*/K_{sp}^*)$ is the Langelier constant (see also Table 9.2) and $f_{La}$ is the summarized activity coefficient of the Langelier equation. The relationship between $f_{La}$ and the activity coefficient of a univalent ion, $\gamma_1$, can be found by considering equations (9.13) and (9.14):

$$f_{La} = \frac{f_{sp}}{f_{a2}} = \frac{\gamma_1^8}{\gamma_1^3} = \gamma_1^5 \qquad \log f_{La} = 5 \log \gamma_1 \tag{9.23}$$

For the determination of $\gamma_1$, see Chapter 3, Section 3.8.

To find an expression for the pH, equation (9.22) has to be written in logarithmic form:

$$pH_{eq} = -\log K_{La} - \log f_{La} - \log c(HCO_3^-) - \log c(Ca^{2+}) \tag{9.24}$$

Equation (9.24) can be used to find the equilibrium pH that is related to the measured concentrations of the calcium and hydrogencarbonate ions. This pH has to then be compared with the measured pH to assess the calcite saturation state in the given water sample. The difference between measured and calculated (equilibrium) pH is referred to as the saturation index, $SI$:

$$SI = pH_{meas} - pH_{eq} \tag{9.25}$$

A negative *SI* (measured pH lower than equilibrium pH) indicates calcite-dissolving water whereas a positive *SI* (measured pH higher than equilibrium pH) indicates calcite-precipitating water (see also Figure 9.1). If *SI* = 0, the water is in the state of equilibrium (i.e., saturated with calcite).

---

**Example 9.2**

For a given water sample, the following data are known: $c(Ca^{2+}) = 3.5$ mmol/L, $c(HCO_3^-) = 5.3$ mmol/L (measured as acid neutralizing capacity to pH = 4.3), electrical conductivity $\kappa_{25} = 96$ mS/m, temperature $\vartheta = 10$ °C, and $pH_{meas} = 6.9$. What is the saturation state of this water with respect to calcite?

**Solution:**

At first, $\log K_{La}$ and $\log f_{La}$ have to be determined. The value of $\log K_{La}$ for 10 °C, calculated from $\log K_{a2}^*$ and $\log K_{sp}^*$, is given in Table 9.2 ($\log K_{La} = -2.082$). To calculate $\log f_{La}$, we need the ionic strength, *I*. According to the available data, the ionic strength has to be calculated approximately by using the electrical conductivity (see Chapter 3):

$$I(mol/L) \approx \frac{\kappa_{25}(mS/m)}{6\,200} = 0.0155$$

The logarithm of the activity coefficient of a univalent ion, $\log \gamma_1$, can be found from the Güntelberg equation:

$$\log \gamma_1 = -0.5 \frac{\sqrt{\dfrac{I}{mol/L}}}{1+1.4\sqrt{\dfrac{I}{mol/L}}} = -0.5\frac{0.124}{1+1.4\times 0.124} = -0.053$$

According to equation (9.23), $\log f_{La}$ is:

$$\log f_{La} = 5 \log \gamma_1 = -0.265$$

Now, the equilibrium pH can be calculated by using equation (9.24):

$$pH_{eq} = -\log K_{La} - \log f_{La} - \log c(HCO_3^-) - \log c(Ca^{2+})$$

Here, it has to be considered that the concentrations have to be inserted in mol/L according to the conventions made for the formulation of the laws of mass action and the related units of the equilibrium constants:

$$\log c(HCO_3^-) = \log (5.3\times 10^{-3}) = -2.276 \qquad \log c(Ca^{2+}) = \log (3.5\times 10^{-3}) = -2.456$$

Thus, we find the equilibrium pH to be:

$$pH_{eq} = 2.082 + 0.265 + 2.276 + 2.456 = 7.079 \approx 7.1$$

Finally, the equilibrium pH has to be compared with the measured pH ($pH_{meas} = 6.9$):

$$SI = pH_{meas} - pH_{eq} = 6.9 - 7.1 = -0.2$$

Since *SI* is negative, the considered water is calcite-dissolving.

---

To determine the calcite dissolution capacity (or inversely precipitation capacity) which is a measure of the distance from the equilibrium state, a numerical simulation of the calcite dissolution (or precipitation) on the basis of equation (9.4) has to be carried out by means of an iteration procedure. After each iteration step, the *SI* is calculated to prove if the equilibrium state is achieved. In the iteration procedure, it has to be taken into account that during dissolution or precipitation of calcium carbonate not only the concentrations of calcium, hydrogencarbonate, and $CO_{2(aq)}$ change their values, but also the pH increases (dissolution) or decreases (precipitation) as can be derived from Figure 9.1.

To calculate the changes of the relevant concentrations during the simulated dissolution or precipitation, we have to know the initial values. Whereas the concentrations of $Ca^{2+}$ and $HCO_3^-$ and also the pH are typically known (see Example 9.2), the related initial concentration of $CO_{2(aq)}$ has to be calculated from the known data by means of equation (9.8). Without the knowledge of the $CO_{2(aq)}$ concentration, the pH after the simulated dissolution or precipitation could not be determined. However, this pH (the new "measured" pH) is required as reference value to calculate *SI*.

If, for example, the water is initially calcite-dissolving, the dissolution of a small amount of calcite is assumed and the resulting concentration increase for calcium and hydrogencarbonate and the concentration decrease of $CO_{2(aq)}$ are calculated under consideration of the stoichiometric ratios given in equation (9.4). With the new $CO_{2(aq)}$ and $HCO_3^-$ concentrations, the new pH can be calculated by equation (9.8). This is the pH that would be measured for the given new concentrations. With the new calcium and hydrogencarbonate concentrations, the related equilibrium pH can be calculated by means of the Langelier equation. Finally the new *SI* can be computed. If the *SI* is still negative, dissolution of a higher amount of calcite has to be simulated. This iteration is done as long as the saturation index is zero. The related dissolved amount of calcite is the calcite dissolution capacity. For calcite-precipitating waters, the calcite precipitation capacity can be calculated in an analogous manner but with the simulation of $CaCO_3$ precipitation.

It has to be noted that for an exact calculation the change of the ionic strength and the resulting changes in $f_{a1}$ and $f_{La}$ have to be considered in each iteration step. The iteration procedure is demonstrated in detail in Example 9.3.

---

**Example 9.3**
In this example, the first iteration step for determining the calcite dissolution capacity is shown. The water data are taken from Example 9.2 (calcite-dissolving water): $c(Ca^{2+}) = 3.5$ mmol/L, $c(HCO_3^-) = 5.3$ mmol/L, pH = 6.9 ($a(H^+) = 1.259 \times 10^{-7}$ mol/L), log $K_{La} = -2.082$, $I = 0.0155$ mol/L, $\gamma_1 = 0.885$. Furthermore, we need the acidity constant for the first dissociation step of $CO_{2(aq)}$, which is $K_{a1}^* = 3.428 \times 10^{-7}$ mol/L at the given temperature of 10 °C (Table 9.2).

**Solution:**
We start with the calculation of $c(CO_2)$ for the given initial conditions by means of equation (9.20):

$$c(CO_2) = \frac{a(H^+) f_{a1}}{K^*_{a1}} c(HCO_3^-) = \frac{(1.259 \times 10^{-7} \text{ mol/L}) (0.885) (5.3 \times 10^{-3} \text{ mol/L})}{3.428 \times 10^{-7} \text{ mol/L}}$$

$$c(CO_2) = 1.723 \times 10^{-3} \text{ mol/L}$$

Note that $f_{a1}$ equals $\gamma_1$ (equation (9.12)). Now we want to simulate the dissolution of 0.2 mmol/L $CaCO_3$. In this case, according to equation (9.4), the concentration of calcium and hydrogencarbonate is increased by 0.2 mmol/L and 0.4 mmol/L, respectively. The new values are $c(Ca^{2+}) = 3.7$ mmol/L, ($3.7 \times 10^{-3}$ mol/L) and $c(HCO_3^-) = 5.7$ mmol/L ($5.7 \times 10^{-3}$ mol/L). Simultaneously, the $CO_2$ concentration is decreased by 0.2 mmol/L to 1.523 mmol/L ($1.523 \times 10^{-3}$ mol/L). The increased concentrations of $Ca^{2+}$ and $HCO_3^-$ result in an increase of the ionic strength by $6 \times 10^{-4}$ mol/L.

$$\Delta I = 0.5 \sum_i \Delta c_i z_i^2 = 0.5 \left[ (2 \times 10^{-4} \text{ mol/L}) \, 4 + (4 \times 10^{-4} \text{ mol/L}) \, 1 \right] = 6 \times 10^{-4} \text{ mol/L}$$

The new ionic strength is therefore:

$$I_{new} = I_{old} + \Delta I = 1.55 \times 10^{-2} \text{ mol/L} + 6 \times 10^{-4} \text{ mol/L} = 0.0161 \text{ mol/L}$$

The related activity coefficient, $\gamma_1$, is found by means of the Güntelberg equation:

$$\log \gamma_1 = -0.5 \frac{\sqrt{\frac{I}{\text{mol/L}}}}{1 + 1.4 \sqrt{\frac{I}{\text{mol/L}}}} = -0.5 \frac{0.127}{1 + (1.4)(0.127)} = -0.054$$

$$\gamma_1 = 10^{-0.054} = 0.883$$

Now, we can calculate the new pH that would be measured after the dissolution. To do that, we can use the equation given previously, but in rearranged form and with the new data:

$$a(H^+) = \frac{K^*_{a1} c(CO_2)}{f_{a1} c(HCO_3^-)} = \frac{(3.428 \times 10^{-7} \text{ mol/L}) (1.523 \times 10^{-3} \text{ mol/L})}{(0.883) (5.7 \times 10^{-3} \text{ mol/L})}$$

$$a(H^+) = 1.037 \times 10^{-7} \text{mol/L}$$

$$pH = -\log a(H^+) = -\log(1.037 \times 10^{-7}) = 6.98$$

Now, we can apply the Langelier equation to find the equilibrium pH. With:

$$\log f_{La} = 5 \log \gamma_1 = 5(-0.054) = -0.27$$

we obtain:

$$pH_{eq} = -\log K_{La} - \log f_{La} - \log c(HCO_3^-) - \log c(Ca^{2+})$$

$$pH_{eq} = 2.082 + 0.27 + 2.244 + 2.432 = 7.028 \approx 7.03$$

This is the equilibrium pH that is related to the new $Ca^{2+}$ and $HCO_3^-$ concentrations after dissolution of 0.2 mmol/L $CaCO_3$. Finally, we can calculate the new saturation index:

$$SI = pH_{meas} - pH_{eq} = 6.98 - 7.03 = -0.05$$

As a result of the simulated dissolution, the saturation index has changed from −0.2 (Example 9.2) to −0.05, but the equilibrium ($SI = 0$) is still not reached. To find the exact calcite dissolution capacity, the procedure has to be repeated with a slightly higher amount of $CaCO_3$.

## 9.5 Outlook: Assessment of the Calcite Saturation State Under Consideration of Complex Formation

Some major ions in natural waters are able to form ion pairs, each consisting of a cation and an anion. This specific form of complex formation has to be considered in cases where a very exact calculation of the saturation state should be carried out. In particular, the ions $Ca^{2+}$, $Mg^{2+}$, $HCO_3^-$, $CO_3^{2-}$, and $SO_4^{2-}$ are able to form significant concentrations of ion pairs. The respective reaction equations, here formulated for the complex dissociation, are:

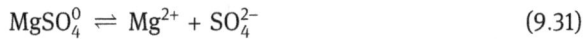

$$CaHCO_3^+ \rightleftharpoons Ca^{2+} + HCO_3^- \tag{9.26}$$

$$CaCO_3^0 \rightleftharpoons Ca^{2+} + CO_3^{2-} \tag{9.27}$$

$$CaSO_4^0 \rightleftharpoons Ca^{2+} + SO_4^{2-} \tag{9.28}$$

$$MgHCO_3^+ \rightleftharpoons Mg^{2+} + HCO_3^- \tag{9.29}$$

$$MgCO_3^0 \rightleftharpoons Mg^{2+} + CO_3^{2-} \tag{9.30}$$

$$MgSO_4^0 \rightleftharpoons Mg^{2+} + SO_4^{2-} \tag{9.31}$$

The charge number zero in the formulas is used to indicate the neutral character of the respective ion pairs in aqueous solution and should distinguish them from solid sulfates and carbonates with the same composition.

To assess the saturation state under consideration of the complex formation reactions, it is necessary to solve a set of equations consisting of the laws of mass action for the reactions described in equations (9.1)–(9.3) and (9.26)–(9.31), the electroneutrality condition, and the material balance equations for the involved species, in particular:

$$c(Ca_{total}) = c(Ca^{2+}) + c(CaHCO_3^+) + c(CaCO_3^0) + c(CaSO_4^0) \tag{9.32}$$

$$c(Mg_{total}) = c(Mg^{2+}) + c(MgHCO_3^+) + c(MgCO_3^0) + c(MgSO_4^0) \tag{9.33}$$

$$c(SO_{4,\,total}) = c(SO_4^{2-}) + c(CaSO_4^0) + c(MgSO_4^0) \tag{9.34}$$

$$c(DIC) = c(CO_2) + c(HCO_3^-) + c(CO_3^{2-}) + c(CaHCO_3^+) + c(CaCO_3^0)$$
$$+ c(MgHCO_3^+) + c(MgCO_3^0) \tag{9.35}$$

Under certain conditions (high phosphate concentration and/or high pH), the complex formation of $Ca^{2+}$ and $Mg^{2+}$ with $H_2PO_4^-$, $HPO_4^{2-}$, and $PO_4^{3-}$ as well as the formation of the hydroxo complexes $Ca(OH)^+$ and $Mg(OH)^+$ may also be relevant. In this case, the set of equations has to be further extended by the respective complex formation equilibrium relationships and the acid/base equilibrium relationships of the phosphoric acid. The additional species have to be introduced into the material balance equations and an additional balance has to be established for the phosphate species.

The solution of such extensive sets of equations can only be found by numerical methods, which requires appropriate software.

## 9.6 Special Case: Fixed $CO_2$ Partial Pressure

In the previous sections, the calco-carbonic equilibrium was considered in a general manner. This means that the concentration of $CO_2$ was allowed to vary. Some reasons for such variations were discussed in Section 9.1. In this general case, we have an infinite number of different equilibrium states, each of them defined by a specific ratio of the concentrations of $CO_{2(aq)}$, $HCO_3^-$, $CO_3^{2-}$, $Ca^{2+}$, and $H^+$. As shown in Section 9.3, we can illustrate these different equilibrium states by the Tillmans curve (Figure 9.1). The concentration distribution of the constituents of the calco-carbonic equilibrium at each point of the curve can be calculated by means of equations (9.8)–(9.10).

Let us now consider a situation where the $CO_2$ concentration is fixed. As we can derive from the Tillmans curve, the equilibrium state is given in this case by only one specific point on the curve. This means that under this condition the concentrations of all other constituents of the calco-carbonic equilibrium are also uniquely fixed. Such a situation can be found, for instance, in surface water that is in equilibrium with atmospheric $CO_2$ and also with solid calcium carbonate. Due to the constant $CO_2$ partial pressure in the atmosphere, the $CO_2$ concentration in the liquid phase is also fixed. Accordingly, the concentration distribution of all species can be calculated by combining equations (9.8)–(9.10) with Henry's law for $CO_{2(g)}$ (Chapter 6) as will be shown below. For the sake of simplification, ideal conditions are assumed.

The elementary reactions and the overall reaction are:

$$CO_{2(g)} \rightleftharpoons CO_{2(aq)} \qquad\qquad H \qquad\qquad (9.36)$$

$$CO_{2(aq)} + H_2O \rightleftharpoons H^+ + HCO_3^- \qquad\qquad K_{a1} \qquad\qquad (9.37)$$

$$H^+ + CO_3^{2-} \rightleftharpoons HCO_3^- \qquad\qquad 1/K_{a2} \qquad\qquad (9.38)$$

$$CaCO_{3(s)} \rightleftharpoons Ca^{2+} + CO_3^{2-} \qquad\qquad K_{sp} \qquad\qquad (9.39)$$

$$CaCO_{3(s)} + CO_{2(g)} + H_2O \rightleftharpoons Ca^{2+} + 2\,HCO_3^- \qquad H\,K_{a1}\,K_{sp}/K_{a2} \qquad (9.40)$$

For 25 °C, the overall equilibrium constant is:

$$K_{overall} = \frac{H\,K_{a1}\,K_{sp}}{K_{a2}} = \frac{\left(3.342 \times 10^{-2}\,\frac{mol}{L \cdot bar}\right)\left(4.406 \times 10^{-7}\,\frac{mol}{L}\right)\left(3.304 \times 10^{-9}\,\frac{mol^2}{L^2}\right)}{4.688 \times 10^{-11}\,\frac{mol}{L}}$$

$$= 1.038 \times 10^{-6}\,mol^3/(L^3 \cdot bar) \qquad\qquad (9.41)$$

Assuming that the concentration ratio of $Ca^{2+}$ and $HCO_3^-$ corresponds to the stoichiometric ratio of the overall reaction:

$$c(Ca^{2+}) = 0.5\,c(HCO_3^-) \qquad (9.42)$$

we can calculate the species distribution by using the laws of mass action related to equations (9.36)–(9.40). It has to be noted that, in principle, the calculation can be carried out in the same manner for any other $Ca^{2+}/HCO_3^-$ ratio.

---

**Example 9.4**
Calculate the pH and the species distribution of $Ca^{2+}$, $HCO_3^-$, $CO_3^{2-}$, and $CO_{2(aq)}$ in a water that is at 25 °C in equilibrium with solid calcium carbonate and with atmospheric air (415 ppm $CO_2$, $p(CO_2) = $ 0.000415 bar) by using the equilibrium constants given in equation (9.41). Ideal conditions and a stoichiometric ratio of $Ca^{2+}$ and $HCO_3^-$ are assumed.

**Solution:**
At first, the law of mass action for the overall reaction (equation (9.40)) is formulated under consideration of the stoichiometric ratio of $Ca^{2+}$ and $HCO_3^-$:

$$K_{overall} = \frac{c(Ca^{2+})\,c^2(HCO_3^-)}{p(CO_2)} = \frac{0.5\,c(HCO_3^-)\,c^2(HCO_3^-)}{p(CO_2)} = \frac{0.5\,c^3(HCO_3^-)}{p(CO_2)}$$

Rearranging this equation and introducing the given data yields:

$$c(HCO_3^-) = \sqrt[3]{2\,p(CO_2)\,K_{overall}} = \sqrt[3]{2(0.000415\,\text{bar})(1.038 \times 10^{-6}\,\text{mol}^3/(L^3 \cdot \text{bar}))}$$

$$c(HCO_3^-) = \sqrt[3]{8.615 \times 10^{-10}\,\text{mol}^3/L^3} = 9.515 \times 10^{-4}\,\text{mol/L}$$

The CO$_2$ concentration can be calculated by means of Henry's law:

$$c(CO_2) = H\,p(CO_2) = (3.342 \times 10^{-2}\,\text{mol}/(L \cdot \text{bar}))(0.000415\,\text{bar}) = 1.387 \times 10^{-5}\,\text{mol/L}$$

Now, we can calculate the pH and the carbonate concentration by using the equilibrium relationships of the carbonic acid system:

$$c(H^+) = \frac{c(CO_2)\,K_{a1}}{c(HCO_3^-)} = \frac{(1.387 \times 10^{-5}\,\text{mol/L})(4.406 \times 10^{-7}\,\text{mol/L})}{9.515 \times 10^{-4}\,\text{mol/L}} = 6.423 \times 10^{-9}\,\text{mol/L}$$

$$pH = -\log c(H^+) = -\log(6.423 \times 10^{-9}) = 8.19$$

$$c(CO_3^{2-}) = \frac{K_{a2}\,c(HCO_3^-)}{c(H^+)} = \frac{(4.688 \times 10^{-11}\,\text{mol/L})(9.515 \times 10^{-4}\,\text{mol/L})}{6.423 \times 10^{-9}\,\text{mol/L}} = 6.945 \times 10^{-6}\,\text{mol/L}$$

Finally, the calcium concentration can be calculated from the solubility product:

$$c(Ca^{2+}) = \frac{K_{sp}}{c(CO_3^{2-})} = \frac{3.304 \times 10^{-9}\,\text{mol}^2/L^2}{6.945 \times 10^{-6}\,\text{mol/L}} = 4.757 \times 10^{-4}\,\text{mol/L}$$

Alternatively, the calcium concentration can be found from the hydrogencarbonate concentration by means of the stoichiometric condition $c(Ca^{2+}) = 0.5\,c(HCO_3^-)$. The calcium concentration can then be used to calculate the carbonate concentration by means of the solubility product.

---

The model calculation shown in Example 9.4 can be used to demonstrate the impact of a further increase of the atmospheric $CO_2$ concentration on the pH of ocean water. To simplify the calculation we want to restrict our consideration to the calco-carbonic equilibrium and neglect the influence of further acid/base systems, such as boric acid/borate or silicic acid/silicate, on the pH. Moreover, we want to assume that the water in the upper layers of the oceans is in equilibrium with the partial pressure of $CO_2$ in the air. To apply an equation for the overall reaction comparable to equation (9.41) we have to consider the specific conditions for seawater:

- The solubility of gases decreases with increasing salinity of the water. This effect is referred to as the salting-out effect. As a consequence, the Henry constants for seawater are lower than those for pure water (or water of low salinity).
- Due to its high salt content, seawater cannot be considered an ideal solution. In principle, we could calculate activity coefficients to derive conditional constants from the thermodynamic constants. For this, however, we would have to apply complex models (e.g., Pitzer model) because the simple methods for calculating activity coefficients (Güntelberg equation, Davies equation, Debye-Hückel equation) are not useful for the high ionic strength of seawater ($\approx 0.71$ mol/L). Therefore we want to apply literature data here.

Conditional equilibrium constants (including Henry constants) for the seawater salinity of 3.5% were published by Millero 2001 (Table 9.3). Using these data, we can calculate the pH of seawater for different $CO_2$ concentrations in the atmosphere (Example 9.5). Note that the constants in the table are based on molality as concentration measure. For simplification we use these constants with molar concentrations which causes only a minor error (see also Chapter 3, Section 3.2).

**Table 9.3:** Henry constants of $CO_2$ and conditional constants of the calco-carbonic equilibrium for seawater with a salinity of 3.5% (Millero 2001).

| Temperature (°C) | log $H$ | log $K_{sp}$ | log $K_{a1}^*$ | log $K_{a2}^*$ |
|---|---|---|---|---|
| 0 | −1.202 | −6.365 | −6.101 | −9.376 |
| 5 | −1.283 | −6.363 | −6.046 | −9.277 |
| 10 | −1.358 | −6.363 | −5.994 | −9.182 |
| 15 | −1.427 | −6.363 | −5.943 | −9.090 |
| 20 | −1.489 | −6.364 | −5.894 | −9.001 |
| 25 | −1.547 | −6.367 | −5.847 | −8.916 |
| 30 | −1.599 | −6.371 | −5.802 | −8.833 |

**Example 9.5**

Calculate the pH and the carbonate concentration of seawater ($\vartheta = 20$ °C) in equilibrium with atmospheric CO$_2$ concentrations of 415 ppm and 600 ppm ($p_{total} = 1$ bar). The respective partial pressures of CO$_2$ in the atmosphere are 0.000415 bar and 0.0006 bar. Ideal conditions and a stoichiometric ratio of Ca$^{2+}$ and HCO$_3^-$ ($c(Ca^{2+}) = 0.5\ c(HCO_3^-)$) are assumed.

**Solution:**

With the constants given in Table 9.3, we can find the overall equilibrium constant for the reaction

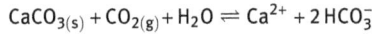

$$CaCO_{3(s)} + CO_{2(g)} + H_2O \rightleftharpoons Ca^{2+} + 2\,HCO_3^-$$

at 20 °C (see equation (9.41)):

$$K_{overall} = \frac{H\,K_{a1}\,K_{sp}}{K_{a2}} = \frac{\left(3.243 \times 10^{-2}\,\frac{mol}{L \cdot bar}\right)\left(1.276 \times 10^{-6}\,\frac{mol}{L}\right)\left(4.325 \times 10^{-7}\,\frac{mol^2}{L^2}\right)}{9.977 \times 10^{-10}\,\frac{mol}{L}}$$

$$= 1.794 \times 10^{-5}\,mol^3/(L^3 \cdot bar)$$

Under the simplifying assumption that the ratio of Ca$^{2+}$ and HCO$_3^-$ equals the stoichiometric ratio ($c(Ca^{2+}) = 0.5\ c(HCO_3^-)$), we get:

$$K_{overall} = \frac{c(Ca^{2+})\,c^2(HCO_3^-)}{p(CO_2)} = \frac{0.5\,c^3(HCO_3^-)}{p(CO_2)}$$

With the given data, we can calculate the hydrogencarbonate concentration for $p(CO_2) = 0.000415$ bar:

$$c(HCO_3^-) = \sqrt[3]{2\,p(CO_2)\,K_{overall}} = \sqrt[3]{2(0.000415\ bar)(1.794 \times 10^{-5}\ mol^3/(L^3 \cdot bar))}$$

$$c(HCO_3^-) = \sqrt[3]{1.489 \times 10^{-8}\ mol^3/L^3} = 2.460 \times 10^{-3}\ mol/L$$

The CO$_2$ concentration can be found by means of Henry's law:

$$c(CO_2) = H\,p(CO_2) = (3.243 \times 10^{-2}\ mol/(L \cdot bar))(0.000415\ bar) = 1.346 \times 10^{-5}\ mol/L$$

Now, we can calculate the pH:

$$c(H^+) = \frac{c(CO_2)\,K_{a1}}{c(HCO_3^-)} = \frac{(1.346 \times 10^{-5}\ mol/L)(1.276 \times 10^{-6}\ mol/L)}{2.460 \times 10^{-3}\ mol/L} = 6.982 \times 10^{-9}\ mol/L$$

$$pH = -\log c(H^+) = -\log(6.982 \times 10^{-9}) = 8.16$$

For the carbonate concentration, we get:

$$c(CO_3^{2-}) = \frac{K_{a2}\,c(HCO_3^-)}{c(H^+)} = \frac{(9.977 \times 10^{-10}\ mol/L)(2.460 \times 10^{-3}\ mol/L)}{6.982 \times 10^{-9}\ mol/L} = 3.515 \times 10^{-4}\ mol/L$$

The same calculation for $p(CO_2) = 0.0006$ bar gives:

$$c(HCO_3^-) = \sqrt[3]{2\,p(CO_2)\,K_{overall}} = \sqrt[3]{2(0.0006\ bar)(1.794 \times 10^{-5}\ mol^3/(L^3 \cdot bar))}$$

$$c(HCO_3^-) = \sqrt[3]{2.153 \times 10^{-8}\ mol^3/L^3} = 2.782 \times 10^{-3}\ mol/L$$

$$c(CO_2) = H\,p(CO_2) = (3.243 \times 10^{-2}\ mol/(L \cdot bar))(0.0006\ bar) = 1.946 \times 10^{-5}\ mol/L$$

$$c(H^+) = \frac{c(CO_2)\,K_{a1}}{c(HCO_3^-)} = \frac{(1.946 \times 10^{-5}\ mol/L)(1.276 \times 10^{-6}\ mol/L)}{2.782 \times 10^{-3}\ mol/L} = 8.926 \times 10^{-9}\ mol/L$$

$$pH = -\log c(H^+) = -\log(8.926 \times 10^{-9}) = 8.05$$

$$c(CO_3^{2-}) = \frac{K_{a2}\,c(HCO_3^-)}{c(H^+)} = \frac{(9.977 \times 10^{-10}\ mol/L)(2.782 \times 10^{-3}\ mol/L)}{8.926 \times 10^{-9}\ mol/L} = 3.110 \times 10^{-4}\ mol/L$$

---

Despite the simplified calculation, the example shows clearly the effects to be expected if the $CO_2$ concentration in the atmosphere continues to rise: decrease of pH, increase of the aqueous $CO_2$ concentration, and decrease of the carbonate concentration. These effects are referred to as ocean acidification.

## 9.7 Problems

**9.1.** As an example for the establishment of a Tillmans curve, we want to calculate the equilibrium concentration of $CO_2$ and the pH of water that contains 1.4 mmol/L $Ca^{2+}$ and 3.8 mmol/L $HCO_3^-$. The ionic strength is 10 mmol/L and the temperature is 10 °C. The required equilibrium constants at 10 °C are given as $\log K_T = 4.383$ and $\log K_{a1}^* = -6.465$.

**9.2.** Assess the calcite saturation state for water that is characterized by the following data: $\vartheta = 10$ °C, $pH_{meas} = 8.2$, $c(Ca^{2+}) = 1.2$ mmol/L, $c(HCO_3^-) = 1.7$ mmol/L, $\kappa_{25} = 45$ mS/m. The logarithm of the Langelier constant for 10 °C is $\log K_{La} = -2.082$.

**9.3.** Take the data from Problem 9.2 and calculate the $CO_2$ concentration that is related to the measured pH and hydrogencarbonate concentration (equilibrium between $CO_2$ and $HCO_3^-$ is assumed). Then calculate the concentration of $CO_2$ in the state of the calco-carbonic equilibrium by the Tillmans equation and compare the results. Additional to the data in Problem 9.2, the following equilibrium constants are given: $K_{a1} = 3.428 \times 10^{-7}$ mol/L, $K_{a2} = 3.251 \times 10^{-11}$ mol/L, $K_{sp} = 3.926 \times 10^{-9}$ mol$^2$/L$^2$.

**9.4.** Calculate the pH and the species distribution of $Ca^{2+}$, $HCO_3^-$, $CO_3^{2-}$, and $CO_{2(aq)}$ in seepage water that is at 10 °C in equilibrium with solid calcium carbonate and with a soil atmosphere in which the partial pressure of $CO_2$ is by a factor of hundred higher than that of the atmosphere ($p(CO_2)_{atm} = 0.000415$ bar). Ideal conditions are assumed. The respective equilibrium constants for the elementary reactions at 10 °C are: $H = 5.247 \times 10^{-2}$ mol/(L $\cdot$ bar), $K_{a1} = 3.428 \times 10^{-7}$ mol/L, $K_{a2} = 3.251 \times 10^{-11}$ mol/L, $K_{sp} = 3.926 \times 10^{-9}$ mol$^2$/L$^2$. The overall equilibrium constant for the reaction $CaCO_{3(s)} + CO_{2(g)} + H_2O \rightleftharpoons Ca^{2+} + 2\,HCO_3^-$ at 10 °C is $K_{overall} = 2.172 \times 10^{-6}$ mol$^3$/(L$^3 \cdot$ bar).

# 10 Redox Equilibria

## 10.1 Introduction

Redox reactions belong, besides acid/base reactions, to the most important processes in aquatic systems. The term "redox" is a combination of the words reduction and oxidation. Oxidation and reduction are reverse reactions and the term "redox" indicates that they are always coupled. An oxidation is defined as a loss of electrons, whereas reduction means gain of electrons. A number of water constituents are able to take part in redox reactions. This means that they can be reduced or oxidized or, in other words, they can change their oxidation state. The change of the oxidation state is typically connected with a change in the properties and in many cases also with a change in the composition of the species. The oxidized and the related reduced species form a redox couple (also referred to as redox pair). Examples of redox couples frequently occurring in aquatic systems are $O_2/H_2O$, $NO_3^-/NH_3$, $NO_3^-/NO_2^-$, $NO_3^-/N_2$, $MnO_2/Mn^{2+}$, $Fe(OH)_3/Fe^{2+}$, $SO_4^{2-}/H_2S$, $SO_4^{2-}/HS^-$, $CO_2/CH_4$, and $CO_2/C_xH_yO_z$. The given examples demonstrate that a number of major and minor water constituents are part of redox couples, which underlines the relevance of redox reactions in aquatic systems. As we can further see, some species are constituents of different redox couples. This is possible when the considered element occurs in more than just two different oxidation states, such as nitrogen, which can occur in aquatic systems as $NO_3^-$, $NO_2^-$, $N_2$, or $NH_3/NH_4^+$.

To formulate redox reaction equations and to carry out equilibrium calculations, we need a specific parameter that characterizes the oxidation state of a considered atom in a given species. This parameter is known as the oxidation number (or oxidation state). It can be estimated in different ways as will be shown in the next section. For a considered redox couple, the lower value of the oxidation number characterizes the reduced form, whereas the higher value characterizes the oxidized form. If necessary, the oxidation number is written above the symbol of the atom, for instance $\overset{+5}{N}O_3^-$ and $\overset{-3}{N}H_3$, where +5 is the oxidation number of nitrogen in the oxidized form (nitrate) and −3 is the oxidation number of nitrogen in the reduced form (ammonia). In spelled-out species names, the oxidation state is frequently given as Roman numerals in brackets; for instance, iron(III) hydroxide or iron(II) hydroxide.

## 10.2 Estimation of Oxidation Numbers (Oxidation States)

For atoms and for ions, which consist only of one sort of atom, the estimation of the oxidation number is very simple, because the charge of the species is directly linked to the number of electrons gained or lost. Accordingly, the oxidation number equals

https://doi.org/10.1515/9783110758788-010

the charge of the species. For instance, metallic iron (Fe) has neither gained nor lost electrons when compared to its basic state and, therefore, the oxidation number is 0. In the case of $Fe^{2+}$, two electrons are lost compared to Fe and the oxidation number is +2. In reverse, the anion chloride, $Cl^-$, has gained one electron when compared to the atom Cl and the charge as well as the oxidation number is therefore −1.

For compounds consisting of two or more atoms, the oxidation number is the charge that results from a hypothetical cleavage of the bonds (hypothetical dissociation of the compound into ions). For compounds with only ionic bonds, for instance salts like NaCl, this hypothetical bond cleavage is identical to the real dissociation reaction:

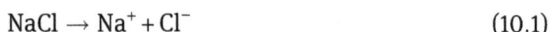

$$NaCl \rightarrow Na^+ + Cl^- \tag{10.1}$$

In this case, the oxidation numbers equal the ion charges after the dissociation (here +1 for $Na^+$ and −1 for $Cl^-$).

In the case of compounds with covalent bonds, the oxidation numbers are found following a hypothetical dissociation that is carried out in such a manner that the more electronegative partner of the bond formally receives all the bonding electrons. The electronegativity is a property that describes the tendency of an atom in a compound to attract electrons or to shift the electron density towards itself. There are different scales of electronegativity in use (Table 10.1). Irrespective of the slightly different numerical values in the different scales, there are some general tendencies with respect to the position of the element in the periodic table of elements. As a rule, the electronegativity of the main group elements increases from left to right along a period, and decreases with increasing atomic number within a group. The d-block elements (transition metals) show a comparable general trend, but with minor variation and some exceptions.

**Table 10.1:** Electronegativities of elements frequently forming covalent bonds.

| Element | Electronegativity (Pauling scale) | Electronegativity (Allen scale) |
|---|---|---|
| F | 3.98 | 4.19 |
| O | 3.44 | 3.61 |
| Cl | 3.16 | 2.87 |
| N | 3.04 | 3.07 |
| Br | 2.96 | 2.69 |
| S | 2.58 | 2.59 |
| C | 2.55 | 2.54 |
| H | 2.20 | 2.30 |
| P | 2.19 | 2.25 |
| Si | 1.90 | 1.92 |

The formal cleavage of bonds is carried out irrespective of whether this reaction is really possible or not. In the case of $H_2O$, for instance, the real dissociation leads to $H^+$ and $OH^-$, whereas the hypothetical dissociation carried out by the described method leads to $2\,H^+$ and $O^{2-}$. Example 10.1 illustrates this method of estimating oxidation numbers.

---

**Example 10.1**
What are the oxidation numbers of all atoms within the compounds $CO_2$, $H_2O$, and $O_2$?

**Solution:**
a. $CO_2$:
   The formal dissociation reaction is:

   $$O=C=O \rightarrow C^{4+} + 2\,O^{2-}$$

   There are eight bonding electrons in the molecule, two from each oxygen atom and four from the carbon atom. Oxygen is more electronegative than carbon and formally receives the four bonding electrons from carbon (two for each oxygen atom); the resulting oxidation numbers are therefore:
   C: + 4, O: –2.
b. $H_2O$:
   The formal dissociation reaction is:

   $$H\text{–}O\text{–}H \rightarrow 2\,H^+ + O^{2-}$$

   There are four bonding electrons in the molecule, one from each hydrogen atom and two from oxygen. Oxygen is more electronegative than hydrogen and formally receives the two bonding electrons from the hydrogen atoms; the resulting oxidation numbers are therefore:
   H: + 1, O: –2.
c. $O_2$:
   The formal dissociation reaction is:

   $$O=O \rightarrow O+O$$

   The $O_2$ molecule consists of the same type of atom. Therefore, no bonding partner is more electronegative than the other and the partners retain their bonding electrons. Consequently, the oxidation numbers of both oxygen atoms are 0.

---

The described general method to estimate oxidation numbers is laborious, particularly in the case of species with complex structure, and, furthermore, requires the knowledge of the electronegativities of all atoms in the compound. Therefore, a simplified procedure for determination of oxidation numbers (*ON*) is often used in practice. This method is based on the application of a number of rules in a defined order of priority. These rules have been derived from the general method described before and read as follows:

Rule 1: Elements and compounds consisting of equal atoms have the oxidation
   number $ON = 0$.
Rule 2: For ions originating from a single atom, the oxidation number equals the
   ion charge.
Rule 3: The sum of the oxidation numbers in a neutral molecule is 0, whereas in
   ions, the sum of the oxidation numbers equals the ion charge.
Rule 4: Fluorine in compounds has the oxidation number $ON = -1$.
Rule 5: Metals in compounds have positive oxidation numbers. Alkali metals (first
   group of the periodic table of elements) have the oxidation number $ON = +1$
   and alkaline earth metals (second group of the periodic table of elements)
   have the oxidation number $ON = +2$.
Rule 6: Hydrogen in compounds has the oxidation number $ON = +1$.
Rule 7: Oxygen in compounds has the oxidation number $ON = -2$.

It has to be noted that in the case of conflicts, the rule with the lower ordinal number invalidates the rule with the higher number. As a consequence, exceptions can
be found in particular for rules 6 and 7. Typical examples for such exceptions are
metal hydrides with $ON = -1$ for hydrogen and peroxo compounds with $ON = -1$ for
oxygen.

---

**Example 10.2**
What are the oxidation numbers of the atoms in $O_2$, $Br^-$, $SO_4^{2-}$, $Fe(OH)_2$, $HClO_3$, LiH, and $H_2O_2$?

**Solution:**
$O_2$: $ON(O) = 0$ (rule 1)
$Br^-$: $ON(Br) = -1$ (rule 2)
$SO_4^{2-}$: $ON(O) = -2$ (rule 7), $ON(S) =$ ion charge $- 4 \times ON(O) = +6$ (rule 3 with $\Sigma\,ON =$ ion charge)
$Fe(OH)_2$: $ON(O) = -2$ (rule 7), $ON(H) = +1$ (rule 6), $ON(Fe) = 0 - 2 \times [ON(O) + ON(H)] = +2$ (rule 3 with $\Sigma\,ON = 0$)
$HClO_3$: $ON(H) = +1$ (rule 6), $ON(O) = -2$ (rule 7), $ON(Cl) = 0 - [ON(H) + 3 \times ON(O)] = +5$ (rule 3 with $\Sigma\,ON = 0$)
LiH: $ON(Li) = +1$ (rule 5), $ON(H) = 0 - ON(Li) = -1$ (rule 3 with $\Sigma\,ON = 0$). Note that rules 3 and 5 invalidate rule 6 (metal hydride as an exception to rule 6).
$H_2O_2$: $ON(H) = +1$ (rule 6), $ON(O) = (0 - 2 \times ON(H))/2 = -1$ (rule 3 with $\Sigma\,ON = 0$). Note that rules 3 and
6 invalidate rule 7 (peroxo compound as an exception to rule 7).

---

## 10.3 Redox Equilibria: Definitions and Basic Concepts

With respect to the quantitative description of redox equilibria, we have to distinguish between half-reactions and complete redox reactions. Half-reactions describe
the electron transfer between the components of a redox couple according to:

$$\text{Ox} + n_e\, e^- \rightleftharpoons \text{Red} \qquad\qquad (10.2)$$

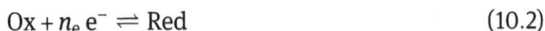

where Ox is the oxidant, Red is the reductant and $n_e$ is the number of the transferred electrons. The oxidant (or oxidizing agent) gains electrons and will be reduced during the reaction, whereas the reductant (or reducing agent) loses electrons and will be oxidized. Accordingly, the reaction from left to right is a reduction (gain of electrons, decrease of oxidation state), whereas the reaction from right to left is an oxidation (loss of electrons, increase of oxidation state). Oxidant and reductant are also referred to as electron acceptor and electron donor, respectively.

Although half-reactions with free electrons do not reflect the real situation in water where free electrons do not occur, this formal approach provides the basis for the definition of the master variable redox intensity, which gives information about the redox state and the ratio of oxidants and reductants in a considered aqueous system (Section 10.4).

A complete redox reaction describes the electron transfer from one redox couple to another. Therefore, a complete redox system consists of two redox couples where the electrons released in one half-reaction are accepted by the other half-reaction according to:

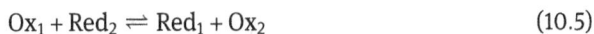

$$\text{Ox}_1 + n_e\, e^- \rightleftharpoons \text{Red}_1 \qquad\qquad (10.3)$$

$$\text{Red}_2 \rightleftharpoons \text{Ox}_2 + n_e\, e^- \qquad\qquad (10.4)$$

_____

$$\text{Ox}_1 + \text{Red}_2 \rightleftharpoons \text{Red}_1 + \text{Ox}_2 \qquad\qquad (10.5)$$

Equation (10.5) describes the overall redox reaction that contains no more free electrons. The mathematical treatment of such complete redox reactions will be subject of Section 10.6.

## 10.4 Half-Reactions

### 10.4.1 Writing Equations for Redox Half-Reactions

In this section, a short guide to setting up reaction equations for redox couples (redox half-reactions) will be given. Complete redox reactions are discussed separately in Section 10.6. To establish equations for redox half-reactions, the following steps have to be carried out:

Step 1: Determination of the oxidation numbers of the components of the redox couple according to the rules given in Section 10.2.

Step 2: Calculation of the number of electrons that are exchanged between the components of the redox couple. The number of exchanged electrons is given as the difference between the oxidation numbers.

Step 3: Writing the reaction equation under consideration of the rule that the component with the higher oxidation number (the oxidant) and the electrons are located on the left-hand side of the equation and the component with the lower oxidation number (the reductant) is located on the right-hand side of the equation. This convention is the basis for the definition of the equilibrium constant and the standard redox intensity (Section 10.4.2). In the case of simple redox couples where the components are derived from a single type of atom only (e.g., $Fe^{3+}/Fe^{2+}$), the reaction equation found in this step is already the final equation. In other cases where the components are of complex composition (e.g., $NO_3^-/NH_4^+$, $SO_4^{2-}/HS^-$), an additional step 4 is necessary, where the equation has to be balanced with respect to the oxygen and hydrogen atoms.

Step 4: Introduction of protons and water into the raw equation found in step 3 in order to balance the equation. Addition of protons, $H^+$, on the side with an oxygen excess and $H_2O$ on the other side. A possible need for protons in the species on the other side has to be taken into account.

After establishing the reaction equation, it is reasonable to check the material and charge balance. The number of the different atoms and the total charge must be the same on each side of the equation.

The principle of setting up reaction equation for redox half-reactions is demonstrated for two different cases in Example 10.3.

---

**Example 10.3**

What are the reaction equations for the redox couples a) $Fe^{3+}/Fe^{2+}$ and b) $NO_3^-/NH_4^+$?

**Solution:**

a. According to rule 2 given in Section 10.2, the oxidation numbers of $Fe^{3+}$ and $Fe^{2+}$ are +3 and +2, respectively. The number of the electrons exchanged in the redox reaction equals the difference of the oxidation numbers, here $3 - 2 = 1$. Following the convention that the component with the higher oxidation number (oxidant) is written on the left-hand side of the equation, we find:

$$Fe^{3+} + e^- \rightleftharpoons Fe^{2+}$$

To check the material and the charge balance, we have to compare the number of Fe atoms and the sum of charges on each side:

    Left-hand side: 1 Fe; charge: $+3 + (-1) = +2$

    Right-hand side: 1 Fe; charge: +2

b. According to the rules 3, 6, and 7, the oxidation number of N is +5 in $NO_3^-$ and -3 in $NH_4^+$. The number of exchanged electrons is $(+5) - (-3) = 8$ and the raw equation (without balancing) reads:

$$NO_3^- + 8\,e^- \rightleftharpoons NH_4^+$$

Balancing the equation with respect to O and H atoms requires the addition of 10 protons on the left-hand side (6 for the formation of 3 $H_2O$ with the three O atoms and 4 for $NH_4^+$). On the right-hand side we have to add 3 $H_2O$. The resulting reaction equation is:

$$NO_3^- + 10\,H^+ + 8\,e^- \rightleftharpoons NH_4^+ + 3\,H_2O$$

To check the material and charge balance, we have to compare the number of N atoms, the number of H atoms, the number of O atoms, and the sum of charges on each side:

Left-hand side: 1 N, 10 H, 3 O; charge: (−1) + (+10) + (−8) = +1
Right-hand side: 1 N, 10 H, 3 O; charge: +1

## 10.4.2 Law of Mass Action and Redox Intensity

The law of mass action related to the reaction:

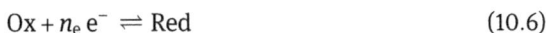

$$Ox + n_e\,e^- \rightleftharpoons Red \qquad (10.6)$$

is given by:

$$K^* = \frac{a(Red)}{a(Ox)\,a^{n_e}(e^-)} \qquad (10.7)$$

For the negative logarithm of the activity of the electrons, a new parameter is introduced that is referred to as redox intensity, pe:

$$pe = -\log a(e^-) \qquad (10.8)$$

Writing equation (10.7) in logarithmic form and applying the definition of the redox intensity gives after rearrangement:

$$pe = \frac{1}{n_e}\log K^* + \frac{1}{n_e}\log\frac{a(Ox)}{a(Red)} \qquad (10.9)$$

After introducing the standard redox intensity, $pe^0$, according to:

$$pe^0 = \frac{1}{n_e}\log K^* \qquad (10.10)$$

we finally find:

$$pe = pe^0 + \frac{1}{n_e}\log\frac{a(Ox)}{a(Red)} \qquad (10.11)$$

From this equation, we can derive that the parameter pe is directly related to the activities of the components of the redox couple. For a given $pe^0$ (or equilibrium constant), the value of pe determines the activity ratio of oxidant and reductant in a considered aqueous system.

It has to be noted that equation (10.11) was derived on the basis of equation (10.6), which describes the simplest case of a redox system. An example of such a simple redox system is:

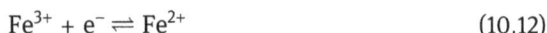

$$Fe^{3+} + e^- \rightleftharpoons Fe^{2+} \tag{10.12}$$

Here, the redox intensity is given according to equation (10.11) (with $n_e = 1$) by:

$$pe = pe^0 + \log\frac{a(Fe^{3+})}{a(Fe^{2+})} \tag{10.13}$$

However, the redox half-reactions are frequently more complex and also include, besides the oxidant and the reductant, water and water-related species ($H_2O$, $H^+$, $OH^-$), for instance:

$$Fe(OH)_{3(s)} + 3\,H^+ + e^- \rightleftharpoons Fe^{2+} + 3\,H_2O \tag{10.14}$$

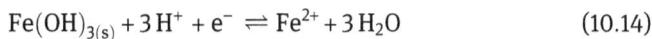

The law of mass action is given here by:

$$K^* = \frac{a(Fe^{2+})}{a^3(H^+)\,a(e^-)} \tag{10.15}$$

and after introducing the definitions of $pe^0$ and pe, we find:

$$pe = pe^0 + \log\frac{a^3(H^+)}{a(Fe^{2+})} \tag{10.16}$$

Two points are noteworthy. First, not only the reductant and the oxidant but other species also taking part in the reaction have to be considered in the formulation of the equation for pe, because the latter is a derivate of the law of mass action. Second, according to the rules for establishing the law of mass action (Chapter 5), $H_2O$ and solids (even if they are reductant or oxidant) will not be considered in the equation.

To generalize equation (10.11), we can write:

$$pe = pe^0 + \frac{1}{n_e}\log\frac{\Pi\,a^v(Ox)}{\Pi\,a^v(Red)} \tag{10.17}$$

where $\Pi\,a^v(Ox)$ and $\Pi\,a^v(Red)$ are the products of the activities of all relevant species on the oxidant side and on the reductant side of the reaction equation, with the respective stoichiometric factors, $v$, as exponents of the activities.

The standard redox intensities, $pe^0$, are well known for all relevant redox couples. Selected values are listed in Table 10.2. Note that the equilibrium constants and therefore also the standard redox intensities generally depend on the temperature. The data given in Table 10.2 are valid for 25 °C. For the sake of simplification, all examples here and in the following sections use the data for 25 °C. Furthermore, it has to be noted that all equilibrium constants and standard redox intensities given in the table refer to the notation of the half-reaction with the oxidant and the electrons on the left-hand side and the reductant on the right-hand side.

Table 10.2 also includes standard redox potentials that are often used instead of the standard redox intensities to characterize the equilibria conditions of half-reactions. The relationship between redox intensity and redox potential is provided

**Table 10.2:** Standard redox intensities ($pe^0$), equilibrium constants ($\log K^*$), standard potentials ($E_H^0$), and standard redox intensities at pH = 7 ($pe^0$(pH = 7)) of selected half-reactions (all data for 25 °C).

| Half-reaction | $pe^0$ | $\log K^* = n_e\,pe^0$ | $E_H^0$ (Volt) | $pe^0$(pH = 7) |
|---|---|---|---|---|
| $O_{2(aq)} + 4\,H^+ + 4\,e^- \rightleftharpoons 2\,H_2O$ | 21.48 | 85.92 | 1.267 | 14.48 |
| $2\,NO_3^- + 12\,H^+ + 10\,e^- \rightleftharpoons N_{2(g)} + 6\,H_2O$ | 21.05 | 210.50 | 1.242 | 12.65 |
| $MnO_{2(s)} + 4\,H^+ + 2\,e^- \rightleftharpoons Mn^{2+} + 2\,H_2O$ | 20.80 | 41.60 | 1.227 | 6.80 |
| $O_{2(g)} + 4\,H^+ + 4\,e^- \rightleftharpoons 2\,H_2O$ | 20.75 | 83.00 | 1.224 | 13.75 |
| $2\,NO_3^- + 12\,H^+ + 10\,e^- \rightleftharpoons N_{2(aq)} + 6\,H_2O$ | 20.73 | 207.31 | 1.223 | 12.33 |
| $Fe(OH)_{3(s)} + 3\,H^+ + e^- \rightleftharpoons Fe^{2+} + 3\,H_2O$ | 16.30 | 16.30 | 0.962 | −4.70 |
| $NO_2^- + 8\,H^+ + 6\,e^- \rightleftharpoons NH_4^+ + 2\,H_2O$ | 15.14 | 90.84 | 0.893 | 5.81 |
| $NO_3^- + 10\,H^+ + 8\,e^- \rightleftharpoons NH_4^+ + 3\,H_2O$ | 14.90 | 119.20 | 0.879 | 6.15 |
| $NO_3^- + 2\,H^+ + 2\,e^- \rightleftharpoons NO_2^- + H_2O$ | 14.15 | 28.30 | 0.835 | 7.15 |
| $Fe^{3+} + e^- \rightleftharpoons Fe^{2+}$ | 13.00 | 13.00 | 0.767 | 13.00 |
| $CH_2O + 4\,H^+ + 4\,e^- \rightleftharpoons CH_{4(g)} + H_2O$ | 6.94 | 27.76 | 0.409 | −0.06 |
| $SO_4^{2-} + 8\,H^+ + 6\,e^- \rightleftharpoons S_{(s)} + 4\,H_2O$ | 6.03 | 36.18 | 0.356 | −3.30 |
| $SO_4^{2-} + 10\,H^+ + 8\,e^- \rightleftharpoons H_2S_{(g)} + 4\,H_2O$ | 5.25 | 42.00 | 0.310 | −3.50 |
| $SO_4^{2-} + 10\,H^+ + 8\,e^- \rightleftharpoons H_2S_{(aq)} + 4\,H_2O$ | 5.12 | 40.96 | 0.302 | −3.63 |
| $N_{2(g)} + 8\,H^+ + 6\,e^- \rightleftharpoons 2\,NH_4^+$ | 4.68 | 28.08 | 0.276 | −4.65 |
| $SO_4^{2-} + 9\,H^+ + 8\,e^- \rightleftharpoons HS^- + 4\,H_2O$ | 4.25 | 34.00 | 0.251 | −3.63 |
| $S_{(s)} + 2\,H^+ + 2\,e^- \rightleftharpoons H_2S_{(g)}$ | 2.89 | 5.78 | 0.171 | −4.11 |
| $CO_{2(g)} + 8\,H^+ + 8\,e^- \rightleftharpoons CH_{4(g)} + 2\,H_2O$ | 2.87 | 22.96 | 0.169 | −4.13 |
| $S_{(s)} + 2\,H^+ + 2e^- \rightleftharpoons H_2S_{(aq)}$ | 2,40 | 4,80 | 0.142 | −4.60 |
| $6\,CO_{2(aq)} + 24\,H^+ + 24\,e^- \rightleftharpoons C_6H_{12}O_6 + 6\,H_2O$ | 0.17 | 4.08 | 0.010 | −6.83 |
| $2\,H^+ + 2\,e^- \rightleftharpoons H_{2(g)}$ | 0.00 | 0.00 | 0.000 | −7.00 |
| $6\,CO_{2(g)} + 24\,H^+ + 24\,e^- \rightleftharpoons C_6H_{12}O_6 + 6\,H_2O$ | −0.20 | −4.80 | −0.012 | −7.20 |
| $CO_{2(aq)} + 4\,H^+ + 4\,e^- \rightleftharpoons CH_2O + H_2O$ | −0.83 | −3.32 | −0.049 | −7.83 |
| $CO_{2(g)} + 4\,H^+ + 4\,e^- \rightleftharpoons CH_2O + H_2O$ | −1.20 | −4.80 | −0.071 | −8.20 |

in Section 10.4.3. For redox reactions, where gases are involved, there are two possible ways to write the half-reaction, leading to different values of $pe^0$. This case is discussed in more detail in Section 10.4.4.

Further, we can see from Table 10.2 that protons are involved in many redox half-reactions. For some applications, it may be convenient to normalize $pe^0$ to a fixed pH, for instance pH = 7. Standard redox intensities at pH = 7 are also included in Table 10.2. The calculation of pH-normalized standard redox intensities will be demonstrated by using the reaction:

$$NO_3^- + 10\,H^+ + 8\,e^- \rightleftharpoons NH_4^+ + 3\,H_2O \tag{10.18}$$

as an example. The redox intensity of this half-reaction is given by:

$$pe = pe^0 + \frac{1}{8}\log\frac{a^{10}(H^+)\,a(NO_3^-)}{a(NH_4^+)} \tag{10.19}$$

Separating the proton activity from the activity ratio of oxidant and reductant and introducing the pH gives:

$$pe = pe^0 + \frac{10}{8}\log a(H^+) + \frac{1}{8}\log\frac{a(NO_3^-)}{a(NH_4^+)} = pe^0 - \frac{10}{8}pH + \frac{1}{8}\log\frac{a(NO_3^-)}{a(NH_4^+)} \tag{10.20}$$

Now, a standard redox intensity at pH = 7, $pe^0(pH = 7)$, can be defined by combining the first two terms of the right-hand side of equation (10.20):

$$pe^0(pH = 7) = pe^0 - \frac{10}{8}pH = pe^0 - \frac{10}{8}7 \tag{10.21}$$

and with $pe^0 = 14.9$, we receive $pe^0(pH = 7) = 6.15$ for our example (see Table 10.2).

Applying this approach to any redox reaction:

$$Ox + n_e\,e^- + n_p\,H^+ \rightleftharpoons Red \tag{10.22}$$

gives the general definition of a pH-normalized standard redox intensity:

$$pe^0(pH) = pe^0 - \frac{n_p}{n_e}pH \tag{10.23}$$

where $n_p$ is the number of the protons in the reaction equation and $n_e$ is the number of the transferred electrons. Some applications of pH-normalized standard redox intensities are discussed in Sections 10.4.5 and 10.6.

From equations (10.11) or (10.17), we can see that the redox intensity is related to the activities of the oxidant and the reductant. This is what makes the parameter redox intensity significant. Obviously, it is an indicator of the redox milieu. It indicates if we have an oxidizing milieu with predominance of the oxidant or a reducing milieu with predominance of the reductant. In this respect, the redox intensity has the same significance for redox systems as the pH for acid/base systems, which indicates acidic or basic milieu and determines the distribution between an acid and its conjugate base. There are some further analogies: The definition equations for pe and pH have the same mathematical form, both parameters are dimensionless, and equation (10.9) is analogous to equation (7.107) (with exception of the number of electrons in equation (10.9)).

In water chemistry, the parameters pe and pH are often referred to as master variables due to their relevance for speciation in aqueous systems and for assessing the milieu conditions. In the most general case, the speciation can depend on both parameters. The speciation of water constituents as a function of pe and pH will be discussed in more detail in Section 10.5.

Equation (10.17) can be used to solve two practical problems. First, the equation can be used to calculate the redox intensity in a considered water on the basis of analytical data, i.e., on the basis of the activities (or concentrations) of the redox system (Example 10.4). Second, the equation allows, inversely, finding of the species distribution for a given redox intensity (Example 10.5).

**Example 10.4**
The following data were measured in a surface water: pH = 8, $c(Mn^{2+}) = 1 \times 10^{-7}$ mol/L. What is the redox intensity in this water if we assume ideal conditions ($a = c$)? The relevant half-reaction is:

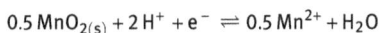

$$0.5\,MnO_{2(s)} + 2\,H^+ + e^- \rightleftharpoons 0.5\,Mn^{2+} + H_2O$$

with log $K^* = 20.8$.

**Solution:**
The standard redox intensity for the given system is:

$$pe^0 = \frac{1}{n_e}\log K^* = 20.8$$

Since only one electron is transferred, $pe^0$ equals log $K^*$. The general equation for the redox intensity reads:

$$pe = pe^0 + \frac{1}{n_e}\log\frac{\Pi a^v(Ox)}{\Pi a^v(Red)}$$

If we assume ideal conditions, we can use the concentrations instead of the activities. Water and the solid $MnO_2$ have not to be considered in the equation (see Chapter 5, Section 5.3). Therefore, the redox intensity equation for the given system is:

$$pe = pe^0 + \log\frac{c^2(H^+)}{c^{0.5}(Mn^{2+})}$$

Further, we obtain:

$$pe = pe^0 + 2\log c(H^+) - 0.5\log c(Mn^{2+}) = pe^0 - 2\,pH - 0.5\log c(Mn^{2+})$$

and, after introducing the given data, we find:

$$pe = 20.8 - 16 - 0.5(-7) = 8.3$$

---

**Example 10.5**

What is the ratio of the concentrations of sulfate ($SO_4^{2-}$) and hydrogensulfide ($HS^-$) at pH = 8 and pe = −4? The half-reaction for the redox couple $SO_4^{2-}/HS^-$ is:

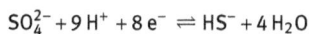

$$SO_4^{2-} + 9H^+ + 8e^- \rightleftharpoons HS^- + 4H_2O$$

and the equilibrium constant is given as log $K^*$ = 34. Ideal conditions ($a = c$) are assumed.

**Solution:**

The standard redox intensity related to log $K^*$ = 34 is:

$$pe^0 = \frac{1}{n_e} \log K^* = \frac{34}{8} = 4.25$$

and the equation for the redox intensity reads:

$$pe = pe^0 + \frac{1}{n_e} \log \frac{\Pi a^v(Ox)}{\Pi a^v(Red)}$$

$$pe = pe^0 + \frac{1}{8} \log \frac{c(SO_4^{2-}) c^9(H^+)}{c(HS^-)}$$

$$= pe^0 + \frac{9}{8} \log c(H^+) + \frac{1}{8} \log \frac{c(SO_4^{2-})}{c(HS^-)}$$

After rearranging this equation, we can find the concentration ratio:

$$\log \frac{c(SO_4^{2-})}{c(HS^-)} = 8 \left[ pe - pe^0 - \frac{9}{8} \log c(H^+) \right] = 8\,pe - 8\,pe^0 + 9\,pH$$

$$\log \frac{c(SO_4^{2-})}{c(HS^-)} = -32 - 34 + 72 = 6$$

$$\frac{c(SO_4^{2-})}{c(HS^-)} = 10^6$$

Obviously, $SO_4^{2-}$ is the dominant species under the given conditions.

　　Note: The redox equation can also be written in the form:

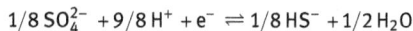

$$1/8\,SO_4^{2-} + 9/8\,H^+ + e^- \rightleftharpoons 1/8\,HS^- + 1/2\,H_2O$$

where only one electron is exchanged. In this case, the standard redox intensity remains constant, whereas log $K^*$ is reduced by a factor of 8 (according to the division of the reaction equation by 8) and equals $pe^0$. The redox intensity, pe, is now given by:

$$pe = pe^0 + \log \frac{c^{1/8}(SO_4^{2-}) c^{9/8}(H^+)}{c^{1/8}(HS^-)}$$

which obviously leads to the same equation as found in the case before:

$$pe = pe^0 + \frac{9}{8}\log c(H^+) + \frac{1}{8}\log\frac{c(SO_4^{2-})}{c(HS^-)}$$

Thus, the kind of writing the reaction equation is optional.

---

### 10.4.3 Redox Intensity Versus Redox Potential

As shown in the previous section, the redox intensity as a master variable in redox systems can be calculated on the basis of analytical data. However, the redox intensity is also related to a directly measurable quantity, the redox potential.

The relation between the redox potential and the activities of the oxidant and the reductant is given by the Nernst equation:

$$E_H = E_H^0 + \frac{2.303\,R\,T}{n_e\,F}\log\frac{\Pi\,a^v(Ox)}{\Pi\,a^v(Red)} \tag{10.24}$$

where $E_H^0$ is the standard redox potential, $R$ is the universal gas constant (8.3145 J/(mol·K)), $T$ is the absolute temperature, $n_e$ is the number of the transferred electrons, and $F$ is the Faraday constant ($F = 96\ 485$ C/mol). To quantify potentials, a reference state is needed. By definition, the potentials $E_H$ and $E_H^0$ are potentials relative to the standard hydrogen electrode (proton activity of 1 mol/L and hydrogen partial pressure of 1 bar) as a reference. This reference electrode corresponds to the half-reaction:

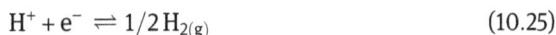

$$H^+ + e^- \rightleftharpoons 1/2\,H_{2(g)} \tag{10.25}$$

for which it holds that $\log K^* = pe^0 = 0$.

The redox potential can be measured as the potential difference between an inert working electrode (e.g., platinum) and a reference electrode. For practical reasons, not the standard hydrogen electrode but other reference electrodes with stable potentials that are easier to handle (e.g., Ag/AgCl electrode or calomel electrode) are typically used. To make the measured values comparable, they are converted into the potential relative to the hydrogen electrode by adding the potential difference between the applied reference electrode and the standard hydrogen electrode. This difference must therefore be known for the applied electrode. For redox potential measurements, commercial devices are available, referred to as redox potential meters or ORP (oxidation reduction potential) meters.

If we compare equation (10.24) with equation (10.17), we can easily derive a relationship between the redox intensity and the measurable redox potential:

$$pe = \frac{F}{2.303\,R\,T}E_H \tag{10.26}$$

The same relationship holds for the standard values, $pe^0$ and $E_H^0$:

$$pe^0 = \frac{F}{2.303\,R\,T}E_H^0 \tag{10.27}$$

Given that $1\,C = 1\,A\cdot s$ and $1\,J = 1\,W\cdot s = 1\,V\cdot A\cdot s$, we can find the conversion factor at 25 °C as follows:

$$pe = \frac{96\,485\,A\cdot s/mol}{2.303\,(8.3145\,V\cdot A\cdot s/(mol\cdot K))\,(298.15\,K)} = \frac{1}{0.059\,V}E_H \tag{10.28}$$

and

$$pe^0 = \frac{1}{0.059\,V}E_H^0 \tag{10.29}$$

In this manner, redox intensities can be estimated from measured redox potentials. Often, the redox potentials themselves are used as parameters to characterize the redox state of ground or surface waters. This is acceptable because both approaches are equivalent. For equilibrium calculations, however, the dimensionless redox intensity is better suited than a potential given in volts. Moreover, the link of the standard redox intensity to the law of mass action and the equilibrium constant is more direct. The analogy to the other master variable, the pH, and the conceptual similarity in treating acid/base and redox reactions are further arguments for using redox intensities instead of redox potentials.

In practice, the easily measurable redox potential is often not interpreted critically enough. It has to be pointed out that in many cases redox potential measurements are not very exact. There are different reasons for possible inaccuracies. A number of relevant species show slow electrode kinetics or are even more or less inactive, for instance $O_2$, $SO_4^{2-}$, and $NO_3^-$. Poisoning of the electrode surface is a further error source in redox potential measurements. Consequently, measured redox potentials seldom exactly match the related values calculated on the basis of analytical data. If possible, the analytical determination of the relevant redox species with subsequent calculation of pe should be preferred.

Independent of the parameter used, it has to be noted that redox equilibria are often not fully established in natural waters due to the high number of redox systems and the slow electron transfer processes in the aqueous phase. This is especially true for water bodies with strongly varying concentrations of redox-relevant species.

### 10.4.4 Special Case: Redox Reactions with Dissolved Gases

Many dissolved gases, such as $O_2$, $N_2$, $CO_2$, $H_2S$, and $CH_4$, play an important role as constituents of redox systems in natural waters. The laws of mass action of the respective half-reactions can be formulated in two different ways with respect to the quantification of the gas concentration: (i) by using the partial pressure above the considered liquid phase according to the convention made in Chapter 5 (Section 5.3), or (ii) with the molar liquid-phase concentration for the dissolved gas. Although the redox reactions under consideration occur in the liquid phase, the application of the partial pressure is not only formal, but may be appropriate in cases where the water is in contact with the gas phase and the gas-phase partial pressure influences the liquid-phase concentration according to Henry's law. Both approaches are equivalent and the related constants can be converted by means of the Henry constant. The conversion will be demonstrated below with the oxygen/water system as an example.

If we start with a reaction equation where $O_2$ is considered formally as a gas and then add the gas/water partitioning equilibrium (according to Henry's law in inverse form, see Chapter 6), we receive the reaction equation with oxygen as dissolved species:

$$O_{2(g)} + 4\,H^+ + 4\,e^- \rightleftharpoons 2\,H_2O \tag{10.30}$$

$$O_{2(aq)} \rightleftharpoons O_{2(g)} \tag{10.31}$$

$$\overline{\phantom{O_{2(aq)} + 4\,H^+ + 4\,e^- \rightleftharpoons 2\,H_2O}}$$

$$O_{2(aq)} + 4\,H^+ + 4\,e^- \rightleftharpoons 2\,H_2O \tag{10.32}$$

In Chapter 5, we have seen that in the case where reaction equations are added, the constants have to be multiplied or the logarithms of the constants have to be added. With $pe_1^0 = 20.75$ ($\log K_1 = 83$) for the first reaction and $\log K_2 = \log (1/H) = 2.95$ for the second reaction (see Table 6.1 in Chapter 6), we find the logarithm of the constant for the third reaction according to:

$$\log K_3 = \log K_1 + \log K_2 = \log K_1 + \log (1/H) = 83 + 2.95 = 85.95 \tag{10.33}$$

which is equivalent to $pe_3^0 = 21.48$. Accordingly, the pe functions related to the equations (10.30) and (10.32) (with $a = c$) are:

$$pe_1 = pe_1^0 + \frac{1}{4}\log [p(O_2)\,c^4(H^+)] = 20.75 + \frac{1}{4}\log p(O_2) - pH \tag{10.34}$$

$$pe_3 = pe_3^0 + \frac{1}{4}\log [c(O_2)\,c^4(H^+)] = 21.48 + \frac{1}{4}\log c(O_2) - pH \tag{10.35}$$

In practice, both types of formulation of redox processes with gases can be used. However, it has to be considered that, depending on the type of formulation, different constants are valid. To avoid confusion, the subscripts "aq" (aqueous solution) and "g" (gas phase) should be always used in the reaction equations.

### 10.4.5 Crossover Points Between Predominance Areas of Reduced and Oxidized Species

In Section 10.4.2 we discussed some analogies between acid/base and redox systems. In both cases, we have two different species that form a couple that is connected by a proton exchange (acid/base couple) or by an electron exchange (redox couple). The laws of mass action of these proton or electron transfer reactions provide the basis for calculating the crossover point between the predominance areas of the coupled species. For acid/base systems $(HA/A^-)$, it is a simple matter to find the crossover point. According to equation (7.107) (Chapter 7) we can write:

$$pH = pK_a^* - \log\frac{a(HA)}{a(A^-)} = pK_a^* + \log\frac{a(A^-)}{a(HA)} = pK_a^* + \log\frac{a(base)}{a(acid)} \qquad (10.36)$$

Defining the crossover point, $pH_c$, as that pH where the activities of acid and base are just the same, we have:

$$pH_c = pK_a^* \qquad (10.37)$$

Below this pH, the acid predominates, whereas above this pH, the base is the predominant species (see also Section 7.6 in Chapter 7).

For redox systems, we can find the crossover point in analogous manner from equation (10.11) by setting $a(Ox) = a(Red)$:

$$pe_c = pe^0 \qquad (10.38)$$

However, equation (10.38) can only be used for simple redox reactions with both oxidant and reductant present in the aqueous phase and without influence of pH. For instance, for the system $Fe^{2+}/Fe^{3+}$ with $pe_c = pe^0 = 13$ (see Table 10.2), the reduced species, $Fe^{2+}$, predominates at $pe < 13$ and the oxidized species, $Fe^{3+}$, predominates at $pe > 13$.

If protons occur in the redox reaction, the crossover point depends on pH and has therefore to be assigned to a specified pH value (Example 10.6).

---

**Example 10.6**
What is the pe where the activity of $SO_4^{2-}$ equals the activity of dissolved $H_2S$ at pH = 6 (crossover point)?

**Solution:**
From Table 10.2, we find:

$$SO_4^{2-} + 10\,H^+ + 8\,e^- \rightleftharpoons H_2S_{(aq)} + 4\,H_2O \qquad pe^0 = 5.12$$

The related pe function reads:

$$pe = pe^0 + \frac{1}{8} \log \frac{a(SO_4^{2-}) \, a^{10}(H^+)}{a(H_2S_{(aq)})} = pe^0 + \frac{10}{8} \log a(H^+) + \frac{1}{8} \log \frac{a(SO_4^{2-})}{a(H_2S_{(aq)})}$$

With $pH = -\log a(H^+)$ and setting $a(SO_4^{2-}) = a(H_2S_{(aq)})$, we find the crossover point at $pH = 6$ to be:

$$pe_c(pH = 6) = pe^0 - \frac{10}{8} pH = 5.12 - \frac{10}{8} 6 = -2.38$$

Generalizing the equation for $pe_c$ derived in the example under consideration of equation (10.23), we can write for the crossover point at a given pH:

$$pe_c(pH) = pe^0(pH) = pe^0 - \frac{n_p}{n_e} pH \qquad (10.39)$$

or for the neutral point (pH = 7) as the reference value:

$$pe_c(pH = 7) = pe^0(pH = 7) = pe^0 - \frac{n_p}{n_e} 7 \qquad (10.40)$$

where $n_p$ is the number of the protons in the reaction equation and $n_e$ is the number of the transferred electrons.

If water or solids are involved in the redox process, they do not occur in the law of mass action and in the related pe function. This means that only one component (oxidant or reductant) of the redox couple occur in the pe function and the definition of the crossover point cannot be done by setting $a(\text{oxidant}) = a(\text{reductant})$. In this case, the crossover point has to be defined by setting the activity of that partner that occurs in the pe function to an appropriate value. Typically, a very small concentration value is used to define the crossover point, because this point reflects a situation where the dissolved species loses its relevance and the other partner of the redox couple is considered to be predominant. This approach is demonstrated in Example 10.7.

**Example 10.7**
What is the crossover point of the system $Mn^{2+}/MnO_{2(s)}$ at $pH = 7$ if we define the boundary between the predominance areas of $Mn^{2+}$ and $MnO_2$ by $c(Mn^{2+}) = 1 \times 10^{-6}$ mol/L? Ideal conditions with $a = c$ are assumed.

**Solution:**
The half-reaction (see Table 10.2) is:

$$0.5 \, MnO_{2(s)} + 2 \, H^+ + e^- \rightleftharpoons 0.5 \, Mn^{2+} + H_2O \qquad pe^0 = 20.8$$

and the pe equation reads:

$$pe = pe^0 + \log \frac{c^2(H^+)}{c^{0.5}(Mn^{2+})}$$

or

$$pe = pe^0 - 2\,pH - 0.5\,\log c(Mn^{2+})$$

For $c(Mn^{2+}) = 1 \times 10^{-6}$ mol/L and pH = 7, we obtain:

$$pe_c(pH = 7) = 20.8 - 14 + 3 = 9.8$$

If pe in a considered water is lower than the crossover point, we can expect $Mn^{2+}$ in concentrations higher than $1 \times 10^{-6}$ mol/L. This is, according to our definition, the predominance area of $Mn^{2+}$. For pe higher than the crossover point, the concentration of $Mn^{2+}$ is lower than $1 \times 10^{-6}$ mol/L and, by definition, $MnO_2$ predominates.

### 10.4.6 Speciation as a Function of pe

In the previous section we discussed the estimation of the crossover point. In systems where both oxidant and reductant are present in dissolved form, the crossover point is that redox intensity where both components occur with equal activities. In the same manner, we can also determine the redox intensities for other activity ratios in such systems. In doing so, we can find the complete speciation with dependence on pe. This calculation is analogous to that shown for acid/base systems in Section 7.6 (Chapter 7).

First, we want to consider a simple redox system without the influence of pH. For such a system, we can derive the following relationship between the activity ratio and pe from equation (10.11):

$$\frac{a(Ox)}{a(Red)} = 10^{n_e(pe - pe^0)} \tag{10.41}$$

Introducing the fraction of the oxidant in the given oxidant/reductant system, $f(Ox)$,

$$f(Ox) = \frac{a(Ox)}{a(Ox) + a(Red)} \tag{10.42}$$

and combining this definition with equation (10.41) leads, after some rearrangements, to:

$$f(Ox) = \frac{1}{1 + 10^{n_e(pe^0 - pe)}} \tag{10.43}$$

Equation (10.43) has the same mathematical form as equation (7.110) for the fraction of a base in an acid/base system (Chapter 7). The fraction of the related reductant is:

$$f(\text{Red}) = 1 - f(\text{Ox}) \tag{10.44}$$

Knowing $pe^0$, we can calculate the fractions of oxidant and reductant as a function of pe. As an example, the speciation of the $Fe^{2+}/Fe^{3+}$ system is shown in Figure 10.1. It has to be noted that the simple $Fe^{2+}/Fe^{3+}$ system is only relevant under strongly acidic conditions, because $Fe^{3+}$ forms hydroxo complexes at higher pH values, which affects the redox equilibrium (see Section 10.5.7).

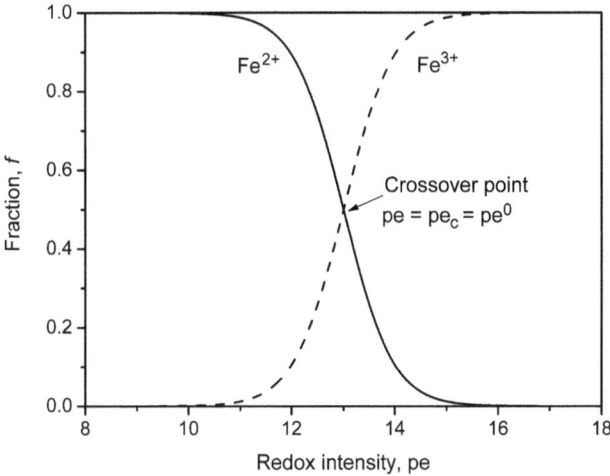

**Figure 10.1:** Speciation in the system $Fe^{2+}/Fe^{3+}$ with dependence on pe ($pe^0 = 13$).

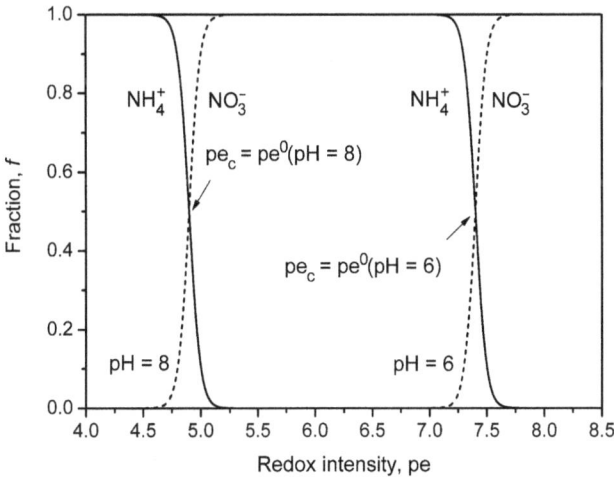

**Figure 10.2:** Speciation in the system $NH_4^+/NO_3^-$ with dependence on pe at two different pH values. $pe^0(pH = 6) = 7.4$, $pe^0(pH = 8) = 4.7$.

The same calculation can also be carried out for systems that are influenced by the pH. To do this, we have to replace $pe^0$ by $pe^0(pH)$ in equation (10.43):

$$f(Ox) = \frac{1}{1 + 10^{n_e[pe^0(pH) - pe]}} \qquad (10.45)$$

The standard redox intensity for a given pH can be found from equation (10.23). Figure 10.2 shows the speciation in the system $NH_4^+/NO_3^-$ for two different pH values. Here, the speciation depends not only on pe but also on pH.

### 10.4.7 Water as a Redox System

Water itself represents a redox system, because the hydrogen in the water molecule ($ON$: +1) can be reduced to $H_2$ ($ON = 0$) and the oxygen in the water molecule ($ON = -2$) can be oxidized to $O_2$ ($ON = 0$). These reactions would lead to a decomposition of water. Therefore, the redox conditions related to these reactions represent the boundary conditions for the stability of liquid water.

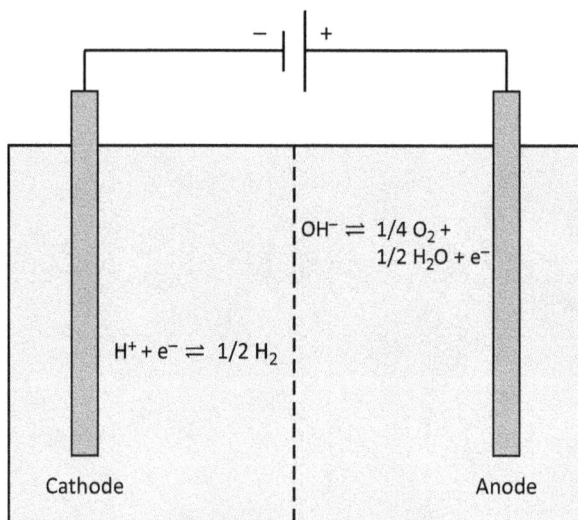

**Figure 10.3:** Decomposition of water by electrolysis.

To illustrate the decomposition, we want to consider the electrolysis of water, schematically shown in Figure 10.3. The relevant reactions are the autoprotolysis (self-ionization) of water (equation (10.46)) and the reactions on the cathode (equation (10.47)) and on the anode (equation (10.48)). Combining these reactions leads to the overall reaction equation of water decomposition (equation (10.49)):

$$H_2O \rightleftharpoons H^+ + OH^- \tag{10.46}$$

$$H^+ + e^- \rightleftharpoons 1/2\,H_{2(g)} \qquad \text{(cathode)} \tag{10.47}$$

$$OH^- \rightleftharpoons 1/4\,O_{2(g)} + 1/2\,H_2O + e^- \qquad \text{(anode)} \tag{10.48}$$

$$1/2\,H_2O \rightleftharpoons 1/4\,O_{2(g)} + 1/2\,H_{2(g)} \tag{10.49}$$

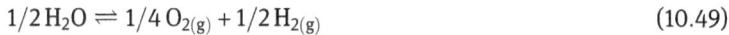

To bring equation (10.48) into a form that is consistent with the definition of a redox half-reaction (oxidant on the left-hand side, see equation (10.2) and Table 10.2) and includes protons instead of OH⁻, we have to add equations (10.48) and (10.46) and to write the resulting equation:

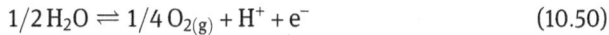

$$1/2\,H_2O \rightleftharpoons 1/4\,O_{2(g)} + H^+ + e^- \tag{10.50}$$

in reverse form:

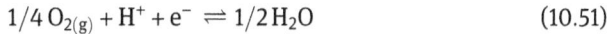

$$1/4\,O_{2(g)} + H^+ + e^- \rightleftharpoons 1/2\,H_2O \tag{10.51}$$

Given that the decomposition of water is described by equations (10.47) and (10.51), the related redox intensities define the stability limits of liquid water or, in other words, the highest and lowest redox intensities that are possible in liquid water. To quantify these limits, assumptions regarding the partial pressures of the decomposition products oxygen and hydrogen have to be made. Here, we will assume 1 bar as the maximum partial pressure of the decomposition products $H_2$ and $O_2$. The standard redox intensities are $pe^0 = 20.75$ for the oxygen/water half-reaction and $pe^0 = 0$ for the proton/hydrogen half-reaction (see Table 10.2). Under these conditions, we find for the $O_2/H_2O$ half-reaction (equation (10.51)):

$$pe = pe^0 + \log a(H^+) + \frac{1}{4}\log p(O_2) \tag{10.52}$$

$$pe = pe^0 - pH = 20.75 - pH \qquad \text{for} \quad p(O_2) = 1\,\text{bar} \tag{10.53}$$

and, in the same manner, we obtain for the $H^+/H_2$ half-reaction (equation (10.47)):

$$pe = pe^0 + \log a(H^+) - \frac{1}{2}\log p(H_2) \tag{10.54}$$

$$pe = pe^0 - pH = -pH \qquad \text{for} \quad p(H_2) = 1\,\text{bar} \tag{10.55}$$

In both cases, the redox intensity depends on the pH. By using equations (10.53) and (10.55) we can construct a pe–pH diagram that shows the stability area of liquid water (Figure 10.4). In natural waters, we can only find redox intensities that are located between both boundary lines.

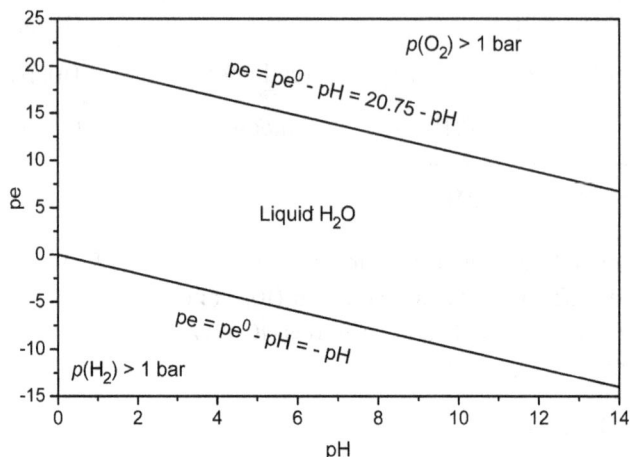

**Figure 10.4:** Stability diagram of water.

---

**Example 10.8**

What is the maximum redox intensity in water at pH = 7? Compare this value with the redox inten-sity of a surface water that is in equilibrium with the atmospheric oxygen ($p(O_2) = 0.209$ bar). The half-reaction for the redox couple $O_{2(g)}/H_2O$ is given in equation (10.51).

**Solution:**

The maximum redox intensity can be found by setting $p(O_2) = 1$ bar (see equation (10.53)):

$$pe = pe^0 - pH = 20.75 - pH$$

$$pe = 20.75 - 7 = 13.75$$

For the water saturated with oxygen, we have to apply equation (10.52) with $p(O_2) = 0.209$ bar, the partial pressure of oxygen in the air (see Table 6.3 in Chapter 6):

$$pe = pe^0 + \log a(H^+) + \frac{1}{4}\log p(O_2)$$

$$pe = 20.75 - 7 - 0.17 = 13.58$$

The redox intensity of the water saturated with atmospheric oxygen at pH = 7 is close to the maxi-mum redox intensity at this pH. Obviously, we have a strongly oxidizing milieu in oxygen-saturated water. This underlines the relevance of dissolved oxygen as oxidant in natural waters.

---

## 10.5 Construction of pe–pH Diagrams

### 10.5.1 Introduction

Many elements, such as C, S, N, Fe, or Mn, may occur in aquatic systems in different forms (species), depending on the master variables pe and pH of the water. The conditions, with respect to pH and pe, under which the different species predominate can be illustrated by means of pe–pH diagrams. These diagrams are also referred to as Pourbaix or predominance area diagrams. They include lines that represent the crossover points of the respective acid/base and redox systems as a function of pe and pH. In other words, the lines illustrate the boundaries between the predominance areas of the different species. A simple form of a Pourbaix diagram was already used to demonstrate the stability area of liquid water (Figure 10.4). Instead of pe, the redox potential, $E_H$, can also be used in predominance area diagrams to characterize the redox state (Section 10.4.3).

In the following sections, the step-by-step construction of a pe–pH diagram is demonstrated under consideration of the different types of equilibria that may be relevant. For the sake of simplification, we want to assume ideal conditions with $c = a$ and $K = K^*$ and a temperature of 25 °C in all cases.

## 10.5.2 Boundary Lines for Pure Acid/Base Systems

According to equation (7.107) (Chapter 7) or equation (10.36) (Section 10.4.5), we can describe the relationship between pH and the species distribution by:

$$pH = pK_a + \log \frac{c(\text{base})}{c(\text{acid})} \tag{10.56}$$

If we assume that the boundary between the predominance areas of the acid and the conjugate base is given by $c(\text{acid}) = c(\text{base})$ (crossover point), we find the following simple equation for the boundary line:

$$pH = pK_a \tag{10.57}$$

Obviously, the boundary lines between acids and their conjugate bases are independent of pe. Consequently, they are vertical lines (parallel to the pe axis) in the pe–pH diagram. At $pH < pK_a$, the acid predominates, whereas at $pH > pK_a$, the base predominates.

## 10.5.3 Boundary Lines for Complex Acid/Base Systems

Sometimes, the acid/base equilibria are coupled with further reactions, for instance precipitation. As an example, we want to consider the coupling of the iron(II) carbonate precipitation with the dissociation of $HCO_3^-$:

$$HCO_3^- \rightleftharpoons H^+ + CO_3^{2-} \tag{10.58}$$

$$Fe^{2+} + CO_3^{2-} \rightleftharpoons FeCO_{3(s)} \tag{10.59}$$

$$\overline{Fe^{2+} + HCO_3^- \rightleftharpoons FeCO_{3(s)} + H^+} \tag{10.60}$$

Equation (10.60) describes the formation of iron(II) carbonate in a pH range where $HCO_3^-$ is the dominant carbonate species. It can be used to find the boundary line between the stability areas of $Fe^{2+}$ and $FeCO_{3(s)}$ in the pe–pH diagram. Since protons occur in the reaction equation, the formation of $FeCO_{3(s)}$ depends on the pH. The pH function can be found from the respective law of mass action:

$$K = \frac{c(H^+)}{c(Fe^{2+})\,c(HCO_3^-)} \tag{10.61}$$

$$c(H^+) = K\,c(Fe^{2+})\,c(HCO_3^-) \tag{10.62}$$

$$pH = -\log K - \lg c(Fe^{2+}) - \log c(HCO_3^-) \tag{10.63}$$

Since $FeCO_{3(s)}$ is a solid, it does not occur in the law of mass action and consequently no concentration ratio can be used to define the boundary between the stability areas of the iron species. Furthermore, this boundary line additionally depends on the hydrogencarbonate concentration. The only possible way to construct the boundary line is to define appropriate concentrations for $Fe^{2+}$ and $HCO_3^-$. Since the location of the boundary line depends on the assumed concentrations, these concentrations have to be exactly specified in the legend of the diagram. For carbonate species, typically a total concentration (DIC, dissolved inorganic carbon) is defined, because the fractions of the different carbonate species (here hydrogencarbonate) depend on pH. The respective hydrogencarbonate concentration has to be calculated from $c(DIC)$ for each pH (Chapter 7, Section 7.7). Since pH is the wanted parameter and unknown at the beginning of the calculation, an iteration procedure has to be applied to find the boundary line.

### 10.5.4 Boundary Lines for Pure Redox Systems with Oxidant and Reductant in Dissolved Form

In analogy to the acid/base system, we can start with the pe equation:

$$pe = pe^0 + \frac{1}{n_e}\log\frac{c(Ox)}{c(Red)} \tag{10.64}$$

and assume $c(Ox) = c(Red)$ as a boundary (crossover point, see Section 10.4.5) between the predominance areas. In so doing, we obtain:

$$pe = pe^0 \qquad (10.65)$$

as the equation for the boundary line. The boundary lines for pure redox systems are independent on pH and run horizontally in the diagram (parallel to the pH axis). The system $Fe^{2+}/Fe^{3+}$ is a typical example for such simple redox systems.

### 10.5.5 Boundary Lines for pH-Dependent Redox Systems with Oxidant and Reductant in Dissolved Form

As an example for this case, we want to consider the redox equilibrium between the iron(III) hydroxo complex $FeOH^{2+}$ and the $Fe^{2+}$ ion:

$$FeOH^{2+} + H^+ + e^- \rightleftharpoons Fe^{2+} + H_2O \qquad (10.66)$$

The respective pe equation reads:

$$pe = pe^0 + \log\frac{c(FeOH^{2+})\, c(H^+)}{c(Fe^{2+})} \qquad (10.67)$$

With $c(FeOH^{2+}) = c(Fe^{2+})$ as condition for the boundary line, we receive:

$$pe = pe^0 - pH \qquad (10.68)$$

In this special case, the boundary line in the diagram is a transversal line with the slope −1. Generally, the boundary lines in pH-dependent redox systems are transversal lines with a slope that depends on the stoichiometric factors of the protons and electrons in the half-reaction. A general equation for the boundary line can be derived from equation (10.39) that describes the pH dependence of the crossover point:

$$pe = pe_c(pH) = pe^0 - \frac{n_p}{n_e}pH \qquad (10.69)$$

### 10.5.6 Boundary Lines for pH-Dependent Redox Systems Where Only One Partner Occurs in Dissolved Form

This case occurs when a solid is a reaction partner in the redox reaction. Since a solid compound is not considered in the law of mass action and in the related pe equation, the condition $c(Ox) = c(Red)$ cannot be used to define the boundary line. Instead of that, an absolute concentration for the dissolved component that occurs in the pe equation has to be defined. The system $Mn^{2+}/MnO_{2(s)}$ is a typical example. In analogy to the calculation of the crossover point in Section 10.4.5 (Example 10.7), the equation for the boundary line between the predominance areas can be found as follows.

The half-reaction is:

$$1/2\,MnO_{2(s)} + 2\,H^+ + e^- \rightleftharpoons 1/2\,Mn^{2+} + H_2O \tag{10.70}$$

and the pe equation reads:

$$pe = pe^0 + \log \frac{c^2(H^+)}{c^{0.5}(Mn^{2+})} \tag{10.71}$$

or

$$pe = pe^0 - 2\,pH - 0.5 \log c(Mn^{2+}) \tag{10.72}$$

Here, an appropriate concentration has to be defined to construct the boundary line. For instance, with $c(Mn^{2+}) = 1 \times 10^{-6}$ mol/L, we find the following equation for the boundary line:

$$pe = pe^0 - 2\,pH + 3 \tag{10.73}$$

For all pe lower than the boundary line, we can expect $Mn^{2+}$ in concentrations higher than $1 \times 10^{-6}$ mol/L. This is, according to our definition, the predominance area of $Mn^{2+}$. For pe higher than the boundary line, the concentration of $Mn^{2+}$ is lower than $1 \times 10^{-6}$ mol/L and, by definition, $MnO_2$ predominates.

Since for such redox systems the location of the boundary line depends on the assumed concentration, this concentration has to be specified in the legend of the diagram.

### 10.5.7 Example: The pe–pH Diagram of Iron

In this section, as an example, the construction of a pe–pH diagram for iron species will be shown. The first step in constructing pe–pH diagrams is to define the species to be considered. If we initially restrict our consideration to the $Fe/H_2O$ system, the following species have to be taken into account:

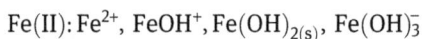

$$Fe(III): Fe^{3+},\ FeOH^{2+},\ Fe(OH)_2^+,\ Fe(OH)_{3(s)},\ Fe(OH)_4^-$$

$$Fe(II): Fe^{2+},\ FeOH^+,\ Fe(OH)_{2(s)},\ Fe(OH)_3^-$$

Besides the ions $Fe^{2+}$ and $Fe^{3+}$, the hydroxo complexes and the hydroxides are particularly relevant within the stability area of liquid water. Metallic iron is not stable under these pe and pH conditions.

The next step is to formulate the relevant pH-dependent reactions and the related pH functions. These equations together with the respective equilibrium constants are given in Table 10.3. The given reaction equations represent the successive deprotonation of the hydrated ferric ($Fe^{3+}$) and ferrous ($Fe^{2+}$) ions as well as the formation of the respective hydroxides. Since the iron species concentrations occur in some of the pH functions, the total iron concentration, for which the diagram is to be established, has to be defined. For our example, we want to assume a total iron concentration of $c(Fe_{total}) = 1 \times 10^{-5}$ mol/L. Now, we can calculate the pH values for the vertical boundary lines between the predominance areas of the species. These values are also given in Table 10.3.

**Table 10.3:** pe–pH diagram of iron: pH functions.

| No. | Reaction | pH function | p$K_a$, log $K$ | pH of the boundary line at $c(Fe_{total}) = 1 \times 10^{-5}$ mol/L |
|---|---|---|---|---|
| 1 | $Fe^{3+} + H_2O \rightleftharpoons$ $FeOH^{2+} + H^+$ | pH = p$K_a$ | p$K_a$ = 2.2 | 2.2 |
| 2 | $FeOH^{2+} + H_2O \rightleftharpoons$ $Fe(OH)_2^+ + H^+$ | pH = p$K_a$ | p$K_a$ = 3.6 | 3.6 |
| 3 | $Fe(OH)_2^+ + H_2O \rightleftharpoons$ $Fe(OH)_{3(s)} + H^+$ | pH = $-\log K$ $-\log c(Fe(OH)_2^+)$ | log $K$ = 2.5 | 2.5 |
| 4 | $Fe(OH)_{3(s)} + H_2O \rightleftharpoons$ $Fe(OH)_4^- + H^+$ | pH = $-\log K$ $+\log c(Fe(OH)_4^-)$ | log $K$ = $-18.4$ | 13.4 |
| 5 | $Fe^{2+} + H_2O \rightleftharpoons$ $FeOH^+ + H^+$ | pH = p$K_a$ | p$K_a$ = 9.1 | 9.1 |
| 6 | $FeOH^+ + H_2O \rightleftharpoons$ $Fe(OH)_{2(s)} + H^+$ | pH = $-\log K$ $-\log c(FeOH^+)$ | log $K$ = $-2.6$ | 7.6 |
| 7 | $Fe(OH)_{2(s)} + H_2O \rightleftharpoons$ $Fe(OH)_3^- + H^+$ | pH = $-\log K$ $+\log c(Fe(OH)_3^-)$ | log $K$ = $-18.9$ | 13.9 |
| 2 + 3 | $FeOH^{2+} + 2H_2O \rightleftharpoons$ $Fe(OH)_{3(s)} + 2H^+$ | pH = $-0.5 \log K$ $-0.5 \log c(FeOH^{2+})$ | log $K$ = $-1.1$ | 3.05 |
| 5 + 6 | $Fe^{2+} + 2H_2O \rightleftharpoons$ $Fe(OH)_{2(s)} + 2H^+$ | pH = $-0.5 \log K$ $-0.5 \log c(Fe^{2+})$ | log $K$ = $-11.7$ | 8.35 |

According to the successive deprotonation (reactions 1–4 and 5–7 in Table 10.3), we would expect a successive increase of the boundary pH for the Fe(III) species as well as for the Fe(II) species. However, inspection of the data shows that, in contrast to the expectation, the pH of the boundary line $Fe(OH)_2^+/Fe(OH)_{3(s)}$ is lower than the pH of the boundary line $FeOH^{2+}/Fe(OH)_2^+$. The same is true for the boundary lines $FeOH^+/Fe(OH)_{2(s)}$ and $Fe^{2+}/FeOH^+$. This means that there are no predominance areas for $Fe(OH)_2^+$ and $FeOH^+$ at the given iron concentration. We would have a different situation if the iron concentration would be lower, for instance $1 \times 10^{-7}$ mol/L. In this case, we would find a pH of 4.5 from equation 3 in Table 10.3 and a pH of 9.6 from equation 6 which means that we can expect a predominance area of $Fe(OH)_2^+$ between pH = 3.6 and pH = 4.5 and a narrow predominance area of $FeOH^+$ between pH = 9.1 and pH = 9.6.

In our case, where no predominance areas of $Fe(OH)_2^+$ and $FeOH^+$ exist, we have to add the reactions 2 and 3 and also the reactions 5 and 6 to find the boundary lines $Fe(OH)^{2+}/Fe(OH)_{3(s)}$ and $Fe^{2+}/Fe(OH)_{2(s)}$, respectively. This will be shown by example for the reactions 2 and 3 (Table 10.3):

$$FeOH^{2+} + H_2O \rightleftharpoons Fe(OH)_2^+ + H^+ \qquad \log K = -3.6 \qquad (10.74)$$

$$Fe(OH)_2^+ + H_2O \rightleftharpoons Fe(OH)_{3(s)} + H^+ \qquad \log K = 2.5 \qquad (10.75)$$

$$FeOH^{2+} + 2H_2O \rightleftharpoons Fe(OH)_{3(s)} + 2H^+ \qquad \log K = -1.1 \qquad (10.76)$$

In the same manner, we can combine reactions 5 and 6 to find the boundary line $Fe^{2+}/Fe(OH)_{2(s)}$. The combined reaction equations (2 + 3 and 5 + 6) and the related pH functions are also listed in Table 10.3 (last two rows).

The next step in the construction of the pe–pH diagram is to draw the vertical lines (pH functions) as shown in Figure 10.5.

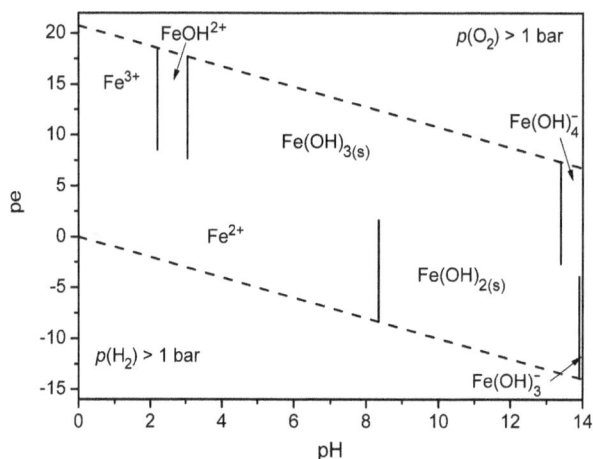

**Figure 10.5:** pH functions in the pe–pH diagram of iron ($c(Fe_{total}) = 1 \times 10^{-5}$ mol/L).

Now, we have to find the boundary lines between the reduced and the oxidized species. Which boundaries have to be considered can be derived from Figure 10.5.

From Figure 10.5 and Table 10.3, we can also find the relevant pH ranges of the respective pe lines. The reaction equations and pe functions are given in Table 10.4. In principle, only the standard redox intensity of the reaction 8 has to be known. All other constants can be derived on the basis of the solubility products of the hydroxides (see Table 8.1 in Chapter 8), the water dissociation constant, and the constants of the hydrolysis reactions given in Table 10.3. Example 10.9 demonstrates such a calculation.

**Table 10.4:** pe–pH diagram of iron: pe functions.

| No. | Reaction | pe function | pe⁰ |
|-----|----------|-------------|-----|
| 8 | $Fe^{3+} + e^- \rightleftharpoons Fe^{2+}$ | $pe = pe^0$ | $pe^0 = 13$ |
| 9 | $FeOH^{2+} + H^+ + e^- \rightleftharpoons Fe^{2+} + H_2O$ | $pe = pe^0 - pH$ | $pe^0 = 15.2$ |
| 10 | $Fe(OH)_{3(s)} + 3\,H^+ + e^- \rightleftharpoons Fe^{2+} + 3\,H_2O$ | $pe = pe^0 - 3\,pH - \log c(Fe^{2+})$ | $pe^0 = 16.3$ |
| 11 | $Fe(OH)_{3(s)} + H^+ + e^- \rightleftharpoons Fe(OH)_{2(s)} + H_2O$ | $pe = pe^0 - pH$ | $pe^0 = 4.6$ |
| 12 | $Fe(OH)_4^- + 2\,H^+ + e^- \rightleftharpoons Fe(OH)_{2(s)} + 2\,H_2O$ | $pe = pe^0 - 2\,pH + \log c(Fe(OH)_4^-)$ | $pe^0 = 23$ |
| 13 | $Fe(OH)_4^- + H^+ + e^- \rightleftharpoons Fe(OH)_3^- + H_2O$ | $pe = pe^0 - pH$ | $pe^0 = 4.1$ |

**Example 10.9**

What are the standard redox intensity and the pe function of the reaction:

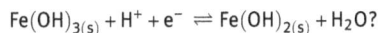

$$Fe(OH)_{3(s)} + H^+ + e^- \rightleftharpoons Fe(OH)_{2(s)} + H_2O?$$

The required data are given in Table 8.1 and 10.3.

**Solution:**

If the standard redox intensity for a redox reaction is unknown, we can derive it from the standard redox intensities of other reactions after combining these reactions in such a manner that the reaction of interest results as the overall reaction. The redox equation in this example can be considered an overall reaction consisting of the following elementary reactions: dissolution of $Fe(OH)_3$ and $Fe(OH)_2$, redox reaction $Fe^{3+}/Fe^{2+}$ and water dissociation. Accordingly, the overall equilibrium constant can be found from the solubility products of the hydroxides, the standard redox intensity of the redox system $Fe^{3+}/Fe^{2+}$, and the dissociation constant of water. Note that in cases where the reaction equation has to be written in reverse direction compared to the definition of the equilibrium constant, the logarithm of the constant receives an opposite sign.

$$Fe(OH)_{3(s)} \rightleftharpoons Fe^{3+} + 3\,OH^- \qquad \log K = -pK_{sp} = -38.7$$

$$Fe^{2+} + 2\,OH^- \rightleftharpoons Fe(OH)_{2(s)} \qquad \log K = pK_{sp} = 16.3 \;\;(\text{reverse reaction!})$$

$$Fe^{3+} + e^- \rightleftharpoons Fe^{2+} \qquad \log K = pe^0 = 13$$

$$H^+ + OH^- \rightleftharpoons H_2O \qquad \log K = pK_w = 14 \;\;(\text{reverse reaction!})$$

---

$$Fe(OH)_{3(s)} + H^+ + e^- \rightleftharpoons Fe(OH)_{2(s)} + H_2O \qquad \log K = pe^0 = -38.7 + 16.3 + 13 + 14 = 4.6$$

The respective pe function reads:

$$pe = pe^0 + \log c(H^+) = pe^0 - pH = 4.6 - pH$$

Under consideration of all pe functions, we can now complete the pe–pH diagram (Figure 10.6).

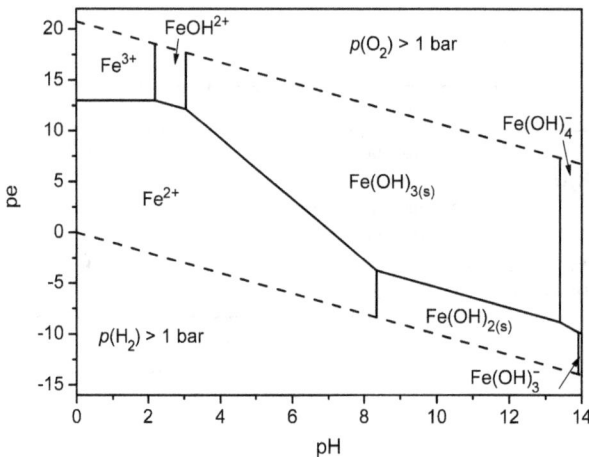

**Figure 10.6:** Complete pe–pH diagram of iron ($c(Fe_{total}) = 1 \times 10^{-5}$ mol/L).

It has to be noted that Figure 10.6 shows the predominance area diagram of iron in its simplest form, because it considers, besides the ions $Fe^{2+}$ and $Fe^{3+}$, only the hydroxo complexes and the hydroxides. However, iron is also able to form compounds with other water constituents, for instance with carbonate or sulfide ($FeCO_{3(s)}$ or $FeS_{(s)}$). In particular the iron(II) carbonate, $FeCO_{3(s)}$, is relevant due to the relatively high concentration of inorganic carbon in most aquatic systems. Therefore, we will take $FeCO_{3(s)}$ as an example to demonstrate how the diagram can be extended to consider further species.

At first we have to think about the possible position of the $FeCO_{3(s)}$ predominance area. According to the species distribution in the carbonic acid system (Section 7.7 in Chapter 7), we can expect that the formation of $FeCO_{3(s)}$ is favored at higher pH values, where enough carbonate is available. On the other hand, we see from a comparison of the solubility products of $FeCO_{3(s)}$ ($pK_{sp} = 10.5$) and $Fe(OH)_{2(s)}$ ($pK_{sp} = 16.3$) that $FeCO_{3(s)}$ is much more soluble than $Fe(OH)_{2(s)}$. Consequently, $Fe(OH)_{2(s)}$ will preferentially precipitate at very high pH values (high $OH^-$ concentrations). Thus, we can expect that the predominance area of $FeCO_{3(s)}$ is located between the areas of $Fe^{2+}$ and $Fe(OH)_{2(s)}$. Accordingly, we have to find the new boundary lines $Fe^{2+}/FeCO_{3(s)}$, $FeCO_{3(s)}/Fe(OH)_{2(s)}$ and $Fe(OH)_{3(s)}/FeCO_{3(s)}$. Again, we can combine other basic equations to find the required equilibrium constants for the pH and pe functions (Example 10.10).

---

**Example 10.10**

What is the equilibrium constant of the reaction:

$$FeCO_{3(s)} + 2H_2O \rightleftharpoons Fe(OH)_{2(s)} + H^+ + HCO_3^- ?$$

The following data are given: $pK_{sp}(FeCO_{3(s)}) = 10.5$, $pK_{sp}(Fe(OH)_{2(s)}) = 16.3$, $pK_a(HCO_3^-) = 10.3$ and $pK_w = 14$.

**Solution:**

| | |
|---|---|
| $FeCO_{3(s)} \rightleftharpoons Fe^{2+} + CO_3^{2-}$ | $\log K = -pK_{sp} = -10.5$ |
| $CO_3^{2-} + H^+ \rightleftharpoons HCO_3^-$ | $\log K = pK_a = 10.3$ (reverse reaction!) |
| $Fe^{2+} + 2OH^- \rightleftharpoons Fe(OH)_{2(s)}$ | $\log K = pK_{sp} = 16.3$ (reverse reaction!) |
| $2H_2O \rightleftharpoons 2H^+ + 2OH^-$ | $\log K = -2pK_w = -28$ |

---

$FeCO_{3(s)} + 2H_2O \rightleftharpoons Fe(OH)_{2(s)} + H^+ + HCO_3^-$    $\log K = -10.5 + 10.3 + 16.3 - 28 = -11.9$

---

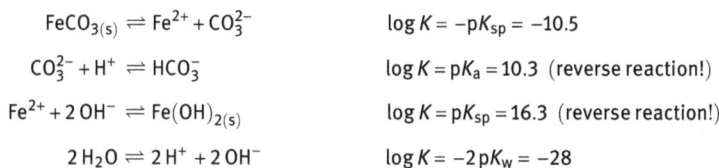

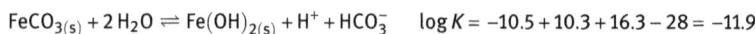

The complete set of additional equations, necessary to consider $FeCO_{3(s)}$ in the diagram, is listed in Table 10.5. Since the hydrogencarbonate concentration occurs in the pH and pe equations, we have to define a respective concentration for which the calculation of the boundary lines will be done. Due to the pH dependence of the species distribution in the carbonic acid system, it is necessary to define a constant DIC concentration and to calculate the hydrogencarbonate concentration for the considered pH as shown in Chapter 7 (equation (7.129)). For our calculation, the DIC concentration is set to $1 \times 10^{-3}$ mol/L. For the pH functions, in general, an iterative procedure starting with $c(HCO_3^-) = c(DIC)$ is necessary, because the pH is the wanted parameter and therefore unknown at the beginning of the calculation. In our special case, however, the hydrogencarbonate concentration is not much different from the DIC concentration because the expected pH values of the boundary lines (Table 10.5) are close to the pH where the fraction of hydrogencarbonate reaches its maximum (pH = 8.35, see Figure 7.4 in Chapter 7). Thus, approximately, the DIC concentration can be set equal to the $HCO_3^-$ concentration for the whole pH range of the predominance area of $FeCO_{3(s)}$.

The complete diagram, including the predominance area of $FeCO_{3(s)}$, is shown in Figure 10.7.

We can derive from Figure 10.7 that under oxidizing conditions iron(III) hydroxide is the predominant species over a broad range of pH, whereas under reducing conditions $Fe^{2+}$ predominates under acidic, neutral and weakly basic conditions. At higher pH values, iron(II) carbonate and iron(II) hydroxide are the predominant species

**Table 10.5:** pe–pH diagram of iron: additional equations necessary to consider $FeCO_{3(s)}$.

| No. | Reaction | pH and pe functions | log $K$, pe$^0$ | pH of the boundary line at $c(Fe_{total}) = 1 \times 10^{-5}$ mol/L, $c(DIC) = 1 \times 10^{-3}$ mol/L |
|---|---|---|---|---|
| 14 | $FeCO_{3(s)} + H^+ \rightleftharpoons$ $Fe^{2+} + HCO_3^-$ | $pH = \log K - \log c(Fe^{2+})$ $- \log c(HCO_3^-)$ | $\log K = -0.2$ | 7.8 |
| 15 | $FeCO_{3(s)} + 2 H_2O \rightleftharpoons$ $Fe(OH)_{2(s)} + H^+ + HCO_3^-$ | $pH = -\log K +$ $\log c(HCO_3^-)$ | $\log K = -11.9$ | 8.9 |
| 16 | $Fe(OH)_{3(s)} + 2 H^+ + HCO_3^-$ $+ e^- \rightleftharpoons FeCO_{3(s)} + 3 H_2O$ | $pe = pe^0 - 2 pH +$ $\log c(HCO_3^-)$ | $pe^0 = 16.5$ | – |

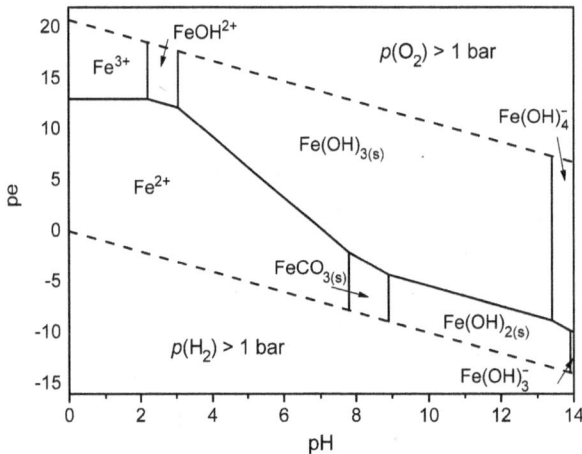

**Figure 10.7:** pe–pH diagram of iron, extended by the predominance area of iron(II) carbonate ($c(Fe_{total}) = 1 \times 10^{-5}$ mol/L, $c(DIC) = 1 \times 10^{-3}$ mol/L).

under reducing conditions. It can be further seen that $Fe^{3+}$ is predominant only at very low pH values and under strongly oxidizing conditions. Furthermore, we can conclude that a higher concentration of $Fe^{2+}$ in water is an indicator for reducing conditions. Such reducing conditions can be found, for instance, in groundwaters or at the bottom of water bodies where an oxygen deficit exists. If such waters come into contact with air, the redox intensity will strongly increase due to the dissolution of oxygen (see also Example 10.8) and, as a result, $Fe^{2+}$ will be oxidized to solid $Fe(OH)_{3(s)}$. This oxidation process is also utilized in drinking water treatment to remove high $Fe^{2+}$ concentrations from raw water (e.g., groundwater) in order to avoid uncontrolled precipitation of iron hydroxide in vessels and tubes during water treatment and distribution. During the so-called deironing process, the raw water is aerated to increase the redox intensity and to oxidize $Fe^{2+}$ to $Fe(OH)_{3(s)}$. The solid oxidation product is subsequently removed by filtration.

## 10.5.8 Example: The pe–pH Diagram of Sulfur

As a second example, the pe–pH diagram of sulfur will be considered in this section. The major sulfur species in aqueous solutions are $HSO_4^-$ and $SO_4^{2-}$, both with the oxidation number 6 for sulfur, elemental S with the oxidation number 0, and $H_2S$, $HS^-$, and $S^{2-}$, all with the oxidation number –2 for sulfur. Since $H_2SO_4$ is a very strong acid, it is completely dissociated in water and has not to be considered here. The respective reaction equations together with the pe and pH functions for the sulfur system are listed in Table 10.6. In this table, the $pK_a$ of $HS^-$ is given as 17. However, this value is very uncertain. Values in the range of 13–19 have been reported in the literature. If we assume a value of 17, there is no predominance area for $S^{2-}$ because the (formal) boundary between $HS^-$ and $S^{2-}$ would be at pH = $pK_a$ = 17 which is above the upper limit of the conventional pH scale. At this point, it is necessary to remember that the boundary lines are not sharp boundaries, but indicate only the crossover from the area where one species predominates to the area where another species predominates. Consequently, $S^{2-}$ also occurs at lower pH values but in a much lower concentration in comparison to the predominant species $HS^-$ (Example 10.11). Otherwise, the formation of metal sulfides under reducing conditions at pH ≪ 14 could not be explained.

---

**Example 10.11**
What is the concentration ratio of sulfide ($S^{2-}$) and hydrogensulfide ($HS^-$) at pH = 12, if the $pK_a$ of $HS^-$ is assumed to be 17?

**Solution:**
According to equation (10.36), we can write:

$$pH = pK_a + \log \frac{c(S^{2-})}{c(HS^-)}$$

and

$$\frac{c(S^{2-})}{c(HS^-)} = 10^{pH-pK_a} = 10^{12-17} = 10^{-5}$$

Accordingly, the concentration of sulfide is $10^5$ times lower than that of $HS^-$, but not zero.

---

Following the methods described in Section 10.5.7, the pe–pH diagram of sulfur can be constructed. From this diagram, shown in Figure 10.8, we can derive that sulfate is the predominant species over a broad range of pH and pe. The species $H_2S$ and $HS^-$ predominate only under strongly reducing conditions. Interestingly, there is a narrow area at low pH and medium pe where elemental sulfur is the predominant species. As can be seen from the equations in Table 10.6, a concentration of the sulfur species has to be assumed for the calculation. For the diagram shown in Figure 10.8, a total sulfur concentration of $10^{-3}$ mol/L was assumed which is in the typical order of magnitude found in natural water bodies.

**Table 10.6:** pe–pH diagram of sulfur: pH and pe functions.

| No. | Reaction | pH or pe function | $pK_a$, $pe^0$ |
|---|---|---|---|
| 1 | $HSO_4^- \rightleftharpoons H^+ + SO_4^{2-}$ | $pH = pK_a$ | $pK_a = 2$ |
| 2 | $H_2S_{(aq)} \rightleftharpoons H^+ + HS^-$ | $pH = pK_a$ | $pK_a = 7$ |
| 3 | $HS^- \rightleftharpoons H^+ + S^{2-}$ | $pH = pK_a$ | $pK_a \approx 17$ |
| 4 | $1/2\,S_{(s)} + H^+ + e^- \rightleftharpoons 1/2\,H_2S_{(aq)}$ | $pe = pe^0 - pH - 0.5\,\log c(H_2S)$ | $pe^0 = 2.4$ |
| 5 | $1/2\,S_{(s)} + 1/2\,H^+ + e^- \rightleftharpoons 1/2\,HS^-$ | $pe = pe^0 - 0.5\,pH - 0.5\,\log c(HS^-)$ | $pe^0 = -1.1$ |
| 6 | $1/8\,SO_4^{2-} + 9/8\,H^+ + e^- \rightleftharpoons 1/8\,HS^- + 1/2\,H_2O$ | $pe = pe^0 - \dfrac{9}{8}pH$ | $pe^0 = 4.25$ |
| 7 | $1/6\,SO_4^{2-} + 4/3\,H^+ + e^- \rightleftharpoons 1/6\,S_{(s)} + 2/3\,H_2O$ | $pe = pe^0 - \dfrac{4}{3}pH + \dfrac{1}{6}\log c(SO_4^{2-})$ | $pe^0 = 6.03$ |
| 8 | $1/6\,HSO_4^- + 7/6\,H^+ + e^- \rightleftharpoons 1/6\,S_{(s)} + 2/3\,H_2O$ | $pe = pe^0 - \dfrac{7}{6}pH + \dfrac{1}{6}\log c(HSO_4^-)$ | $pe^0 = 5.7$ |
| 9 | $1/8\,SO_4^{2-} + 5/4\,H^+ + e^- \rightleftharpoons 1/8\,H_2S_{(aq)} + 1/2\,H_2O$ | $pe = pe^0 - \dfrac{5}{4}pH$ | $pe^0 = 5.12$ |

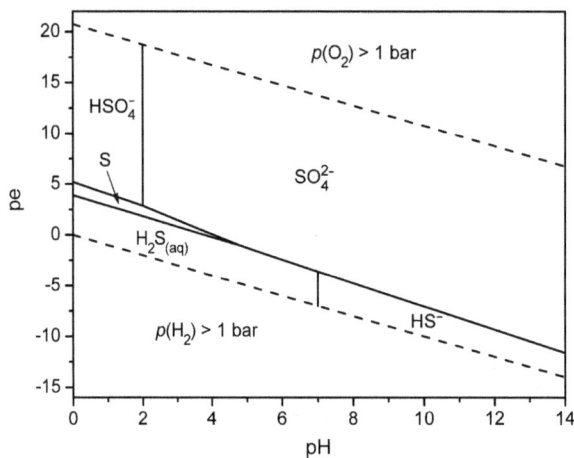

**Figure 10.8:** pe–pH diagram of sulfur ($c(S_{total}) = 0.001$ mol/L).

We can use the sulfur diagram to demonstrate the influence of the assumed sulfur concentration on the location of the boundary lines in the pe–pH diagram. Figure 10.9 shows the pe–pH diagram of sulfur for a sulfur concentration that is 100 times higher than that used for the calculation of the diagram in Figure 10.8. Despite the strong difference between the assumed concentrations, the form of the diagrams differs only slightly. The most obvious difference consists in the size of the predominance area of elemental sulfur, which is larger at higher total sulfur concentrations. We can derive from this comparison that the boundary lines in the pe–pH diagrams are not very sensitive with respect to the concentrations assumed for the diagram construction as long as the concentration differences are not too large. This can be explained by the logarithmic character of the parameters pe and pH.

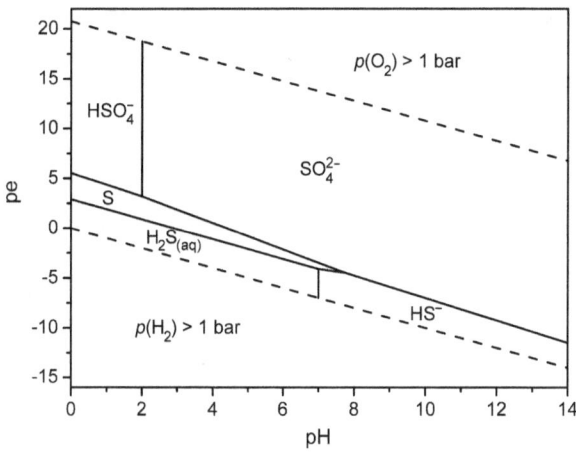

**Figure 10.9:** pe–pH diagram of sulfur ($c(S_{total}) = 0.1$ mol/L).

## 10.6 Complete Redox Reactions

### 10.6.1 Basic Relationships

As already mentioned in Section 10.3, free electrons do not exist in aqueous systems over longer periods of time. Therefore, a complete redox reaction consists of two redox couples between which electrons are exchanged. Accordingly, no more free electrons occur in the complete redox reaction equation:

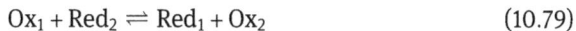

$$Ox_1 + n_e\ e^- \rightleftharpoons Red_1 \tag{10.77}$$

$$Red_2 \rightleftharpoons Ox_2 + n_e\ e^- \tag{10.78}$$

$$Ox_1 + Red_2 \rightleftharpoons Red_1 + Ox_2 \tag{10.79}$$

The law of mass action for the complete redox reaction reads:

$$K^* = \frac{a(Ox_2)\,a(Red_1)}{a(Red_2)\,a(Ox_1)} \tag{10.80}$$

The equilibrium constant of the complete redox reaction is related to the standard redox intensities of the half-reactions. This relationship can be derived from the equilibrium condition and the redox intensity equations for both half-reactions. If the equilibrium is established in the considered water, then this water is characterized by only one single value of pe (remember that the same is true for the other master variable, the pH value). Accordingly, the equilibrium condition can be written as:

$$pe_1 = pe_2 \tag{10.81}$$

Substituting the redox intensities by the respective equations for the half-reactions gives:

$$pe_1^0 + \frac{1}{n_e}\log\frac{a(Ox_1)}{a(Red_1)} = pe_2^0 + \frac{1}{n_e}\log\frac{a(Ox_2)}{a(Red_2)} \tag{10.82}$$

Here, it is assumed that the half-reactions are written in such a manner that the same number of electrons, $n_e$, is transferred. Rearranging equation (10.82) leads to:

$$n_e(pe_1^0 - pe_2^0) = \log\frac{a(Ox_2)}{a(Red_2)} - \log\frac{a(Ox_1)}{a(Red_1)} = \log\frac{a(Ox_2)\,a(Red_1)}{a(Red_2)\,a(Ox_1)} \tag{10.83}$$

The quotient on the right-hand side equals the equilibrium constant (see equation (10.80)). Thus, we receive the following relationship that links the equilibrium constant of the complete redox reaction with the standard redox intensities of the half-reactions:

$$\log K^* = n_e\,(pe_1^0 - pe_2^0) \tag{10.84}$$

If we want to find the equilibrium constant from equation (10.84), we have to know which of the standard redox intensities is $pe_1^0$ and which is $pe_2^0$, because the second is subtracted from the first and to interchange them would change the value of $\log K^*$. The decision which standard redox intensity belongs to which half-reaction is easy if we consider the addition of the half-reactions to the complete redox reaction. As we can see from equations (10.77)–(10.79), one of the half-reactions has to be written in the reverse direction to allow addition of both half-reactions to form an overall reaction. This reverse reaction is that reaction whose standard redox intensity, $pe^0$, has the negative sign in equation (10.84) (i.e., $pe_2^0$). This is in accordance with the general relationship $\log K^*_{forward} = -\log K^*_{reverse}$ that can be derived from equation (5.18) (Chapter 5).

Equation (10.80) together with equation (10.84) can be used for equilibrium calculations as will be shown in Example 10.12.

**Example 10.12**

What is the equilibrium concentration of $Mn^{2+}$ in a water that is saturated with atmospheric oxygen ($p(O_2) = 0.209$ bar) at pH = 6? Ideal conditions are assumed ($a = c$, $K^* = K$).

The relevant half-reactions are:

$$1/4\,O_{2(g)} + H^+ + e^- \rightleftharpoons 1/2\,H_2O \qquad pe^0 = 20.75$$

and

$$MnO_{2(s)} + 4\,H^+ + 2\,e^- \rightleftharpoons Mn^{2+} + 2\,H_2O \qquad pe^0 = 20.8$$

**Solution:**

If we want to write the overall reaction as oxidation of $Mn^{2+}$ with oxygen, we have to multiply the first reaction by two (to have the same number of electrons in both equations) and to write the second equation in the reverse direction:

$$1/2\,O_{2(g)} + 2\,H^+ + 2\,e^- \rightleftharpoons H_2O$$

$$Mn^{2+} + 2\,H_2O \rightleftharpoons MnO_{2(s)} + 4\,H^+ + 2\,e^-$$

$$\overline{\phantom{Mn^{2+} + 2\,H_2O \rightleftharpoons MnO_{2(s)} + 4\,H^+ + 2\,e^-}}$$

$$Mn^{2+} + 1/2\,O_{2(g)} + H_2O \rightleftharpoons MnO_{2(s)} + 2\,H^+$$

The law of mass action for the overall reaction reads:

$$K = \frac{c^2(H^+)}{c(Mn^{2+})\,p^{0.5}(O_2)}$$

and the equilibrium constant can be found from:

$$\log K = n_e\,(pe_1^0 - pe_2^0)$$

with $pe_1^0 = 20.75$ and $pe_2^0 = 20.8$ (subscript 2 refers to the reaction that is written in reverse direction). The number of transferred electrons is $n_e = 2$. Note that $pe^0$, in contrast to the related half-reaction equilibrium constant, does not change its value when the reaction equation is multiplied by an arbitrary factor. The unit of the constant can be derived from the law of mass action:

$$\log K = 2\,(20.75 - 20.8) = -0.1$$

$$K = 10^{-0.1}\,mol/(L \cdot bar^{0.5}) = 0.794\,mol/(L \cdot bar^{0.5})$$

Now, we can calculate the $Mn^{2+}$ concentration by rearranging the law of mass action:

$$c(Mn^{2+}) = \frac{c^2(H^+)}{K\,p^{0.5}(O_2)} = \frac{1 \times 10^{-12}\,mol^2/L^2}{[0.794\,mol/(L \cdot bar^{0.5})]\,(0.457\,bar^{0.5})} = 2.756 \times 10^{-12}\,mol/L$$

We can learn from this example that in an oxygen-saturated water the reduced species $Mn^{2+}$ occurs only in a negligible concentration. The dominant manganese species under these oxidizing conditions is manganese dioxide.

We can use this example to demonstrate that the choice of the number of transferred electrons and the assignment of the indices 1 or 2 to the reactions have no influence on the result. If we set the number of transferred electrons to 1 and write the oxygen half-reaction in reverse form, the equations for the half-reactions read:

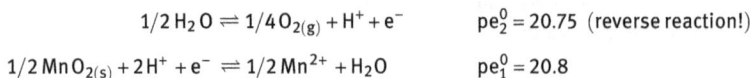

$$1/2\,H_2O \rightleftharpoons 1/4\,O_{2(g)} + H^+ + e^- \qquad pe_2^0 = 20.75 \text{ (reverse reaction!)}$$

$$1/2\,MnO_{2(s)} + 2H^+ + e^- \rightleftharpoons 1/2\,Mn^{2+} + H_2O \qquad pe_1^0 = 20.8$$

$$1/2\,MnO_{2(s)} + H^+ \rightleftharpoons 1/2\,Mn^{2+} + 1/4\,O_{2(g)} + 1/2\,H_2O$$

The related law of mass action is then:

$$K = \frac{c^{0.5}(Mn^{2+})\,p^{0.25}(O_2)}{c(H^+)}$$

and the equilibrium constant, $K$, is given by:

$$\log K = 1\,(20.8 - 20.75) = 0.05$$

$$K = 10^{0.05}\,bar^{0.25} \cdot L^{0.5}/mol^{0.5} = 1.122\,bar^{0.25} \cdot L^{0.5}/mol^{0.5}$$

Finally, we find for the $Mn^{2+}$ concentration:

$$c(Mn^{2+}) = \left(\frac{c(H^+)\,K}{p^{0.25}(O_2)}\right)^2 = \left(\frac{(1\times10^{-6}\,mol/L)\,(1.122\,bar^{0.25}\cdot L^{0.5}/mol^{0.5})}{0.676\,bar^{0.25}}\right)^2$$

$$c(Mn^{2+}) = 2.755 \times 10^{-12}\,mol/L$$

which is nearly the same value as found previously. The small difference between the results is due to round-off errors.

Equation (10.84) can also be used to assess if a considered aquatic system is in the state of redox equilibrium and, if not, in which direction the redox reaction will proceed. As shown in Chapter 5, the equilibrium constant, $K^*$, and the reaction quotient, $Q$, are related to the molar Gibbs energy of the reaction as follows:

$$\Delta_R G = \Delta_R G^0 + RT \ln Q \qquad (10.85)$$

$$\Delta_R G^0 = -RT \ln K^* \qquad (10.86)$$

$$\Delta_R G = -RT \ln K^* + RT \ln Q = RT \ln \frac{Q}{K^*} \qquad (10.87)$$

where $\Delta_R G^0$ is the molar standard Gibbs energy of reaction and $Q$ is the reaction quotient that can be calculated from an expression that has the same form as the law of mass action but includes actual activities (concentrations) instead of equilibrium activities (concentrations). These general relationships are also valid for complete redox reactions and can be used to predict if a redox reaction proceeds spontaneously in the written direction. Given that the condition for a spontaneous reaction is $\Delta_R G < 0$, we can derive that the reaction quotient, $Q$, must be lower than $K^*$ for a spontaneous reaction (see also Table 5.1 in Chapter 5). The probability of meeting this condition is higher the greater the value of $K^*$ or the greater the difference between the standard redox intensities of the half-reactions is. The value of $\Delta_R G^0$ is therefore a good indicator for the probability that a considered redox reactions will occur:

$$\Delta_R G^0 = -R T \ln K^* = -2.303\, R T \log K^* = -2.303\, n_e\, R T (\mathrm{pe}_1^0 - \mathrm{pe}_2^0) \qquad (10.88)$$

The factor 2.303 is used to convert the natural logarithm to the decimal logarithm.

Since redox reactions are often pH-dependent, it may be reasonable to normalize equation (10.88) to a defined pH, typically pH = 7:

$$\Delta_R G^0(\mathrm{pH}=7) = -2.303\, n_e\, R T \left[\mathrm{pe}_1^0(\mathrm{pH}=7) - \mathrm{pe}_2^0(\mathrm{pH}=7)\right] \qquad (10.89)$$

The value of the term $2.303\, R T$ at 25 °C is 5.71 kJ/mol.

Although $\Delta_R G^0$ can be used to get a first impression of the probability of a reaction, for an exact assessment of a considered redox reaction, however, it is necessary to calculate the molar Gibbs energy, $\Delta_R G$, according to equations (10.85) or (10.87) instead of $\Delta_R G^0$, because $\Delta_R G$ also considers the reaction quotient (and therefore the concentrations or activities) and not only the equilibrium constant. Alternatively, we can also use the redox intensities, pe, of the half reactions (which also include the concentrations or activities) to calculate the molar Gibbs energy of the complete redox reaction:

$$\Delta_R G = -2.303\, n_e\, R T (\mathrm{pe}_1 - \mathrm{pe}_2) \qquad (10.90)$$

or, for pH = 7:

$$\Delta_R G(\mathrm{pH}=7) = -2.303\, n_e\, R T \left[\mathrm{pe}_1(\mathrm{pH}=7) - \mathrm{pe}_2(\mathrm{pH}=7)\right] \qquad (10.91)$$

The redox intensities of the half-reactions can be calculated by using equation (10.17). If we apply $\mathrm{pe}^0(\mathrm{pH}=7)$ to calculate $\mathrm{pe}(\mathrm{pH}=7)$, we have to recognize that the proton activity has not to be included in the logarithmic argument of the pe equation because it has already been accounted for in $\mathrm{pe}^0(\mathrm{pH}=7)$.

According to equations (10.90) and (10.91), a comparison of the redox intensities of the half-reactions gives information about the establishment of the equilibrium and, in the case of a non-equilibrium state, about the expected direction of the redox reaction (Table 10.7).

**Table 10.7:** Assessment of redox systems with respect to the establishment of the equilibrium state.

| Condition | Change of Gibbs free energy | Direction of reaction | Change of redox intensities necessary to reach the state of equilibrium |
|---|---|---|---|
| $pe_1 > pe_2$ | $\Delta_R G < 0$ | Reaction proceeds in the direction as written in the complete reaction equation (equation (10.79), from left to right) | $pe_1$ decreases and $pe_2$ increases until $pe_1 = pe_2$ |
| $pe_1 = pe_2$ | $\Delta_R G = 0$ | Equilibrium state, no reaction | no change of $pe_1$ and $pe_2$ |
| $pe_1 < pe_2$ | $\Delta_R G > 0$ | Reaction proceeds in the reverse direction (equation (10.79), from right to left) | $pe_1$ increases and $pe_2$ decreases until $pe_1 = pe_2$ |

**Example 10.13**

$1 \times 10^{-5}$ mol/L $Fe^{2+}$ is introduced into a surface water that is at $\vartheta = 25$ °C and pH $= 7$ in equilibrium with the atmosphere ($p(O_2) = 0.209$ bar). Discuss the fate of $Fe^{2+}$ on the basis of the molar Gibbs energy, $\Delta_R G$, for the redox reaction:

$$Fe^{2+} + 0.25\ O_{2(g)} + 2.5\ H_2O \rightleftharpoons Fe(OH)_{3(s)} + 2\ H^+$$

The universal gas constant is $R = 8.3145$ J/(mol · K) and the product $2.303\ R\ T$ is 5.71 kJ/mol at 25 °C.

**Solution:**

The complete redox system is a combination of the redox couples $Fe^{2+}/Fe(OH)_{3(s)}$ and $O_2/H_2O$ for which the standard redox intensities at pH $= 7$ can be taken from Table 10.2:

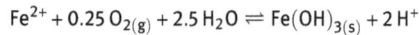
$$0.25\ O_{2(g)} + H^+ + e^- \rightleftharpoons 0.5\ H_2O \qquad pe_1^0(pH = 7) = 13.75$$

$$Fe^{2+} + 3\ H_2O \rightleftharpoons Fe(OH)_{3(s)} + 3\ H^+ + e^- \qquad pe_2^0(pH = 7) = -4.7$$

---

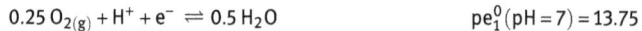
$$Fe^{2+} + 0.25\ O_{2(g)} + 2.5\ H_2O \rightleftharpoons Fe(OH)_{3(s)} + 2\ H^+$$

The molar Gibbs energy of the reaction can be calculated in various ways, for instance by using equation (10.85) together with equation (10.89) or by using equation (10.91).

Option 1:

The molar standard Gibbs energy is calculated by using equation (10.89):

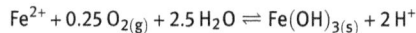
$$\Delta_R G^0(pH = 7) = -2.303\ n_e\ R\ T\left[pe_1^0(pH = 7) - pe_2^0(pH = 7)\right]$$

$$\Delta_R G^0(pH = 7) = (-5.71\ kJ/mol)\ (13.75 + 4.7) = -105.35\ kJ/mol$$

The reaction quotient can be found from an equation similar to the law of mass action by introducing the actual values of the $Fe^{2+}$ concentration and the $O_2$ partial pressure:

$$Q = \frac{1}{c(Fe^{2+})\, p^{0.25}(O_2)} = \frac{1}{(1\times 10^{-5}\, mol/L)\,(0.676\; bar^{0.25})} = 1.479\times 10^5\; L/(mol\cdot bar^{0.25})$$

Here, it has to be noted that $c(H^+)$ does not occur in the equation because it is already accounted for in $\Delta_R G^0(pH=7)$. Finally, the Gibbs energy of the reaction can be calculated according to equation (10.85):

$$\Delta_R G(pH=7) = \Delta_R G^0(pH=7) + R\,T\ln Q = \Delta_R G^0(pH=7) + 2.303\,R\,T\log Q$$

$$\Delta_R G(pH=7) = -105.35\; kJ/mol + (5.71\; kJ/mol)\,(5.17) = -75.83\; kJ/mol$$

The same result is received by using the general (pH-independent) value of $\Delta_R G^0$ (calculated from the $pe^0$ given in the second column of Table 10.2) instead of $\Delta_R G^0(pH=7)$ and considering $c(H^+)$ in the reaction quotient:

$$\Delta_R G^0 = -2.303\, n_e\, R\, T(pe_1^0 - pe_2^0) = -5.71\; kJ/mol\,(20.75-16.30) = -25.41\; kJ/mol$$

$$Q = \frac{c^2(H^+)}{c(Fe^{2+})\, p^{0.25}(O_2)} = \frac{1\times 10^{-14}\; mol^2/L^2}{(1\times 10^{-5}\; mol/L)\,(0.676\; bar^{0.25})} = 1.479\times 10^{-9}\; mol/(L\cdot bar^{0.25})$$

$$\Delta_R G = \Delta_R G^0 + R\,T\ln Q = \Delta_R G^0 + 2.303\,R\,T\log Q$$

$$\Delta_R G = -25.41\; kJ/mol + (5.71\; kJ/mol)\,(-8.83) = -75.83\; kJ/mol$$

Option 2:
To apply equation (10.91), at first, we have to calculate the redox intensities as follows:

$$pe_1(pH=7) = pe_1^0(pH=7) + \frac{1}{n_e}\log\frac{p^{0.25}(O_2)}{1} = 13.75 - 0.17 = 13.58$$

$$pe_2(pH=7) = pe_2^0(pH=7) + \frac{1}{n_e}\log\frac{1}{c(Fe^{2+})} = -4.7 + 5 = 0.3$$

Then, the Gibbs energy can be found by:

$$\Delta_R G(pH=7) = -2.303\, n_e\, R\, T\,[pe_1(pH=7) - pe_2(pH=7)]$$

$$\Delta_R G(pH=7) = (-5.71\; kJ/mol)\,(13.58 - 0.3) = -75.83\; kJ/mol$$

As in option 1, we can alternatively use the general equation for $\Delta_R G$ instead of the specific equation for $pH=7$ if we consider the proton concentration in the logarithmic arguments of the pe equations:

$$pe_1 = pe_1^0 + \frac{1}{n_e}\log\frac{p^{0.25}(O_2)\, c(H^+)}{1} = 20.75 - 7.17 = 13.58$$

$$pe_2 = pe_2^0 + \frac{1}{n_e}\log\frac{c^3(H^+)}{c(Fe^{2+})} = 16.3 - 16 = 0.3$$

$$\Delta_R G = -2.303\, n_e\, R\, T(pe_1 - pe_2) = (-5.71\; kJ/mol)\,(13.28) = -75.83\; kJ/mol$$

Independent of the applied solution method, we find $\Delta_R G = -75.83$ kJ/mol. The negative value of the Gibb energy of the reaction indicates that a spontaneous oxidation of $Fe^{2+}$ to $Fe(OH)_{3(s)}$ can be expected under the given conditions.

### 10.6.2 Redox Reactions Within the Global Carbon Cycle

The two main processes in the global carbon cycle are the production of biomass from inorganic carbon ($CO_2$) through photosynthesis and, as the reverse process, the degradation (mineralization) of organic material to inorganic carbon. Both processes are redox processes that are also relevant for aqueous systems. In carbon dioxide, carbon has the highest possible oxidation number +4, whereas in organic compounds the oxidation number of carbon is lower than +4. The lowest oxidation state of carbon is −4 as can be found in methane, $CH_4$. Although the photosynthesis and most of the degradation processes are mediated by organisms, they can be treated by means of the theoretical principles of chemical redox reactions presented in the previous sections.

Photosynthesis, in aquatic systems performed by aquatic plants and algae, proceeds in a very complex reaction mechanism and requires light as an energy source. It can be described in a very simplified manner by using a carbohydrate (here glucose) as a model compound for biomass as follows:

$$6\,H_2O + 6\,CO_2 \rightleftharpoons C_6H_{12}O_6 + 6\,O_2 \tag{10.92}$$

Here, the oxygen in the water molecule is the reductant that is oxidized from the oxidation state −2 in $H_2O$ to the oxidation state ±0 in $O_2$, whereas the carbon in $CO_2$ is the oxidant that is reduced from the oxidation state +4 in $CO_2$ to the oxidation state ±0 in $C_6H_{12}O_6$. The respective half-reactions are:

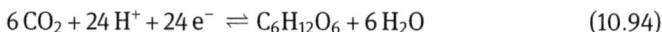

$$6\,O_2 + 24\,H^+ + 24\,e^- \rightleftharpoons 12\,H_2O \tag{10.93}$$

$$6\,CO_2 + 24\,H^+ + 24\,e^- \rightleftharpoons C_6H_{12}O_6 + 6\,H_2O \tag{10.94}$$

To find the overall reaction (equation (10.92)), the first reaction has to be written in the reverse direction.

The degradation of organic material can take place in different ways, biotic or abiotic. Many organisms are able to utilize organic material as an energy source where the energy stored in the organic material is released by oxidation of this material. Humans, animals, and a number of microorganisms use oxygen as an oxidant for the oxidative degradation of organic material to carbon dioxide. This process is known as aerobic respiration. It is the dominant microbial degradation mechanism for organic material, also in aqueous systems, as long as oxygen is available. In the absence of oxygen, specialized microorganisms are able to use other oxidants, such as nitrate (nitrate respiration, denitrification) or sulfate (sulfate respiration, desulfurication). Iron(III) hydroxide and manganese dioxide are further oxidants that can act in chemical but also in biochemical oxidation reactions. The respective half-reactions for the different oxidants are given in Table 10.2. To formulate complete redox reactions, these equations have to be combined with the half reaction for the $CO_2$/glucose redox couple, also given in Table 10.2. The equilibrium constants of the overall reactions can be found by means of equation (10.84) (Example 10.14). As explained in Section 10.4.4, we can use the partial pressures or the molar concentrations of the involved gas components.

---

**Example 10.14**

What is the equilibrium constant of the oxidation of glucose by oxygen (aerobic respiration)?

**Solution:**

The respective half-reactions, taken from Table 10.2 and written as a one-electron transfer, are:

$$1/4\,O_{2(g)} + H^+ + e^- \rightleftharpoons 1/2\,H_2O \qquad\qquad pe^0 = 20.75$$

and

$$1/4\,CO_{2(g)} + H^+ + e^- \rightleftharpoons 1/24\,C_6H_{12}O_6 + 1/4\,H_2O \qquad pe^0 = -0.20$$

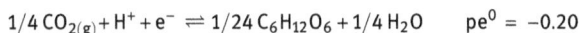

Before both half-reaction can be combined to a complete redox reaction, the second equation has to be written in the reverse direction:

$$1/24\,C_6H_{12}O_6 + 1/4\,H_2O \rightleftharpoons 1/4\,CO_{2(g)} + H^+ + e^-$$

After adding both half reactions, we obtain:

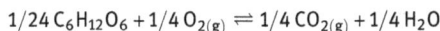

$$1/24\,C_6H_{12}O_6 + 1/4\,O_{2(g)} \rightleftharpoons 1/4\,CO_{2(g)} + 1/4\,H_2O$$

The respective law of mass action is:

$$K^* = \frac{p^{1/4}(CO_2)}{p^{1/4}(O_2)\,a^{1/24}(C_6H_{12}O_6)}$$

and the equilibrium constant can be found from the standard redox intensities:

$$\log K^* = n_e\,(pe_1^0 - pe_2^0) = 1\,(20.75 + 0.20) = 20.95$$

$$K^* = 10^{20.95}\,L^{1/24}/mol^{1/24} = 8.91 \times 10^{20}\,L^{1/24}/mol^{1/24}$$

The high value of the equilibrium constant indicates that the reaction equilibrium lies to the right-hand side and, thus we can expect that glucose is not thermodynamically stable in presence of oxygen.

---

Table 10.8 lists the complete redox reactions for the degradation (mineralization) of organic material by different oxidants that are relevant for aqueous systems. In natural systems, a specific redox sequence is often observed in such a manner that at first the strongest oxidant is utilized followed by the other oxidants in order of decreasing strength. The strongest oxidant is that oxidant that is farthest from the equilibrium with the organic substance and consequently releases the most Gibbs energy during the mineralization reaction. This sequence cannot be seen from the standard Gibbs energies given in Table 10.8 alone because it is determined by the difference between the redox intensities, $\Delta pe$, rather than by the difference of the standard redox intensities, $\Delta pe^0$. To calculate the differences of the redox intensities and the related Gibbs energies, the concentrations (or partial pressures) of the components involved in the redox reaction have to be known (see Example 10.13) or realistic presumptions concerning the concentrations or partial pressures have to be made. To demonstrate the redox sequence, Gibbs energies for the mineralization of the organic model compound glucose with different oxidants, calculated for typical conditions, are shown in Table 10.9 (last column). Indeed, in lake sediments or in the subsurface near infiltration sites the oxidants are consumed in the order $O_2$, $NO_3^-$, $MnO_2$, $Fe(OH)_3$, $SO_4^{2-}$, resulting in specific spatial concentration gradients of characteristic species ($O_2$, $NO_3^-$, $Mn^{2+}$, $Fe^{2+}$, $HS^-$). Generally, the consecutive consumption of oxidants according to the redox sequence is connected with a decreasing redox intensity level in the water.

**Table 10.8:** Mineralization of the organic model compound glucose by different oxidants (data for 25 °C).

| Reaction | log $K^*$ | $\Delta_R G^0 = -2.303$ $RT \log K$ (kJ/mol) |
|---|---|---|
| $1/24\,C_6H_{12}O_6 + 1/5\,NO_3^- + 1/5\,H^+ \rightleftharpoons 1/4\,CO_{2(g)} + 1/10\,N_{2(g)} + 7/20\,H_2O$ | 21.25 | −121.34 |
| $1/24\,C_6H_{12}O_6 + 1/2\,MnO_{2(s)} + H^+ \rightleftharpoons 1/4\,CO_{2(g)} + 1/2\,Mn^{2+} + 3/4\,H_2O$ | 21.00 | −119.91 |
| $1/24\,C_6H_{12}O_6 + 1/4\,O_{2(g)} \rightleftharpoons 1/4\,CO_{2(g)} + 1/4\,H_2O$ | 20.95 | −119.62 |
| $1/24\,C_6H_{12}O_6 + Fe(OH)_{3(s)} + 2\,H^+ \rightleftharpoons 1/4\,CO_{2(g)} + Fe^{2+} + 11/4\,H_2O$ | 16.50 | −94.22 |
| $1/24\,C_6H_{12}O_6 + 1/8\,SO_4^{2-} + 1/8\,H^+ \rightleftharpoons 1/4\,CO_{2(g)} + 1/8\,HS^- + 1/4\,H_2O$ | 4.45 | −25.41 |

**Table 10.9:** Redox sequence in the order of Gibbs energy release at pH = 7, $\Delta_R G$ (pH = 7). All data valid for 25 °C.

| Reaction | $\Delta pe^0$ (pH = 7) | $\Delta_R G^0$ (pH = 7) (kJ/mol) | $\Delta pe$ (pH = 7) | $\Delta_R G$ (pH = 7) (kJ/mol) |
|---|---|---|---|---|
| $1/24\,C_6H_{12}O_6 + 1/4\,O_{2(g)} \rightleftharpoons$ $1/4\,CO_{2(g)} + 1/4\,H_2O$ | 20.95 | −119.62 | 21.46 | −122.54 |
| $1/24\,C_6H_{12}O_6 + 1/5\,NO_3^- + 1/5\,H^+ \rightleftharpoons$ $1/4\,CO_{2(g)} + 1/10\,N_{2(g)} + 7/20\,H_2O$ | 19.85 | −113.34 | 19.74 | −112.72 |
| $1/24\,C_6H_{12}O_6 + 1/2\,MnO_{2(s)} + H^+ \rightleftharpoons$ $1/4\,CO_{2(g)} + 1/2\,Mn^{2+} + 3/4\,H_2O$ | 14.00 | −79.94 | 17.68 | −100.95 |
| $1/24\,C_6H_{12}O_6 + Fe(OH)_{3(s)} + 2\,H^+ \rightleftharpoons$ $1/4\,CO_{2(g)} + Fe^{2+} + 11/4\,H_2O$ | 2.50 | −14.28 | 9.18 | −52.42 |

**Table 10.9** (continued)

| Reaction | $\Delta pe^0$ (pH = 7) | $\Delta_R G^0$ (pH = 7) (kJ/mol) | $\Delta pe$ (pH = 7) | $\Delta_R G$ (pH = 7) (kJ/mol) |
|---|---|---|---|---|
| $1/24\,C_6H_{12}O_6 + 1/8\,SO_4^{2-} + 1/8\,H^+ \rightleftharpoons$ $1/4\,CO_{2(g)} + 1/8\,HS^- + 1/4\,H_2O$ | 3.57 | −20.38 | 4.25 | −24.27 |

$\Delta pe$ (pH = 7) calculated with $c(NO_3^-) = 1 \times 10^{-4}$ mol/L, $c(C_6H_{12}O_6) = 1 \times 10^{-4}$ mol/L, $c(Fe^{2+}) = 1 \times 10^{-6}$ mol/L, $c(Mn^{2+}) = 1 \times 10^{-6}$ mol/L, $c(SO_4^{2-}) = c(HS^-)$, and partial pressures according to the atmospheric conditions ($p(O_2) = 0.209$ bar, $p(CO_2) = 0.000415$ bar, $p(N_2) = 0.781$ bar).

### 10.6.3 Further Oxidation Reactions Mediated by Microorganisms

In water bodies with a sufficient content of dissolved oxygen, reduced species (e.g., methane, sulfide, iron(II), manganese(II), ammonium) are not stable over longer periods of time and will be oxidized. These oxidation reactions are often mediated by microorganisms. The nitrification process that proceeds in two stages with nitrite as an intermediate (nitritation: $NH_4^+ \rightarrow NO_2^-$, nitratation: $NO_2^- \rightarrow NO_3^-$) is a prominent example of such oxidation processes that may occur in natural water bodies. The nitrification process is also utilized in wastewater treatment in combination with a denitrification stage (reaction of nitrate with organic substances under formation of $N_2$; see Table 10.8) in order to remove the nutrient nitrogen from wastewater by the following reaction sequence: $NH_4^+ \rightarrow NO_3^- \rightarrow N_2$. Other relevant reactions are methane oxidation, sulfide oxidation, iron(II) oxidation, and manganese(II) oxidation. For these oxidation reactions, the complete redox equations are listed in Table 10.10 together with the equilibrium constants and the molar standard Gibbs energies.

**Table 10.10:** Aerobic oxidation processes (data for 25 °C).

| Reaction | log $K^*$ | $\Delta_R G^0$ (kJ/mol) | $\Delta_R G^0$ (pH = 7) |
|---|---|---|---|
| $1/8\,CH_{4(g)} + 1/4\,O_{2(g)} \rightleftharpoons 1/8\,CO_{2(g)} + 1/4\,H_2O$ | 17.88 | −102.09 | −102.09 |
| $1/8\,HS^- + 1/4\,O_{2(g)} \rightleftharpoons 1/8\,SO_4^{2-} + 1/8\,H^+$ | 16.50 | −94.22 | −99.24 |
| $Fe^{2+} + 1/4\,O_{2(g)} + 5/2\,H_2O \rightleftharpoons Fe(OH)_{3(s)} + 2\,H^+$ | 4.45 | −25.41 | −105.35 |
| $1/2\,Mn^{2+} + 1/4\,O_{2(g)} + 1/2\,H_2O \rightleftharpoons 1/2\,MnO_2 + H^+$ | −0.05 | 0.29 | −39.68 |
| $1/6\,NH_4^+ + 1/4\,O_{2(g)} \rightleftharpoons 1/6\,NO_2^- + 1/6\,H_2O + 1/3\,H^+$ | 5.61 | −32.03 | −45.34 |
| $1/2\,NO_2^- + 1/4\,O_{2(g)} \rightleftharpoons 1/2\,NO_3^-$ | 6.60 | −37.69 | −37.69 |

## 10.7 Problems

Note: It is assumed that ideal conditions exist in all cases ($a = c$, $K^* = K$). If not stated otherwise, the constants are valid for 25 °C.

**10.1.** The $Fe^{2+}$ concentration in a water with pH = 6 was found to be $c(Fe^{2+}) = 2 \times 10^{-5}$ mol/L. What is the redox intensity of this water? The respective redox half-reaction is:

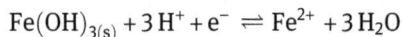

$$Fe(OH)_{3(s)} + 3\,H^+ + e^- \rightleftharpoons Fe^{2+} + 3\,H_2O$$

with $pe^0 = 16.3$.

**10.2.** The standard redox intensity of the reaction:

$$Fe^{3+} + e^- \rightleftharpoons Fe^{2+}$$

is $pe^0 = 13$ (see Table 10.2). What is the related standard redox potential at 25 °C? The universal gas constant is $R = 8.3145$ J/(mol · K) and the Faraday constant is $F = 96\,485$ C/mol.

**10.3.** The standard redox intensity of the half-reaction:

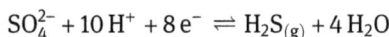

$$SO_4^{2-} + 10\,H^+ + 8\,e^- \rightleftharpoons H_2S_{(g)} + 4\,H_2O$$

is $pe^0 = 5.25$. What is the standard redox intensity of the reaction:

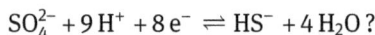

$$SO_4^{2-} + 9\,H^+ + 8\,e^- \rightleftharpoons HS^- + 4\,H_2O\,?$$

The Henry constant of $H_2S$ is $H(H_2S) = 102.2$ mol/(m$^3$ · bar) and the acidity constant of $H_2S$ is $pK_a(H_2S) = 7$.

**10.4.** The redox intensity of a water is pe = 10. What is the expected maximum $Mn^{2+}$ concentration if this water is at pH = 6 in contact with a $MnO_2$-containing sediment? The half-reaction of the $Mn^{2+}/MnO_2$ redox couple is:

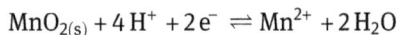

$$MnO_{2(s)} + 4\,H^+ + 2\,e^- \rightleftharpoons Mn^{2+} + 2\,H_2O$$

with $\log K = 41.6$.

**10.5.** Calculate $\Delta pe^0$, $\Delta pe^0(pH = 7)$, $\log K^*$, $\Delta_R G^0$, and $\Delta_R G^0(pH = 7)$ for the reaction:

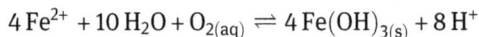

$$4\,Fe^{2+} + 10\,H_2O + O_{2(aq)} \rightleftharpoons 4\,Fe(OH)_{3(s)} + 8\,H^+$$

at the standard temperature 25 °C. The respective half-reactions are:

$$O_{2(aq)} + 4\,H^+ + 4\,e^- \rightleftharpoons 2\,H_2O \qquad\qquad pe^0 = 21.48$$

$$Fe(OH)_{3(s)} + 3\,H^+ + e^- \rightleftharpoons Fe^{2+} + 3\,H_2O \qquad pe^0 = 16.30$$

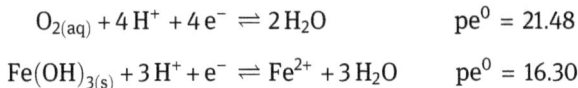

**10.6.** What is the $Mn^{2+}$ equilibrium concentration (in mg/L) in a water with an oxygen content that is one hundredths of the saturation concentration at 25 °C ($c_{sat}\,(O_2) = 0.261$ mmol/L)? The pH of the water is 6 and the molecular weight of manganese is $M(Mn) = 55$ g/mol. The half-reactions to be considered are:

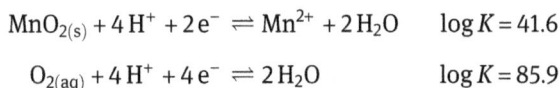

$$MnO_{2(s)} + 4\,H^+ + 2\,e^- \rightleftharpoons Mn^{2+} + 2\,H_2O \qquad \log K = 41.6$$

$$O_{2(aq)} + 4\,H^+ + 4\,e^- \rightleftharpoons 2\,H_2O \qquad\qquad \log K = 85.9$$

**10.7.** In which direction will the reaction:

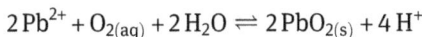

$$2\,Pb^{2+} + O_{2(aq)} + 2\,H_2O \rightleftharpoons 2\,PbO_{2(s)} + 4\,H^+$$

proceed if the following conditions hold: $c(O_2) = 0.261$ mmol/L, pH $= 7$, $\rho^*(Pb^{2+}) = 50\ \mu g/L$, $\vartheta = 25$ °C? The half-reactions and the related standard redox intensities are:

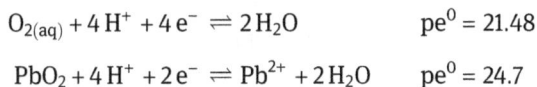

$$O_{2(aq)} + 4\,H^+ + 4\,e^- \rightleftharpoons 2\,H_2O \qquad\qquad pe^0 = 21.48$$

$$PbO_2 + 4\,H^+ + 2\,e^- \rightleftharpoons Pb^{2+} + 2\,H_2O \qquad pe^0 = 24.7$$

and the molecular weight of Pb is $M = 207$ g/mol.

**10.8.** What is the equilibrium concentration of oxygen in a water that contains at pH $= 7.5$ equal concentrations of $SO_4^{2-}$ and $HS^-$? The respective half-reactions are:

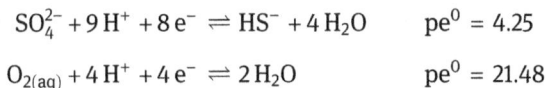

$$SO_4^{2-} + 9\,H^+ + 8\,e^- \rightleftharpoons HS^- + 4\,H_2O \qquad pe^0 = 4.25$$

$$O_{2(aq)} + 4\,H^+ + 4\,e^- \rightleftharpoons 2\,H_2O \qquad\qquad pe^0 = 21.48$$

**10.9.** Is ammonium stable in a water that is at pH $= 7$ saturated with oxygen from air ($p(O_2) = 0.209$ bar)? The respective half-reactions to be considered are:

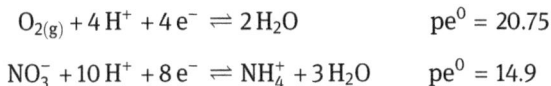

$$O_{2(g)} + 4\,H^+ + 4\,e^- \rightleftharpoons 2\,H_2O \qquad\qquad pe^0 = 20.75$$

$$NO_3^- + 10\,H^+ + 8\,e^- \rightleftharpoons NH_4^+ + 3\,H_2O \qquad pe^0 = 14.9$$

# 11 Complex Formation

## 11.1 Introduction

Metal ions in aqueous systems are able to react with other dissolved species (molecules or ions) to form distinct entities that are referred to as complexes (or more precisely coordination complexes). These complexes are typically readily soluble and their formation has a strong impact on the behavior of the metal ions in other reactions, such as acid/base, precipitation/dissolution, or redox processes. This impact on other reaction equilibria makes complex formation significant for aqueous systems.

Since complex formation determines the behavior of the metal ions in aquatic systems, it is of interest to know which metal species occur in the water under given conditions. The concentration distribution of the different species is also referred to as speciation and can be calculated by means of the laws of mass action for the complex formation together with material balance equations. The speciation of metal ions is discussed in more detail in Section 11.7.

A coordination complex consists of a central metal ion (coordination center) and surrounding ions or neutral molecules that are referred to as ligands or complexing agents. In most cases, the bonds between the central ion and the ligands are coordinate bonds (dative covalent bonds). This means that the ligand provides lone electron pairs that can occupy free orbitals of the central ion. Accordingly, the ligands can be characterized as electron pair donors, whereas the central ions act as electron pair acceptors. Lone electron pairs are provided for instance by atoms such as oxygen, nitrogen, sulfur, or chlorine. These atoms are referred to as ligator atoms and may occur in inorganic and organic ligands.

If a ligand is only able to form one single bond, it is referred to as a monodentate (sometimes also as a unidentate) ligand. Organic molecules or ions that possess two or more functional groups with the above-mentioned ligator atoms are able to form more than one bond with the central ion. Such ligands are called polydentate (or multidentate) ligands. Polydentate ligands are also referred to as chelating agents and the complexes are named chelate complexes. The name chelate comes from the Greek word for crab's claw and should illustrate that the ligand totally surrounds the central ion.

The number of bonds within the complex is called the coordination number. It determines the geometry of the complex. For instance, the coordination number 6 is related to an octahedral geometry whereas the coordination number 4 is related to a quadratic planar or to a tetrahedral geometry.

Figure 11.1 schematically illustrates the structure of complexes with monodentate and polydentate ligands.

https://doi.org/10.1515/9783110758788-011

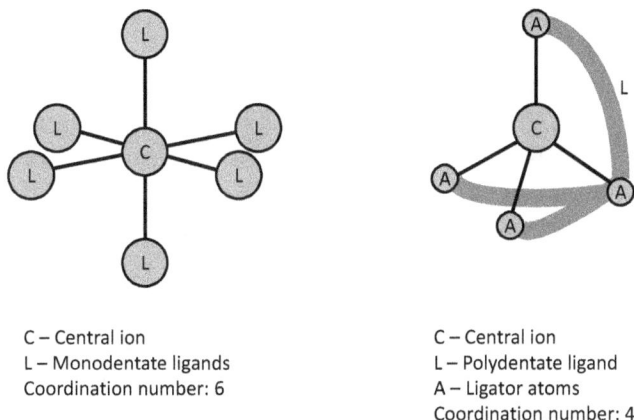

C – Central ion
L – Monodentate ligands
Coordination number: 6

C – Central ion
L – Polydentate ligand
A – Ligator atoms
Coordination number: 4

**Figure 11.1:** Coordination complexes with monodentate and polydentate ligands.

In aqueous systems, metal ions are always surrounded by water molecules. The water molecules in the immediate vicinity of the metal ions constitute the so-called inner hydration sphere. These water molecules act as ligands and form aqua complexes with the metal ions (Section 11.2). Stronger ligands can replace the water molecules in the inner hydration sphere (ligand exchange).

Complexes with water molecules or other ligands that are directly coordinated with the central ions are referred to as inner-sphere complexes. In outer-sphere complexes, the constituents of the complex remain separated by the water molecules of the hydration sphere.

Ion pairs can be considered a specific form of complexes. They consist of a cation and an anion that are held together by electrostatic interactions and behave as one unit in an aqueous solution. Many ion pairs show the typical structure of an outer-sphere complex (so-called solvent-shared ion pairs). The occurrence of ion pairs was already discussed with respect to the calco–carbonic equilibrium (Section 9.5 in Chapter 9).

The name of a complex is formed in the following way: Greek prefix denoting the number of ligands (mono, di, tri, tetra, penta, hexa, etc.) + ligand name + name of the central ion. For neutral ligands, the common name of the molecule is used. Exceptions are water (aqua) and ammonia (ammin). For anionic ligands, the endings -ide, -ate, -ite are replaced by -o, -ato, and -ito, respectively (e.g., hydroxo, chloro, sulfato, carbonato). In cationic complexes, the name of the central ion remains unchanged (e.g., hexaaquairon(III) cation, $Fe(H_2O)_6^{3+}$). In anionic complexes, the ending -ate replaces -um or -ium in the (Latin) element name (e.g., tetrahydroxoaluminate, $Al(OH)_4^-$, or tetrahydroxoferrate, $Fe(OH)_4^-$).

Complexes can be positively charged, negatively charged, or neutral, depending on the charge of the central ion and the number and charge of the ligands. Neutral complexes in aqueous solutions are typically indicated by the superscript "0" to distinguish them from solids with the same composition.

## 11.2 Ligands in Aquatic Systems

Before considering other ligands, we want to discuss the special role of water with respect to the complex formation in aqueous systems. Metal cations in aqueous solutions do not occur as free ions but always as hydrated ions (ions surrounded by water molecules). As already mentioned in Section 11.1, these hydrated metal ions are complexes with water as a ligand (aqua complexes). Typically, the number of water molecules in the first coordination sphere (i.e., the coordination number) is 6. Therefore, in the strict sense, we would have to write $Cu(H_2O)_6^{2+}$ or $Al(H_2O)_6^{3+}$ instead of $Cu^{2+}$ and $Al^{3+}$ for the dissolved ions. Any complex formation with other ligands is therefore a ligand exchange where the water molecules are partly or totally substituted by the other ligands. For the sake of simplification, however, the water ligands are often not written explicitly. The formation of a complex of copper with chloride is therefore formulated usually as:

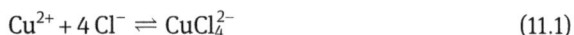

$$Cu^{2+} + 4\,Cl^- \rightleftharpoons CuCl_4^{2-} \tag{11.1}$$

rather than as:

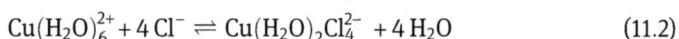

$$Cu(H_2O)_6^{2+} + 4\,Cl^- \rightleftharpoons Cu(H_2O)_2Cl_4^{2-} + 4\,H_2O \tag{11.2}$$

Accordingly, metal ions with only water molecules in the coordination sphere are often referred to as free or uncomplexed metal ions, despite the fact that they are not really free or unbound.

Water molecules bound as ligands can release protons easier than free water molecules due to their polarization by the central ion. Therefore, aqua complexes tend to dissociate under formation of mixed aqua/hydroxo complexes. Consequently, the hydroxo complex formation can also be considered a specific acid/base reaction (also referred to as hydrolysis). This aspect will be discussed in more detail in Section 11.6.

Since water as the solvent is always the excess component is aqueous solutions, only strong complexing agents can replace water in the coordination sphere. According to their frequent occurrence, hydroxide, sulfate, chloride, hydrogencarbonate, and carbonate ions are particularly relevant monodentate ligands in freshwater. In seawater, chloride is the dominant monodentate ligand.

The most important polydentate ligands in natural waters are humic substances. Humic substances possess a large number of functional groups that contains in particular oxygen but also nitrogen and sulfur as potential ligator atoms. Anthropogenic polydentate ligands are introduced into surface or ground waters with wastewater or from diffuse sources. Frequently occurring synthetic complexing agents are aminopolyacetic acids (e.g., ethylenediaminetetraacetic acid, nitrilotriacetic acid) or polyphosphonic acids. Figure 11.2 shows the structures of ethylenediaminetetraacetic acid (EDTA) and nitrilotriacetic acid (NTA). EDTA contains six ligator atoms ($4 \times O$, $2 \times N$), whereas NTA contains four ligator atoms ($3 \times O$, $1 \times N$). In both cases, the anions of the acids act as ligands in complexes. Therefore, the complex formation is coupled to the acid/base reactions of the complexing agents. This has to be taken into account in equilibrium calculations. Figures 11.3 and 11.4 show some possible structures of EDTA and NTA complexes.

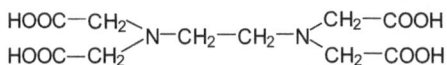

EDTA

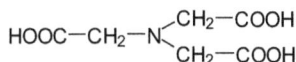

NTA

**Figure 11.2:** Structures of the synthetic complexing agents ethylenediaminetetraacetic acid (EDTA) and nitrilotriacetic acid (NTA).

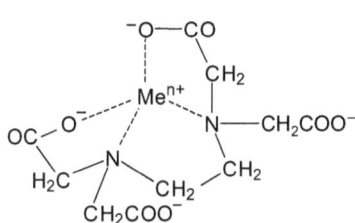

Coordination number: 4
(tetrahedral structure)

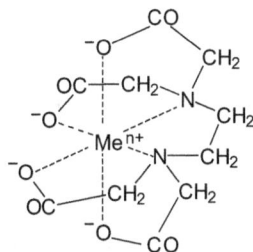

Coordination number: 6
(octahedral structure)

**Figure 11.3:** Possible structures of EDTA complexes.

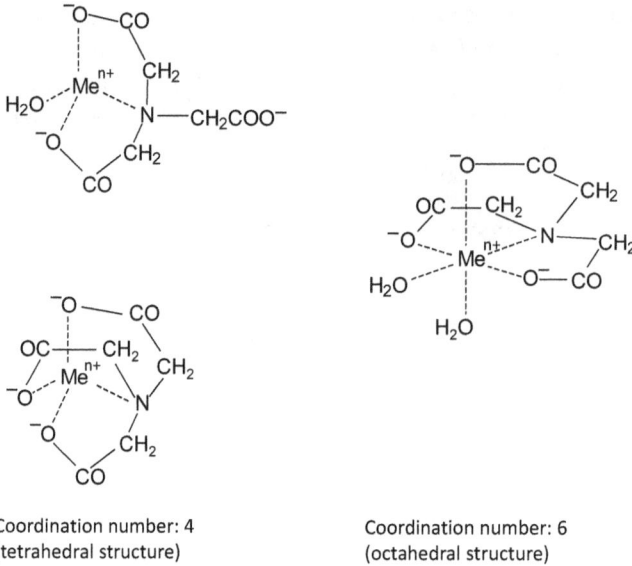

Figure 11.4: Possible structures of NTA complexes.

## 11.3 Equilibrium Relationships and Constants

The law of mass action for complex formation can be formulated in various ways depending on the kind of reaction equation and the types of ligands.

If only one ligand is bound to the central ion as in the case of a chelate complex with a polydentate ligand, the formation of the complex can be written in a general form (with Me = metal ion and L = ligand) as:

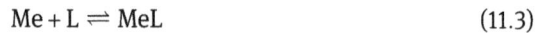

$$Me + L \rightleftharpoons MeL \tag{11.3}$$

and the related law of mass action reads:

$$K_{form}^* = \frac{a(MeL)}{a(Me)\,a(L)} \tag{11.4}$$

where $K_{form}^*$ is the complex formation (or complex stability) constant. The dissociation of the complex is the reverse reaction:

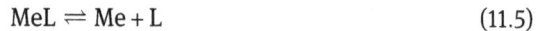

$$MeL \rightleftharpoons Me + L \tag{11.5}$$

and the related law of mass action is given by:

$$K_{diss}^* = \frac{a(Me)\,a(L)}{a(MeL)} \tag{11.6}$$

where $K^*_{diss}$ is the complex dissociation constant that is related to the complex formation constant by:

$$K^*_{diss} = \frac{1}{K^*_{form}} \qquad (11.7)$$

Since complex reactions are typically written as formation reactions, we want to omit the subscript "form" in the following text.

If monodentate ligands are bound to the central ion, two different series of reaction equations and laws of mass actions can be formulated. The first version describes a stepwise addition of ligands until the maximum possible number of ligands is bound:

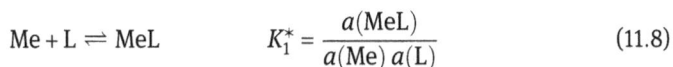

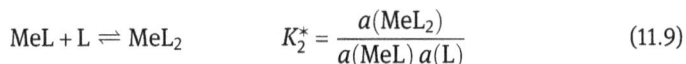

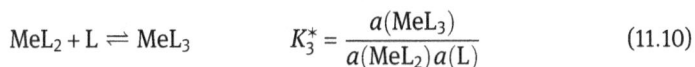

$$Me + L \rightleftharpoons MeL \qquad K^*_1 = \frac{a(MeL)}{a(Me)\,a(L)} \qquad (11.8)$$

$$MeL + L \rightleftharpoons MeL_2 \qquad K^*_2 = \frac{a(MeL_2)}{a(MeL)\,a(L)} \qquad (11.9)$$

$$MeL_2 + L \rightleftharpoons MeL_3 \qquad K^*_3 = \frac{a(MeL_3)}{a(MeL_2)\,a(L)} \qquad (11.10)$$

$$\cdots$$

$$MeL_{n-1} + L \rightleftharpoons MeL_n \qquad K^*_n = \frac{a(MeL_n)}{a(MeL_{n-1})\,a(L)} \qquad (11.11)$$

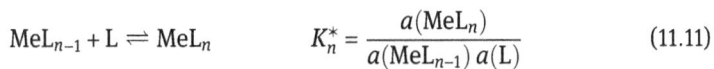

Here, the complex formation constants, $K^*_i\,(i = 1 \ldots n)$, are referred to as stepwise complex formation constants or as individual complex formation constants.

The other opportunity is to formulate overall equations. Thus, for a complex with two ligands we have to write:

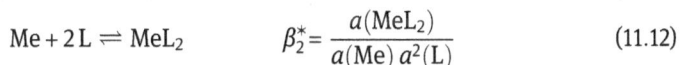

$$Me + 2L \rightleftharpoons MeL_2 \qquad \beta^*_2 = \frac{a(MeL_2)}{a(Me)\,a^2(L)} \qquad (11.12)$$

For three ligands, we obtain:

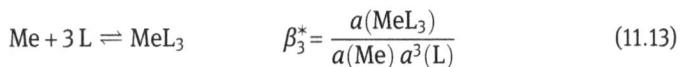

$$Me + 3L \rightleftharpoons MeL_3 \qquad \beta^*_3 = \frac{a(MeL_3)}{a(Me)\,a^3(L)} \qquad (11.13)$$

and for $n$ ligands, we can write:

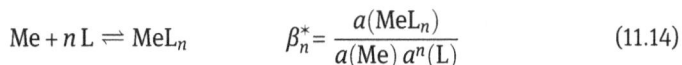

$$Me + nL \rightleftharpoons MeL_n \qquad \beta^*_n = \frac{a(MeL_n)}{a(Me)\,a^n(L)} \qquad (11.14)$$

The overall complex formation constants, $\beta_i$, are related to the individual constants by:

$$\beta_1^* = K_1^* \tag{11.15}$$

$$\beta_2^* = K_1^* \, K_2^* \tag{11.16}$$

$$\beta_3^* = K_1^* \, K_2^* \, K_3^* \tag{11.17}$$

$$\cdots$$

$$\beta_n^* = K_1^* \, K_2^* \, K_3^* \, ... \, K_n^* \tag{11.18}$$

The dissociation constants for the single and overall reactions can be found in analogy to equation (11.7).

As for other reaction equilibria also holds that under ideal conditions (ideal dilute solutions) the thermodynamic constants equal the conditional constants ($K^* = K$, $\beta^* = \beta$).

## 11.4 Strength of Complexation: Monodentate Versus Polydentate Ligands

The strength of complexation can be derived from the value of the complex formation constant or from the related release of free energy. As shown in Chapter 5, Section 5.4, the molar Gibbs energy of a reaction is given by:

$$\Delta_R G = \Delta_R H - T \Delta_R S \tag{11.19}$$

where $\Delta_R H$ is the molar enthalpy of reaction, $\Delta_R S$ is the molar entropy of reaction, and $T$ is the absolute temperature. In the state of equilibrium, the following relationship between the molar standard Gibbs energy of reaction, $\Delta_R G^0$, the molar standard enthalpy of reaction, $\Delta_R H^0$, the molar standard entropy of reaction, $\Delta_R S^0$, and the equilibrium constant, $K^*$, holds:

$$\Delta_R G^0 = \Delta_R H^0 - T \Delta_R S^0 = -R \, T \, \ln K^* \tag{11.20}$$

Accordingly, a strong complexation is characterized by a strongly negative value of $\Delta_R G^0$ and a high value of the equilibrium constant.

Table 11.1 shows complex stability constants for EDTA complexes of heavy metal ions in comparison with constants for complexes of the same metal ions with monodentate ligands. The chelate complexes with EDTA are obviously much stronger than the complexes with monodentate ligands. Such differences are generally found if polydentate and monodentate complexes are compared. As can be derived from equation (11.20), a strong complexation can be a result of a strongly negative value of $\Delta_R H^0$ and/or of a strongly positive value of $\Delta_R S^0$.

**Table 11.1:** Complex stability constants of MeL complexes of different heavy metals.

| Metal ion | L = EDTA log $K^*$ | L = $SO_4^{2-}$ log $K^*$ | L = $CO_3^{2-}$ log $K^*$ | L = $Cl^-$ log $K^*$ |
|---|---|---|---|---|
| $Pb^{2+}$ | 18.00 | 2.75 | 7.24 | 1.60 |
| $Cd^{2+}$ | 16.48 | 2.46 | 5.40 | 1.98 |
| $Ni^{2+}$ | 18.60 | 2.29 | 6.87 | 0.40 |
| $Cu^{2+}$ | 18.78 | 2.31 | 6.73 | 0.43 |
| $Zn^{2+}$ | 16.48 | 2.37 | 5.30 | 0.43 |

The strength of chelate complexes is mainly a result of an entropy effect. Remember that entropy is a measure of the disorder (or randomness) of a system. This entropy effect can be explained in a simplified manner by a comparison of the ligand exchange that takes place if water molecules are replaced by monodentate or polydentate ligands.

For the replacement of water molecules from the coordination sphere of a bivalent metal ion by two monodentate, twofold negatively charged ligands, we can write:

$$Me(H_2O)_6^{2+} + 2L^{2-} \rightleftharpoons Me(H_2O)_4L_2^{2-} + 2H_2O$$
$$\text{3 species} \qquad\qquad\qquad \text{3 species}$$

(11.21)

Since two ligand ions replace exactly two water molecules from the coordination sphere, there is no change in the numbers of both free and bound species. Therefore, no strong entropy effect (change in the degree of disorder) occurs. The same result will be found for any other number of monodentate ligands. On the other hand, if a polydentate ligand (here $L^{4-}$ as an example) replaces water molecules from the coordination sphere, the number of free species increases. Consequently, it can be expected that the entropy that indicates the degree of disorder in the system also increases:

$$Me(H_2O)_6^{2+} + L^{4-} \rightleftharpoons MeL^{2-} + 6H_2O$$
$$\text{2 species} \qquad\qquad \text{7 species}$$

(11.22)

According to equation (11.20), an entropy increase leads to a stronger negative Gibbs energy and a higher value of the equilibrium constant.

Besides the entropy effect, the stronger bond forces in the case of polydentate ligands also contribute to a certain extent to the higher stability of chelate complexes in comparison to complexes with monodentate ligands.

## 11.5 Complex Formation and Solubility

The strong impact of complex formation on other reaction equilibria can be demonstrated impressively by taking the dissolution/precipitation equilibrium as an example. Let us consider the dissolution of a solid consisting of a bivalent metal cation $Me^{2+}$ and a bivalent anion $A^{2-}$. The respective reaction equation is:

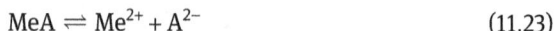

$$MeA \rightleftharpoons Me^{2+} + A^{2-} \tag{11.23}$$

and, for ideal solutions, the corresponding solubility product reads:

$$K_{sp} = c(Me^{2+})\, c(A^{2-}) \tag{11.24}$$

The solubility, $c_s$, is then given by:

$$c_s = c(Me^{2+}) = c(A^{2-}) = \sqrt{K_{sp}} \tag{11.25}$$

as has been shown in Chapter 8, Section 8.3.1.

Now, we want to assume that the metal cation forms a chelate complex with the anion of nitrilotriacetic acid (NTA) as ligand, here denoted as $L^{3-}$:

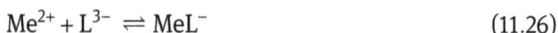

$$Me^{2+} + L^{3-} \rightleftharpoons MeL^- \tag{11.26}$$

$$K_1 = \frac{c(MeL^-)}{c(Me^{2+})\, c(L^{3-})} \tag{11.27}$$

If we take into account that the metal ions released from the solid during dissolution are partly transformed into the complex and the sum of the concentrations of the formed complex and the remaining free metal ions must equal the concentration of the initially released metal ions, the following relationship between the solubility, $c_s$, and the species concentrations in the state of equilibrium holds:

$$c_s = c(Me^{2+}) + c(MeL^-) = c(A^{2-}) \tag{11.28}$$

After introducing the law of mass action for the complex formation, we find:

$$c_s = c(Me^{2+}) + K_1 c(Me^{2+})\, c(L^{3-}) = c(Me^{2+})[1 + K_1 c(L^{3-})] \tag{11.29}$$

$$c(Me^{2+}) = \frac{c_s}{1 + K_1 c(L^{3-})} \tag{11.30}$$

Substituting $c(Me^{2+})$ and $c(A^{2-})$ in the solubility product by equations (11.30) and (11.28), respectively, gives:

$$K_{sp} = \frac{c_s}{1 + K_1 c(L^{3-})}\, c_s \tag{11.31}$$

and the solubility, $c_s$, can be expressed by:

$$c_s = \sqrt{K_{sp}\,[1 + K_1\,c(L^{3-})]}$$ (11.32)

If we compare equation (11.32) with equation (11.25), we can see that the solubility increases the higher the ligand concentration and/or the larger the complex formation constant is. A particularly strong effect of complex formation on the solubility can be expected for chelating ligands due to their large complex formation constants. This effect is of practical interest with respect to the water quality because the introduction of synthetic complexing agents with wastewater into rivers or lakes can lead to a remobilization of hazardous heavy metals from the sediment where they are normally immobilized as hardly soluble hydroxides or sulfides.

## 11.6 Hydrolysis of Hydrated Metal Ions

It was already mentioned in Section 11.2 that water molecules bound as ligands can release protons easier than free water molecules. Therefore, aqua complexes tend to dissociate under formation of mixed aqua/hydroxo complexes. This process is known as hydrolysis and proceeds analogous to a stepwise dissociation of a polyprotic acid:

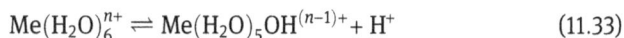

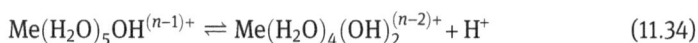

$$Me(H_2O)_6^{n+} \rightleftharpoons Me(H_2O)_5OH^{(n-1)+} + H^+$$ (11.33)

$$Me(H_2O)_5OH^{(n-1)+} \rightleftharpoons Me(H_2O)_4(OH)_2^{(n-2)+} + H^+$$ (11.34)

and so on.

Alternative to this kind of formulating of the hydrolysis reactions, other simplified notations are in use. These notations and the relations between them will be demonstrated below by taking the first dissociation step as an example.

The law of mass action for the first dissociation step according to equation (11.33) reads:

$$K_{a1}^* = \frac{a(H^+)\,a(Me(H_2O)_5OH^{(n-1)+})}{a(Me(H_2O)_6^{n+})}$$ (11.35)

where the subscript "a" is used to denote the constant as an acidity constant. To simplify equations (11.33) and (11.35), we can omit all the water molecules that do not take part in the reaction and write:

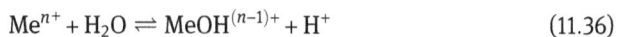

$$Me^{n+} + H_2O \rightleftharpoons MeOH^{(n-1)+} + H^+$$ (11.36)

$$K_{a1}^* = \frac{a(H^+)\,a(MeOH^{(n-1)+})}{a(Me^{n+})}$$ (11.37)

Since $Me^{n+}$ and $MeOH^{(n-1)+}$ are only simplified notations for the species $Me(H_2O)_6^{n+}$ and $Me(H_2O)_5OH^{(n-1)+}$, respectively, the numerical value of the equilibrium constant, $K_{a1}^*$, is identical for both formulations of the law of mass action.

Moreover, the formation of the hydroxo complex $MeOH^{(n-1)+}$ can also be described according to equation (11.8) as a reaction of the central ion $Me^{n+}$ with the $OH^-$ ligand:

$$Me^{n+} + OH^- \rightleftharpoons MeOH^{(n-1)+} \tag{11.38}$$

Then, the law of mass action reads:

$$K_1^* = \beta_1^* = \frac{a(MeOH^{(n-1)+})}{a(Me^{n+})\,a(OH^-)} \tag{11.39}$$

The equilibrium constants, $K_1^*$ and $K_{a1}^*$ are related by the dissociation constant of water:

$$K_w^* = a(H^+)\,a(OH^-) \tag{11.40}$$

$$K_1^*\,K_w^* = \frac{a(MeOH^{(n-1)+})\,a(H^+)\,a(OH^-)}{a(Me^{n+})\,a(OH^-)} = \frac{a(MeOH^{(n-1)+})\,a(H^+)}{a(Me^{n+})} = K_{a1}^* \tag{11.41}$$

Accordingly, the following general relationships for the individual equilibrium constants can be derived:

$$K_n^*\,K_w^* = K_{a,n}^* \tag{11.42}$$

$$\log K_n^* + \log K_w^* = \log K_{a,n}^* \tag{11.43}$$

For the overall constants, we can write:

$$\beta_n^*\,(K_w^*)^n = \beta_{a,n}^* \tag{11.44}$$

$$\log \beta_n^* + n \log K_w^* = \log \beta_{a,n}^* \tag{11.45}$$

Here, $\beta_{a,n}^*$ is the product of the acidity constants, $K_{a1}^*\,K_{a2}^* \ldots K_{a,n}^*$.

---

**Example 11.1**

The overall complex formation constant for the reaction:

$$Cu^{2+} + 4\,OH^- \rightleftharpoons Cu(OH)_4^{2-}$$

is given as $\log \beta_4^* = 16.40$. What is the constant $\beta_{a4}^*$ for the reaction:

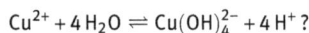

$$Cu^{2+} + 4\,H_2O \rightleftharpoons Cu(OH)_4^{2-} + 4\,H^+\,?$$

The dissociation constant of water is $K_w^* = 1 \times 10^{-14}\ mol^2/L^2$.

**Solution:**

The equation for the second reaction can be received by combining the equation for the first reaction with that of the dissociation of water:

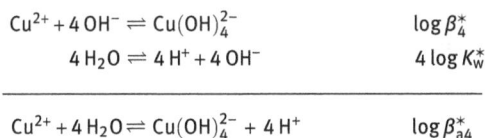

$$Cu^{2+} + 4\,OH^- \rightleftharpoons Cu(OH)_4^{2-} \qquad\qquad \log \beta_4^*$$
$$4\,H_2O \rightleftharpoons 4\,H^+ + 4\,OH^- \qquad\qquad 4\log K_w^*$$

---

$$Cu^{2+} + 4\,H_2O \rightleftharpoons Cu(OH)_4^{2-} + 4\,H^+ \qquad\qquad \log \beta_{a4}^*$$

Accordingly, we can calculate $\log \beta_{a4}^*$ by:

$$\log \beta_{a4}^* = \log \beta_4^* + 4\log K_w^* = 16.40 + 4\,(-14) = -39.60$$

This equation corresponds to equation (11.45) with $n = 4$. Thus, the equilibrium constant is:

$$\beta_{a4}^* = 2.51 \times 10^{-40}\ mol^4/L^4$$

Note that the unit can be derived from the respective law of mass action.

---

## 11.7 Speciation of Metal Ions

### 11.7.1 Introduction

The distribution of an element amongst defined chemical species in a system is referred to as speciation. We have already seen specific forms of speciation as we have discussed the distribution of acid/base pairs with dependence on pH (Section 7.6 in Chapter 7) and the distribution of redox couples with dependence on pe (Section 10.4 in Chapter 10). Here, we want to discuss the speciation of metals as a result of complex formation. In this regard, the pH dependence of speciation is of special interest, because the hydrated metal ions underlie pH-dependent hydrolysis reactions and some of the frequently occurring ligands are involved in acid/base reactions (e.g., hydroxide or carbonate ions). The speciation of metals in aqueous systems is typically found by calculations on the basis of metal and ligand material balances in combination with the equilibrium relationships of the complex formation and other related pH-dependent reactions of the metal and/or the ligands.

In particular, two types of metal speciation are of practical interest. In the first case, the total metal concentration in the system is constant and known, either as an assumed value for the calculation or determined by an analytical method. This case corresponds to a situation where the speciation is not influenced by a solid phase and dissolution/precipitation has not to be considered in the calculation. In the second case, the water is in equilibrium with a solid metal compound that determines the liquid-phase concentrations. In this case, the solubility product has to be considered in addition to the material balance equations and the equilibrium relationships for the liquid phase.

### 11.7.2 Speciation of Dissolved Metal Ions at Constant Total Metal Concentration

To demonstrate the speciation at a constant total metal concentration, we want to consider a metal ion $Me^{2+}$ that forms four different complexes with the ligand $L^-$. Ideal behavior ($\beta_n^* = \beta_n$, $a = c$) is assumed for all species.

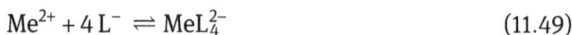

$$Me^{2+} + L^- \rightleftharpoons MeL^+ \tag{11.46}$$

$$Me^{2+} + 2L^- \rightleftharpoons MeL_2^0 \tag{11.47}$$

$$Me^{2+} + 3L^- \rightleftharpoons MeL_3^- \tag{11.48}$$

$$Me^{2+} + 4L^- \rightleftharpoons MeL_4^{2-} \tag{11.49}$$

The related laws of mass action are:

$$\beta_1 = \frac{c(MeL^+)}{c(Me^{2+})\,c(L^-)} \tag{11.50}$$

$$\beta_2 = \frac{c(MeL_2^0)}{c(Me^{2+})\,c^2(L^-)} \tag{11.51}$$

$$\beta_3 = \frac{c(MeL_3^-)}{c(Me^{2+})\,c^3(L^-)} \tag{11.52}$$

$$\beta_4 = \frac{c(MeL_4^{2-})}{c(Me^{2+})\,c^4(L^-)} \tag{11.53}$$

Now, we can establish a material balance for the metal:

$$c(Me_{total}) = c(Me^{2+}) + c(MeL^+) + c(MeL_2^0) + c(MeL_3^-) + c(MeL_4^{2-}) \tag{11.54}$$

and for the ligand:

$$c(L_{total}) = c(L^-) + c(MeL^+) + c(MeL_2^0) + c(MeL_3^-) + c(MeL_4^{2-}) \tag{11.55}$$

If the total concentrations of the metal and the ligand are known, we have, in our example, six unknown concentrations ($Me^+$, $L^-$, and four complexes) and the same number of equations (equations (11.50)–(11.55)). Therefore, the set of equations has a unique solution that can be found by using an appropriate algorithm.

Under certain conditions, equation (11.55) can be omitted, which allows a simpler solution of the set of equations. If the concentration of the metal is much lower than the concentration of the ligand, the concentrations of the complexes in the ligand balance (equation (11.55)) can be neglected (note that the metal concentration limits the concentrations of the complexes) and we can set:

$$c(L_{total}) = c(L^-) \tag{11.56}$$

Consequently, the concentration of the free ligand is no longer unknown and the ligand balance is no longer necessary. This means that the number of the unknowns as well the number of the equations would be reduced by one. A comparable simplification results if we know the free ligand concentration, $c(L^-)$, which is, for instance, the case for the ligand $OH^-$ that is available from the pH. Also in this case, equation (11.55) is dispensable because it is only necessary to find $c(L^-)$ from $c(L_{total})$. In our example, we will make use of this simplification.

If we know the total metal concentration, $c(Me_{total})$, and the ligand concentration, $c(L^-)$, we can calculate the concentration of the free metal ions after substituting the complex concentrations in the metal balance by means of the equilibrium relationships (equations (11.50)–(11.53)) and subsequently factoring out $c(Me^{2+})$:

$$c(Me_{total}) = c(Me^{2+}) + \beta_1 c(Me^{2+}) c(L^-) + \beta_2 c(Me^{2+}) c^2(L^-) + \beta_3 c(Me^{2+}) c^3(L^-)$$
$$+ \beta_4 c(Me^{2+}) c^4(L^-) \tag{11.57}$$

$$c(Me_{total}) = c(Me^{2+}) \left[ 1 + \beta_1 c(L^-) + \beta_2 c^2(L^-) + \beta_3 c^3(L^-) + \beta_4 c^4(L^-) \right] \tag{11.58}$$

$$c(Me^{2+}) = \frac{c(Me_{total})}{1 + \beta_1 c(L^-) + \beta_2 c^2(L^-) + \beta_3 c^3(L^-) + \beta_4 c^4(L^-)} \tag{11.59}$$

After that, the concentrations of the complexes can be found by inserting the concentrations $c(Me^{2+})$ and $c(L^-)$ into the laws of mass action:

$$c(MeL^+) = \beta_1 c(Me^{2+}) c(L^-) \tag{11.60}$$

$$c(MeL_2^0) = \beta_2 c(Me^{2+}) c^2(L^-) \tag{11.61}$$

$$c(MeL_3^-) = \beta_3 c(Me^{2+}) c^3(L^-) \tag{11.62}$$

$$c(MeL_4^{2-}) = \beta_4 c(Me^{2+}) c^4(L^-) \tag{11.63}$$

---

**Example 11.2**

What are the concentrations of $Cu^{2+}$, $CuOH^+$, $Cu(OH)_2^0$, $Cu(OH)_3^-$, and $Cu(OH)_4^{2-}$ at $pH = 7$ if the total concentration of copper is $c(Cu_{total}) = 1 \times 10^{-7}$ mol/L? The overall complex formation constants are $\beta_1 = 1 \times 10^6$ L/mol, $\beta_2 = 2.09 \times 10^{14}$ L$^2$/mol$^2$, $\beta_3 = 1.26 \times 10^{15}$ L$^3$/mol$^3$, $\beta_4 = 2.51 \times 10^{16}$ L$^4$/mol$^4$.

**Solution:**

Since $pOH = 14 - pH = 14 - 7 = 7$, the concentration of the ligand $OH^-$ is $c(OH^-) = 10^{-pOH}$ mol/L = $1 \times 10^{-7}$ mol/L. According to equation (11.59), we can find $c(Cu^{2+})$ from a combination of the copper balance with the equilibrium relationships:

$$c(\text{Cu}^{2+}) = \frac{c(\text{Cu}_{\text{total}})}{1 + \beta_1\, c(\text{OH}^-) + \beta_2\, c^2(\text{OH}^-) + \beta_3\, c^3(\text{OH}^-) + \beta_4\, c^4(\text{OH}^-)}$$

Introducing $c(\text{Cu}_{\text{total}})$ and $c(\text{OH}^-)$, we obtain:

$$\beta_1\, c(\text{OH}^-) = (1 \times 10^6\ \text{L/mol})\,(1 \times 10^{-7}\ \text{mol/L}) = 0.1$$

$$\beta_2\, c^2(\text{OH}^-) = (2.09 \times 10^{14}\ \text{L}^2/\text{mol}^2)\,(1 \times 10^{-14}\ \text{mol}^2/\text{L}^2) = 2.09$$

$$\beta_3\, c^3(\text{OH}^-) = (1.26 \times 10^{15}\ \text{L}^3/\text{mol}^3)\,(1 \times 10^{-21}\ \text{mol}^3/\text{L}^3) = 1.26 \times 10^{-6}$$

$$\beta_4\, c^4(\text{OH}^-) = (2.51 \times 10^{16}\ \text{L}^4/\text{mol}^4)\,(1 \times 10^{-28}\ \text{mol}^4/\text{L}^4) = 2.51 \times 10^{-12}$$

$$c(\text{Cu}^{2+}) = \frac{1 \times 10^{-7}\ \text{mol/L}}{1 + 0.1 + 2.09 + 1.26 \times 10^{-6} + 2.51 \times 10^{-12}} = 3.13 \times 10^{-8}\ \text{mol/L}$$

The complex concentrations can be found from the laws of mass action according to equations (11.60)–(11.63):

$$c(\text{CuOH}^+) = \beta_1\, c(\text{OH}^-)\, c(\text{Cu}^{2+}) = 0.1\,(3.13 \times 10^{-8}\ \text{mol/L}) = 3.13 \times 10^{-9}\ \text{mol/L}$$

$$c(\text{Cu(OH)}_2^0) = \beta_2\, c^2(\text{OH}^-)\, c(\text{Cu}^{2+}) = 2.09\,(3.13 \times 10^{-8}\ \text{mol/L}) = 6.54 \times 10^{-8}\ \text{mol/L}$$

$$c(\text{Cu(OH)}_3^-) = \beta_3\, c^3(\text{OH}^-)\, c(\text{Cu}^{2+}) = (1.26 \times 10^{-6})\,(3.13 \times 10^{-8}\ \text{mol/L}) = 3.94 \times 10^{-14}\ \text{mol/L}$$

$$c(\text{Cu(OH)}_4^{2-}) = \beta_4\, c^4(\text{OH}^-)\, c(\text{Cu}^{2+}) = (2.51 \times 10^{-12})\,(3.13 \times 10^{-8}\ \text{mol/L}) = 7.86 \times 10^{-20}\ \text{mol/L}$$

We can verify the results by adding the concentrations and comparing the sum with the given total concentration:

$$c(\text{Cu}_{\text{total}}) = c(\text{Cu}^{2+}) + c(\text{CuOH}^+) + c(\text{Cu(OH)}_2^0) + c(\text{Cu(OH)}_3^-) + c(\text{Cu(OH)}_4^{2-})$$

$$= 3.13 \times 10^{-8}\ \text{mol/L} + 3.13 \times 10^{-9}\ \text{mol/L} + 6.54 \times 10^{-8}\ \text{mol/L} + 3.94 \times 10^{-14}\ \text{mol/L}$$

$$+ 7.86 \times 10^{-20}\ \text{mol/L} = 9.98 \times 10^{-8}\ \text{mol/L}$$

The small deviation from the given total concentration of $1 \times 10^{-7}$ mol/L is due to round-off errors. From the calculation results, we can conclude that the major species at pH = 7 are $\text{Cu}^{2+}$ and the neutral complex $\text{Cu(OH)}_2^0$.

---

If the ligands are anions of others than very strong acids and only total concentrations (i.e., acid plus related anions) are available, we have to additionally consider the dissociation equilibria. This is particularly relevant in the case of the natural carbonato complexes (see carbonic acid system in Chapter 7, Section 7.7) and also in the case of anthropogenic ligands such as EDTA or NTA where the anions act as chelate ligands (Section 11.2). In such cases, the ligand balance equation has to include the concentrations of the neutral acid and the deprotonated species.

To demonstrate this principle, Example 11.2 is extended in Example 11.3 by the additional consideration of the formation of the carbonato complexes $\text{CuCO}_3^0$ and $\text{Cu(CO}_3)_2^{2-}$.

**Example 11.3**

What are the concentrations of $Cu^{2+}$, $CuOH^+$, $Cu(OH)_2^0$, $Cu(OH)_3^-$, $Cu(OH)_4^{2-}$, $CuCO_3^0$, and $Cu(CO_3)_2^{2-}$ at pH = 7 and $c(DIC) = 1 \times 10^{-3}$ mol/L if the total concentration of copper is $c(Cu_{total}) = 1 \times 10^{-7}$ mol/L? The overall complex formation constants of the hydroxo complexes are $\beta_{hydr,1} = 1 \times 10^6$ L/mol, $\beta_{hydr,2} = 2.09 \times 10^{14}$ L²/mol², $\beta_{hydr,3} = 1.26 \times 10^{15}$ L³/mol³, and $\beta_{hydr,4} = 2.51 \times 10^{16}$ L⁴/mol⁴. The respective constants for the carbonato complexes are $\beta_{carb,1} = 5.37 \times 10^6$ L/mol and $\beta_{carb,2} = 6.76 \times 10^9$ L²/mol². The acidity constants of the carbonic acid are $K_{a1} = 4.4 \times 10^{-7}$ mol/L and $K_{a2} = 4.7 \times 10^{-11}$ mol/L.

**Solution:**

The copper balance reads:

$$c(Cu_{total}) = c(Cu^{2+}) + c(CuOH^+) + c(Cu(OH)_2^0) + c(Cu(OH)_3^-) + c(Cu(OH)_4^{2-}) + c(CuCO_3^0) + c(Cu(CO_3)_2^{2-})$$

After substituting the complex concentrations in the balance by means of the laws of mass action and factoring out the concentration of the copper ions (see Example 11.2), we find:

$$c(Cu^{2+}) = \frac{c(Cu_{total})}{1 + \sum\limits_{n=1}^{4} \beta_{hydr,n}\, c^n(OH^-) + \sum\limits_{n=1}^{2} \beta_{carb,n}\, c^n(CO_3^{2-})}$$

To find $c(Cu^{2+})$, we need the ligand concentrations. Whereas the concentration of $OH^-$ is known from the pH (pOH = $pK_w$ − pH = 14 − 7 = 7, $c(OH^-) = 1 \times 10^{-7}$ mol/L), the carbonate concentration has to be calculated by means of the carbonate species balance:

$$c(DIC) = c(CO_2) + c(HCO_3^-) + c(CO_3^{2-}) + c(CuCO_3^0) + c(Cu(CO_3)_2^{2-})$$

Due to the much lower total concentration of copper in comparison to $c(DIC)$, we can neglect the complex concentrations in the inorganic carbon balance. Under this condition, the concentration of carbonate can be calculated by using equation (7.130) from Chapter 7:

$$c(CO_3^{2-}) = \frac{c(DIC)}{\dfrac{c^2(H^+)}{K_{a1}K_{a2}} + \dfrac{c(H^+)}{K_{a2}} + 1} = \frac{1 \times 10^{-3}\ \text{mol/L}}{\dfrac{1 \times 10^{-14}\ \text{mol}^2/\text{L}^2}{(4.4 \times 10^{-7}\ \text{mol/L})(4.7 \times 10^{-11}\ \text{mol/L})} + \dfrac{1 \times 10^{-7}\ \text{mol/L}}{4.7 \times 10^{-11}\ \text{mol/L}} + 1}$$

$$= 3.83 \times 10^{-7}\,\text{mol/L}$$

Now, we can calculate the terms in the denominator of the rearranged metal balance equation:

$$\sum_{n=1}^{4} \beta_{hydr,n}\, c^n(OH^-) = (1 \times 10^6\ \text{L/mol})(1 \times 10^{-7}\ \text{mol/L}) + (2.09 \times 10^{14}\ \text{L}^2/\text{mol}^2)(1 \times 10^{-14}\ \text{mol}^2/\text{L}^2)$$

$$+ (1.26 \times 10^{15}\ \text{L}^3/\text{mol}^3)(1 \times 10^{-21}\ \text{mol}^3/\text{L}^3) + (2.51 \times 10^{16}\ \text{L}^4/\text{mol}^4)(1 \times 10^{-28}\ \text{mol}^4/\text{L}^4)$$

$$= 0.1 + 2.09 + 1.26 \times 10^{-6} + 2.51 \times 10^{-12} = 2.19$$

$$\sum_{n=1}^{2} \beta_{carb,n}\, c^n(CO_3^{2-}) = (5.37 \times 10^6\ \text{L/mol})(3.83 \times 10^{-7}\ \text{mol/L})$$

$$+ (6.76 \times 10^9\ \text{L}^2/\text{mol}^2)(3.83 \times 10^{-7}\ \text{mol/L})^2 = 2.06 + 0.00099 = 2.06$$

Thus, we find for the copper ion concentration:

$$c(Cu^{2+}) = \frac{1 \times 10^{-7} \, \text{mol/L}}{1 + 2.19 + 2.06} = 1.90 \times 10^{-8} \, \text{mol/L}$$

The complex concentrations are found from the laws of mass action according to:

$$c(CuOH^+) = \beta_{\text{hydr},1} \, c(Cu^{2+}) \, c(OH^-)$$
$$= (1 \times 10^6 \, \text{L/mol}) \, (1.90 \times 10^{-8} \, \text{mol/L}) \, (1 \times 10^{-7} \, \text{mol/L}) = 1.90 \times 10^{-9} \, \text{mol/L}$$

$$c(Cu(OH)_2^0) = \beta_{\text{hydr},2} \, c(Cu^{2+}) \, c^2(OH^-)$$
$$= (2.09 \times 10^{14} \, \text{L}^2/\text{mol}^2) \, (1.90 \times 10^{-8} \, \text{mol/L}) \, (1 \times 10^{-14} \, \text{mol}^2/\text{L}^2) = 3.97 \times 10^{-8} \, \text{mol/L}$$

$$c(Cu(OH)_3^-) = \beta_{\text{hydr},3} \, c(Cu^{2+}) \, c^3(OH^-)$$
$$= (1.26 \times 10^{15} \, \text{L}^3/\text{mol}^3) \, (1.90 \times 10^{-8} \, \text{mol/L}) \, (1 \times 10^{-21} \, \text{mol}^3/\text{L}^3) = 2.39 \times 10^{-14} \, \text{mol/L}$$

$$c(Cu(OH)_4^{2-}) = \beta_{\text{hydr},4} \, c(Cu^{2+}) \, c^4(OH^-)$$
$$= (2.51 \times 10^{16} \, \text{L}^4/\text{mol}^4) \, (1.90 \times 10^{-8} \, \text{L/mol}) \, (1 \times 10^{-28} \, \text{mol}^4/\text{L}^4) = 4.77 \times 10^{-20} \, \text{mol/L}$$

$$c(CuCO_3^0) = \beta_{\text{carb},1} \, c(Cu^{2+}) \, c(CO_3^{2-})$$
$$= (5.37 \times 10^6 \, \text{L/mol}) \, (1.90 \times 10^{-8} \, \text{mol/L}) \, (3.83 \times 10^{-7} \, \text{mol/L}) = 3.91 \times 10^{-8} \, \text{mol/L}$$

$$c(Cu(CO_3)_2^{2-}) = \beta_{\text{carb},2} \, c(Cu^{2+}) \, c^2(CO_3^{2-})$$
$$= (6.76 \times 10^9 \, \text{L}^2/\text{mol}^2) \, (1.90 \times 10^{-8} \, \text{mol/L}) \, (1.47 \times 10^{-13} \, \text{mol}^2/\text{L}^2) = 1.89 \times 10^{-11} \, \text{mol/L}$$

We can derive from the calculation that the major species at pH = 7 are $Cu^{2+}$, $Cu(OH)_2^0$, and $CuCO_3^0$. To verify the results, we have to add the concentrations and to compare the sum with the given total concentration:

$$c(Cu_{\text{total}}) = c(Cu^{2+}) + c(CuOH^+) + c(Cu(OH)_2^0) + c(Cu(OH)_3^-) + c(Cu(OH)_4^{2-}) + c(CuCO_3^0)$$
$$+ c(Cu(CO_3)_2^{2-}) = 1.90 \times 10^{-8} \, \text{mol/L} + 1.90 \times 10^{-9} \, \text{mol/L} + 3.97 \times 10^{-8} \, \text{mol/L}$$
$$+ 2.39 \times 10^{-14} \, \text{mol/L} + 4.77 \times 10^{-20} \, \text{mol/L} + 3.91 \times 10^{-8} \, \text{mol/L} + 1.89 \times 10^{-11} \, \text{mol/L}$$
$$= 9.97 \times 10^{-8} \, \text{mol/L}$$

The small deviation from the given total concentration of $1 \times 10^{-7}$ mol/L is due to round-off errors.

The last example demonstrates that the complexity of speciation calculations strongly increases when more than one ligand has to be considered and the ligands are involved in further reactions. A further increase in complexity will result if we have to consider possible interactions of the ligands with other metal ions that occur in the water. Such calculations are better done by using specialized computer software such as Visual MINTEQ or PHREEQC.

### 11.7.3 Speciation in the Presence of a Solid That Determines the Liquid-Phase Concentrations

A typical example for this type of speciation is the distribution of free (more exactly hydrated) metal ions and hydroxo complexes in the presence of a solid hydroxide. In this case, the solubility of the solid hydroxide is connected with the pH-depending hydroxo complex formation. Accordingly, the solubility product has to be considered together with the complex formation equilibria to find the speciation. As a result of a speciation calculation for such systems, we can find the concentrations of the free metal ion and the hydroxo complex species in the liquid phase as a function of pH. By adding the single concentrations, we can further find a solubility curve that describes the pH dependence of the total metal concentration in the liquid phase in equilibrium with the solid hydroxide.

This type of speciation will be demonstrated by using aluminum hydroxide $(Al(OH)_{3(s)})$ as an example. In aqueous solutions, aluminum can form the following hydroxo complexes: $AlOH^{2+}$, $Al(OH)_2^+$, $Al(OH)_3^0$, and $Al(OH)_4^-$. If the solution is in equilibrium with solid aluminum hydroxide, the solubility product of $Al(OH)_{3(s)}$ and the complex formation equilibrium relationships have to be satisfied simultaneously. The following reactions have to be considered:

$$Al(OH)_{3(s)} \rightleftharpoons Al^{3+} + 3\,OH^- \tag{11.64}$$

$$K_{sp} = c(Al^{3+})\,c^3(OH^-) \qquad pK_{sp} = 34 \tag{11.65}$$

$$Al^{3+} + OH^- \rightleftharpoons AlOH^{2+} \tag{11.66}$$

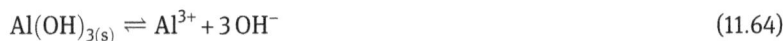

$$\beta_1 = K_1 = \frac{c(AlOH^{2+})}{c(Al^{3+})\,c(OH^-)} \qquad \log\beta_1 = 9 \tag{11.67}$$

$$Al^{3+} + 2\,OH^- \rightleftharpoons Al(OH)_2^+ \tag{11.68}$$

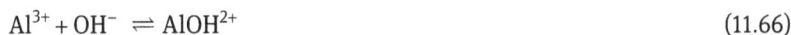

$$\beta_2 = \frac{c(Al(OH)_2^+)}{c(Al^{3+})\,c^2(OH^-)} \qquad \log\beta_2 = 17.9 \tag{11.69}$$

$$Al^{3+} + 3\,OH^- \rightleftharpoons Al(OH)_3^0 \tag{11.70}$$

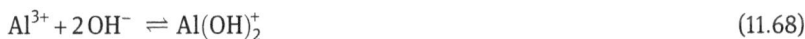

$$\beta_3 = \frac{c(Al(OH)_3^0)}{c(Al^{3+})\,c^3(OH^-)} \qquad \log\beta_3 = 26 \tag{11.71}$$

$$Al^{3+} + 4\,OH^- \rightleftharpoons Al(OH)_4^- \tag{11.72}$$

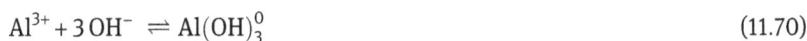

$$\beta_4 = \frac{c(Al(OH)_4^-)}{c(Al^{3+})\,c^4(OH^-)} \qquad \log\beta_4 = 33.8 \tag{11.73}$$

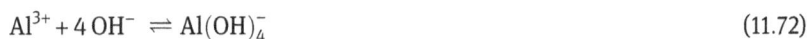

After calculating the $Al^{3+}$ concentration for a given pH from the solubility product, the concentrations of the hydroxo complex species at the same pH can be subsequently computed as shown in Example 11.4.

---

**Example 11.4**

Calculate the concentrations of the species $Al^{3+}$, $AlOH^{2+}$, $Al(OH)_2^+$, $Al(OH)_3^0$, and $Al(OH)_4^-$ that are at pH = 7 in equilibrium with solid $Al(OH)_{3(s)}$. The equilibrium constants are: $K_{sp} = 1 \times 10^{-34}$ mol$^4$/L$^4$, $\beta_1 = 1 \times 10^9$ L/mol, $\beta_2 = 7.94 \times 10^{17}$ L$^2$/mol$^2$, $\beta_3 = 1 \times 10^{26}$ L$^3$/mol$^3$, $\beta_4 = 6.31 \times 10^{33}$ L$^4$/mol$^4$, $K_w = 1 \times 10^{-14}$ mol$^2$/L$^2$.

**Solution:**

At first, the $Al^{3+}$ concentration is calculated from the solubility product of $Al(OH)_{3(s)}$ with:

$$c(OH^-) = \frac{K_w}{c(H^+)} = \frac{1 \times 10^{-14} \text{ mol}^2/L^2}{1 \times 10^{-7} \text{ mol/L}} = 1 \times 10^{-7} \text{ mol/L}$$

according to:

$$c(Al^{3+}) = \frac{K_{sp}}{c^3(OH^-)} = \frac{1 \times 10^{-34} \text{ mol}^4/L^4}{1 \times 10^{-21} \text{ mol/L}} = 1 \times 10^{-13} \text{ mol/L}$$

Then, the other species concentrations can be found from the rearranged laws of mass action:

$$c(AlOH^{2+}) = \beta_1 c(Al^{3+}) c(OH^-) = (1 \times 10^9 \text{ L/mol})(1 \times 10^{-13} \text{ mol/L})(1 \times 10^{-7} \text{ mol/L})$$
$$= 1 \times 10^{-11} \text{ mol/L}$$

$$c(Al(OH)_2^+) = \beta_2 c(Al^{3+}) c^2(OH^-) = (7.94 \times 10^{17} \text{ L}^2/\text{mol}^2)(1 \times 10^{-13} \text{ mol/L})(1 \times 10^{-14} \text{ mol}^2/L^2)$$
$$= 7.94 \times 10^{-10} \text{ mol/L}$$

$$c(Al(OH)_3^0) = \beta_3 c(Al^{3+}) c^3(OH^-) = (1 \times 10^{26} \text{ L}^3/\text{mol}^3)(1 \times 10^{-13} \text{ mol/L})(1 \times 10^{-21} \text{ mol}^3/L^3)$$
$$= 1 \times 10^{-8} \text{ mol/L}$$

$$c(Al(OH)_4^-) = \beta_4 c(Al^{3+}) c^4(OH^-) = (6.31 \times 10^{33} \text{ L}^4/\text{mol}^4)(1 \times 10^{-13} \text{ mol/L})(1 \times 10^{-28} \text{ mol}^4/L^4)$$
$$= 6.31 \times 10^{-8} \text{ mol/L}$$

---

Repeating the calculations shown in Example 11.4 for other pH values gives the complete concentration distribution (Figure 11.5). We can see from this diagram that the lowest solubility of $Al(OH)_3$ is near pH = 6, whereas the solubility is much higher at lower and higher pH values. In the acidic range below pH = 5, the solubility is mainly determined by the hydrated aluminum ion which is, in this pH range, the species with the highest concentration in the liquid phase. The contributions of the other species to the total concentration in the solution is lower (consider the logarithmic scale). At high pH values, in contrast, the solubility is mainly determined by the tetrahydroxoaluminate concentration. The strong pH dependence of the aluminum solubility and in particular the high solubility at low pH values has

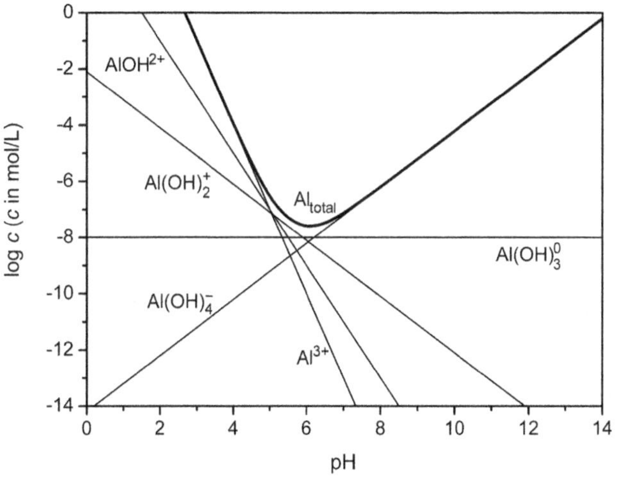

**Figure 11.5:** Aluminum species in equilibrium with solid $Al(OH)_{3(s)}$.

relevance for the mobilization of aluminum from soils as a consequence of acid rain or from sediments as a result of the acidification of lakes.

In Figure 11.5, we can also see that the concentration of the neutral complex $Al(OH)_3^0$, which has the same composition as the solid hydroxide, does not depend on the pH. If $Al(OH)_{3(s)}$ is dissolved under the formation of $Al(OH)_3^0$, there is no consumption or release of $OH^-$ as in the case of the other complex formation reactions. This can be easily shown by adding the dissolution and the complex formation reactions to an overall reaction:

$$Al(OH)_{3(s)} \rightleftharpoons Al^{3+} + 3\,OH^- \qquad\qquad K_{sp} \qquad\qquad (11.74)$$

$$Al^{3+} + 3\,OH^- \rightleftharpoons Al(OH)_3^0 \qquad\qquad \beta_3 \qquad\qquad (11.75)$$

$$Al(OH)_{3(s)} \rightleftharpoons Al(OH)_3^0 \qquad\qquad K = K_{sp}\,\beta_3 \qquad\qquad (11.76)$$

Equation (11.76) provides a simple way to find the pH-independent concentration of the neutral complex. Since the solid does not occur in the law of mass action, we can write:

$$K = c(Al(OH)_3^0) = K_{sp}\,\beta_3 = (1\times 10^{-34}\ \text{mol}^4/\text{L}^4)\,(1\times 10^{26}\ \text{L}^3/\text{mol}^3) = 1\times 10^{-8}\ \text{mol/L}$$

$$(11.77)$$

Solubility diagrams, as shown for $Al(OH)_{3(s)}$, can also be established in the same manner for any other hydroxides. As a further example, the solubility diagram of iron(III) hydroxide is shown in Figure 11.6.

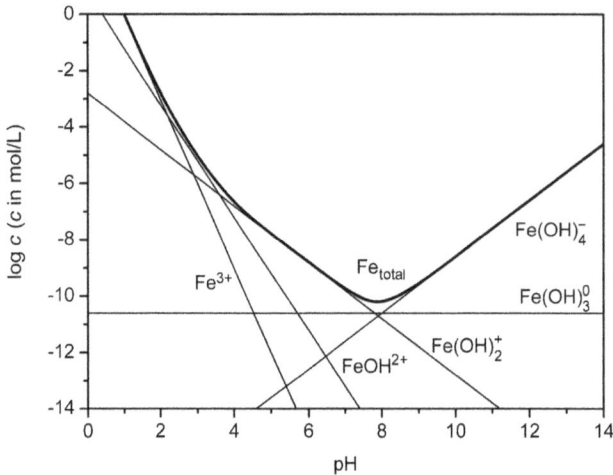

**Figure 11.6:** Iron(III) species in equilibrium with solid $Fe(OH)_{3(s)}$.

## 11.8 Problems

Note: If not stated otherwise, ideal conditions ($a = c$, $K^* = K$, $\beta^* = \beta$) are assumed. The constants are valid for 25 °C.

**11.1.** The individual complex stability constants for $CuCO_3^0$ and $Cu(CO_3)_2^{2-}$ are $K_1^* = 5.37 \times 10^6$ L/mol and $K_2^* = 1.26 \times 10^3$ L/mol. What are the overall complex formation constant and the overall dissociation constant for the dicarbonatocuprate complex?

**11.2.** $Zn^{2+}$ can be removed from water by precipitation as $Zn(OH)_2$. The solubility product of $Zn(OH)_2$ is $pK_{sp} = 17$. Due to the formation of hydroxo complexes, the solubility of Zn is pH-dependent. At higher pH values, the solubility is mainly determined by the formation of the soluble $Zn(OH)_4^{2-}$ complex. What is the limiting pH from which the $Zn(OH)_4^{2-}$ concentration exceeds 2 mg/L zinc ($M(Zn) = 65.4$ g/mol)? The complex formation constant for the tetrahydroxozincate complex is $\beta_4 = 2.5 \times 10^{15}$ L$^4$/mol$^4$.

**11.3.** We want to assume that the total cadmium concentration in a water is composed of the species $Cd^{2+}$, $CdOH^+$, $Cd(OH)_2^0$, and $CdCO_3^0$. What is the fraction $c(Cd^{2+})/c(Cd_{total})$ if the pH is 8 and the carbonate concentration amounts to $c(CO_3^{2-}) = 5 \times 10^{-5}$ mol/L? The following data are given:

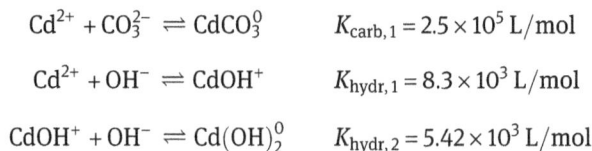

$$Cd^{2+} + CO_3^{2-} \rightleftharpoons CdCO_3^0 \qquad K_{carb,1} = 2.5 \times 10^5 \text{ L/mol}$$

$$Cd^{2+} + OH^- \rightleftharpoons CdOH^+ \qquad K_{hydr,1} = 8.3 \times 10^3 \text{ L/mol}$$

$$CdOH^+ + OH^- \rightleftharpoons Cd(OH)_2^0 \qquad K_{hydr,2} = 5.42 \times 10^3 \text{ L/mol}$$

**11.4.** In a surface water, the total iron(III) concentration was measured to be $1 \times 10^{-5}$ mol/L. Calculate the concentration distribution of the species $Fe^{3+}$, $FeOH^{2+}$, $Fe(OH)_2^+$, $Fe(OH)_3^0$, and $Fe(OH)_4^-$ at pH = 7. The complex formation reactions and the related constants are:

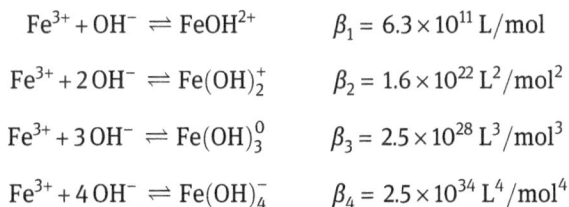

$$Fe^{3+} + OH^- \rightleftharpoons FeOH^{2+} \qquad \beta_1 = 6.3 \times 10^{11} \, L/mol$$

$$Fe^{3+} + 2\,OH^- \rightleftharpoons Fe(OH)_2^+ \qquad \beta_2 = 1.6 \times 10^{22} \, L^2/mol^2$$

$$Fe^{3+} + 3\,OH^- \rightleftharpoons Fe(OH)_3^0 \qquad \beta_3 = 2.5 \times 10^{28} \, L^3/mol^3$$

$$Fe^{3+} + 4\,OH^- \rightleftharpoons Fe(OH)_4^- \qquad \beta_4 = 2.5 \times 10^{34} \, L^4/mol^4$$

**11.5.** Silver chloride, AgCl, has a low solubility. The solubility product is $K_{sp} = 1.8 \times 10^{-10} \, mol^2/L^2$. The solubility increases if ammonia is added to the solution due to the formation of ammin complexes, such as $AgNH_3^+$ and $Ag(NH_3)_2^+$. What is the solubility of AgCl in the absence and presence of 0.5 mol/L ammonia? The relevant complex formation reactions and constants are:

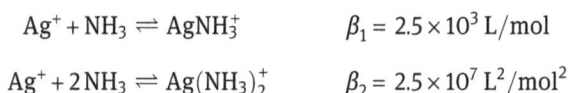

$$Ag^+ + NH_3 \rightleftharpoons AgNH_3^+ \qquad \beta_1 = 2.5 \times 10^3 \, L/mol$$

$$Ag^+ + 2\,NH_3 \rightleftharpoons Ag(NH_3)_2^+ \qquad \beta_2 = 2.5 \times 10^7 \, L^2/mol^2$$

**11.6.** Nitrilotriacetic acid (NTA) is a synthetic complexing agent. Its threefold negatively charged anion acts as a ligand and forms very strong complexes with heavy metal ions, for instance with $Cd^{2+}$. To differentiate between the NTA species resulting from the stepwise dissociation of the triprotonic acid, we want to apply the following notations for the acid and its anions: $H_3NTA$, $H_2NTA^-$, $HNTA^{2-}$, and $NTA^{3-}$. $NTA^{3-}$ is also able to form a complex with $Ca^{2+}$, which is a major constituent of natural waters. Although the $Ca^{2+}$ complex is weaker than the $Cd^{2+}$ complex, $Ca^{2+}$ is a strong competitor for the $NTA^{3-}$ ligand due to its much higher concentration in natural aqueous systems. Furthermore, it has to be considered that $Cd^{2+}$ also forms complexes with $OH^-$ and $CO_3^{2-}$ that occur in all natural waters. Therefore, there is also a competition of these ligands for the $Cd^{2+}$ ions.

What are the equilibrium concentrations of the species $Ca^{2+}$, $Cd^{2+}$, $CaNTA^-$, $CdNTA^-$, $CdCO_3^0$, $CdOH^+$, and $Cd(OH)_2^0$ under the following conditions: $c(NTA_{total}) = 1 \times 10^{-7}$ mol/L, $c(Ca_{total}) = 1 \times 10^{-3}$ mol/L, $c(Cd_{total}) = 1 \times 10^{-9}$ mol/L, $c(DIC) = 1 \times 10^{-3}$ mol/L, pH = 8? Note that of the different species of the involved acids ($H_3NTA$ and carbonic acid) only $HNTA^{2-}$, $NTA^{3-}$, $HCO_3^-$, and $CO_3^{2-}$ are relevant at the given pH value. Furthermore, it holds that $c(CdNTA^-) \ll c(NTA_{total})$, $c(CaNTA^-) \ll c(Ca_{total})$, and $c(CdCO_3^0) \ll c(DIC)$ which simplifies the material balance equations.

Accordingly, the following reactions have to be taken into account:

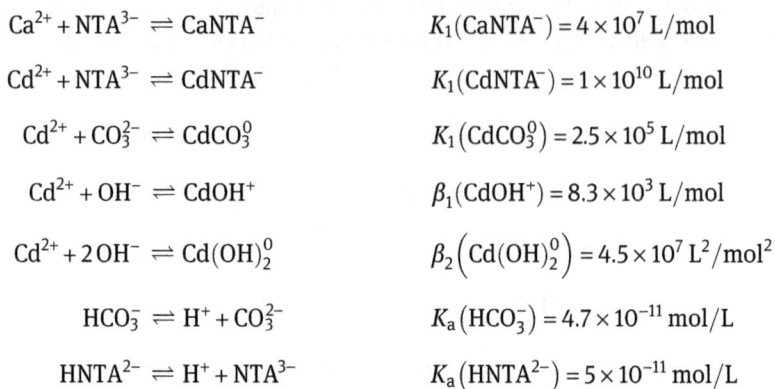

$$Ca^{2+} + NTA^{3-} \rightleftharpoons CaNTA^- \qquad K_1(CaNTA^-) = 4 \times 10^7 \text{ L/mol}$$

$$Cd^{2+} + NTA^{3-} \rightleftharpoons CdNTA^- \qquad K_1(CdNTA^-) = 1 \times 10^{10} \text{ L/mol}$$

$$Cd^{2+} + CO_3^{2-} \rightleftharpoons CdCO_3^0 \qquad K_1(CdCO_3^0) = 2.5 \times 10^5 \text{ L/mol}$$

$$Cd^{2+} + OH^- \rightleftharpoons CdOH^+ \qquad \beta_1(CdOH^+) = 8.3 \times 10^3 \text{ L/mol}$$

$$Cd^{2+} + 2OH^- \rightleftharpoons Cd(OH)_2^0 \qquad \beta_2\left(Cd(OH)_2^0\right) = 4.5 \times 10^7 \text{ L}^2/\text{mol}^2$$

$$HCO_3^- \rightleftharpoons H^+ + CO_3^{2-} \qquad K_a(HCO_3^-) = 4.7 \times 10^{-11} \text{ mol/L}$$

$$HNTA^{2-} \rightleftharpoons H^+ + NTA^{3-} \qquad K_a(HNTA^{2-}) = 5 \times 10^{-11} \text{ mol/L}$$

# 12 Sorption

## 12.1 Introduction

In natural aqueous systems, water is often in contact with solid phases, such as sediments, suspended matter, soil, or aquifer material. In Chapter 8, we have seen that dissolution/precipitation may be an important transfer mechanism between the liquid and the adjacent solid phase. Sorption is another relevant liquid/solid transfer mechanism that also plays an important role in aqueous systems. The term "sorption" is a generic term that describes the accumulation of dissolved species on a solid surface or within the solid material. It includes the surface processes adsorption and ion exchange as well as the uptake within the interior of the solid phase (absorption). The latter is of minor relevance and occurs only in some specific cases (e.g., uptake of dissolved species within solid organic material). Therefore, sorption in aqueous systems is mainly a surface process, where the term "surface" comprises the external surface and – in the case of porous sorbents – also the internal surface. The reverse process (release of accumulated material into the aqueous phase) is referred to as desorption. The receiving solid phase is denoted as sorbent, whereas the substance to be sorbed is called sorbate (Figure 12.1). Natural sorbents are often referred to as geosorbents. Accordingly, the term "geosorption" is sometimes used for natural sorption processes.

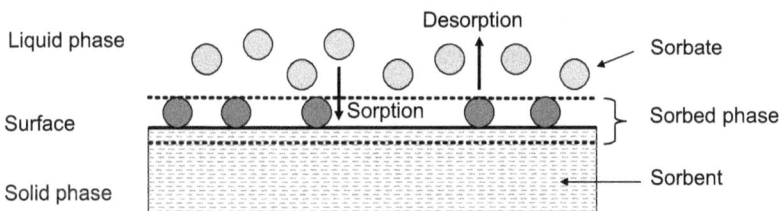

**Figure 12.1:** Fundamental terms used to describe sorption processes.

In principle, sorption can take place at all interfaces where an aqueous phase is in contact with solid material. Table 12.1 gives some typical examples for sorption processes in natural aqueous systems. Sorption as a process that immobilizes dissolved water constituents is a relevant part of the self-purification processes within the water cycle.

Under certain conditions, sorption processes in special environmental compartments can be utilized for water treatment purposes. In drinking water treatment, bank filtration and infiltration of pretreated surface water are typical examples for utilizing the attenuation potential of natural sorption processes. Bank filtration is a pretreatment option in cases where polluted surface water has to be used for drinking water production. In such cases, the raw water is not extracted directly from the river or lake but from extraction wells located at a certain distance from the bank. Due to

https://doi.org/10.1515/9783110758788-012

**Table 12.1:** Examples of solid/liquid interfaces in natural systems where sorption processes are possible.

| Natural solid material acting as sorbent | Liquid phase in contact with the solid |
|---|---|
| Lake and river sediments | Surface water |
| Suspended matter in groundwater and surface water | Groundwater or surface water |
| Soil (vadose zone) | Seepage water, infiltrate |
| Aquifer (saturated zone) | Groundwater, bank filtrate, infiltrate |

the hydraulic gradient between the river or the lake and the extraction wells caused by pumping, the water flows in the direction of the extraction wells. During the subsurface transport, complex attenuation processes take place. Although biodegradation is the most important process, particularly in the first part of the flow-path, sorption onto the aquifer material during the subsurface transport can also contribute to the purification of the raw water, in particular by retardation of nondegradable or poorly degradable solutes. Sorption can also support slow biodegradation processes by extending the retention time in the biologically active zone.

Infiltration of surface water, typically pretreated by engineered processes (e.g., flocculation, sedimentation, filtration), is based on analogous principles. During the contact of the infiltrated water with the soil and the aquifer material, biodegradation and sorption processes can take place, leading to an improvement in the water quality. The infiltrated water is then extracted by extraction wells and further treated by engineered processes.

Natural sorption processes are not only utilized in drinking water treatment but also for reuse of wastewater. In particular in regions with water scarcity, the use of reclaimed wastewater for artificial groundwater recharge becomes increasingly important. In this case, wastewater treated by advanced processes is infiltrated into the subsurface where, in principle, the same attenuation processes as in bank filtration or surface water infiltration take place. Since the degradable water constituents are already removed to a large extent in the wastewater treatment plant, sorption is particularly relevant for the removal of residual nondegradable or poorly degradable solutes.

The sorption of solutes onto aquifer material determines significantly their transport velocity in the subsurface (i.e., in groundwater or during bank filtration and infiltration). More details are discussed in Section 12.6.

## 12.2 Geosorbents

Typically, geosorbents are heterogeneous solids consisting of mineral and organic components. Therefore, different types of interactions with sorbates are possible.

Generally, the following sorbate/sorbent interactions can be distinguished: interactions of inorganic ions with mineral surfaces (electrostatic interactions, ion exchange, surface complex formation) or with solid organic material (complex formation) and interactions of organic solutes with solid organic matter or with mineral surfaces (hydrophobic interactions, van der Waals forces, hydrogen bond formation). Ionized organic species take an intermediate position because the bond forces can include electrostatic interactions and weak intermolecular forces.

The mineral components of geosorbents are mainly oxidic substances and clay minerals. Due to their surface charge (Section 12.4) they sorb preferentially ionic species. In contrast, the organic fractions of the geosorbents are able to sorb especially organic solutes, in particular hydrophobic compounds. The high affinity of hydrophobic solutes to the hydrophobic organic material can be explained by the effect of hydrophobic interactions, which is an entropy-driven effect that induces the hydrophobic solute to leave the aqueous solution and to aggregate with other hydrophobic materials. In accordance with this sorption mechanism, it can be expected that the sorption of organic solutes onto organic sorbent material increases with increasing hydrophobicity (Section 12.5).

## 12.3 Sorption Isotherms

### 12.3.1 General Considerations

Independent of the specific sorption mechanisms, sorption equilibria can be described by sorption isotherms. Sorption isotherms relate the sorbed amount in the state of equilibrium, $q_{eq}$, to the equilibrium concentration in the liquid phase, $c_{eq}$, at constant temperature:

$$q_{eq} = f(c_{eq}) \quad T = \text{constant} \tag{12.1}$$

The sorbed amount, also referred to as the sorbent loading or solid-phase concentration, is defined by:

$$q = \frac{n_{sorbate}}{m_{sorbent}} \tag{12.2}$$

where $n_{sorbate}$ is the sorbed substance amount (in mol or mmol) and $m_{sorbent}$ is the sorbent mass.

Isotherm data can be determined experimentally by adding a defined mass of sorbent, $m_{sorbent}$, to a solution of the sorbate with known volume, $V$, and initial concentration, $c_0$, and determining the residual concentration, $c_{eq}$, after a contact time long enough to establish sorption equilibrium. The sorbed amount related to the equilibrium concentration, $q_{eq}(c_{eq})$ can be calculated by the material balance equation:

$$q_{eq} = \frac{V}{m_{sorbent}} (c_0 - c_{eq}) \tag{12.3}$$

Further isotherm points can be found by varying the sorbent mass or the initial concentration. It has to be noted that this experimental method may fail in the case of very weakly sorbed sorbates. Here, the sorbent mass needed to induce a measurable concentration decrease may be impractically high. In this case, other methods such as column experiments have to be applied (see also Section 12.6).

Instead of the molar liquid-phase and solid-phase concentrations, mass concentrations can also be used to quantify the equilibrium data (e.g., mg/L for the liquid-phase concentration and mg/g for the sorbent loading).

To describe the sorption equilibrium data mathematically, a suitable isotherm equation has to be selected and applied (Section 12.3.2). Although most of the proposed isotherm equations have been derived from a conceptual model, they are typically used as empirical equations to describe measured sorption data without a direct link to a defined sorption mechanism. In principle, sorption isotherm equations can be applied to all kinds of sorbate/sorbent combinations.

For the sorption of ions onto inorganic sorbents an alternative equilibrium model is often used: the model of surface complexation. This model bears analogy to the complex formation in the liquid phase (Chapter 11) and will be discussed in more detail in Section 12.4.

In general, sorption processes are influenced by temperature and pH. However, in the typical temperature range of natural aquatic systems, the temperature dependence of the sorption equilibrium is only weak. In contrast, a strong influence of the pH can be expected in all cases where the sorbent and/or the sorbate are subject to protonation/deprotonation processes.

### 12.3.2 Isotherm Equations

Isotherm equations describe the relationship between the sorbent loading and the liquid-phase concentration of the sorbate in the state of equilibrium. Since we consider only equilibrium data, we will omit the subscript "eq" in the following equations for the sake of simplicity.

If the sorbate concentration is relatively low and the sorption is not very strong, the equilibrium can often be described by a linear relationship:

$$q = K_d \, c \tag{12.4}$$

with $K_d$ as the characteristic isotherm parameter (Figure 12.2). $K_d$ is referred to as the distribution or partition coefficient. Common units are L/g or L/kg. There is obviously an analogy to gas/water partitioning (Chapter 6). Equation (12.4) is therefore also referred to as the Henry isotherm. This equation was found to be well suited in many cases to describing the sorption of organic trace compounds onto geosorbents, in particular on the organic fractions of these sorbents (Section 12.5).

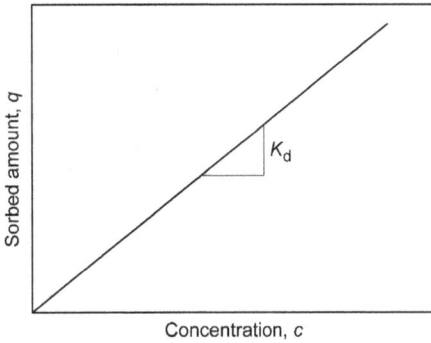

**Figure 12.2:** Henry isotherm.

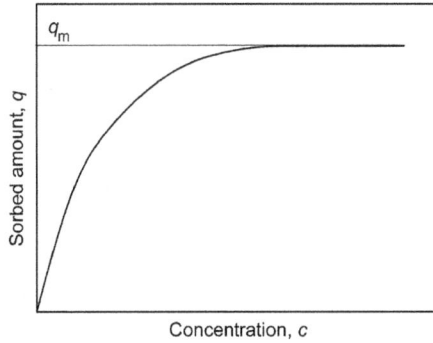

**Figure 12.3:** Langmuir isotherm.

The well-known equations proposed by Langmuir and Freundlich are typical representatives of the group of nonlinear two-parameter isotherms.

The Langmuir isotherm (Figure 12.3) has the form:

$$q = \frac{q_m K_L c}{1 + K_L c} \tag{12.5}$$

where $q_m$ and $K_L$ are the isotherm parameters. The parameter $q_m$, the maximum sorbent loading, has the same unit as the sorbent loading, $q$, and the unit of $K_L$ is the reciprocal of the concentration unit. At low concentrations ($K_L c \ll 1$), equation (12.5) reduces to the linear Henry isotherm:

$$q = q_m K_L c = K_d c \tag{12.6}$$

whereas at high concentrations ($K_L c \gg 1$), a constant saturation value (maximum loading) results:

$$q = q_m = \text{constant} \tag{12.7}$$

While showing plausible limiting cases, the Langmuir isotherm is often not suitable for describing the experimental isotherm data found for aqueous solutions. This might be a consequence of the fact that this theoretically derived isotherm is based on assumptions which are often not fulfilled, in particular mono-layer coverage of the sorbent surface and energetic homogeneity of the sorption sites. On the other hand, the Langmuir isotherm equation was also shown to be applicable in some cases, although these assumptions were obviously not fulfilled.

The Freundlich isotherm is given by:

$$q = K_F c^n \tag{12.8}$$

where $K_F$ and $n$ are the isotherm parameters. The Freundlich isotherm can describe neither the linear range at very low concentrations nor the saturation effect at very high concentrations. By contrast, the medium concentration range of isotherms is often represented very well.

In the Freundlich isotherm, the sorption coefficient, $K_F$, characterizes the strength of sorption. The higher the value of $K_F$ is, the higher is the sorbent loading that can be achieved at a given $n$. The exponent $n$ is related to the energetic heterogeneity of the sorbent surface and determines the curvature of the isotherm. The lower the $n$ value is, the more concave (with respect to the concentration axis) is the isotherm shape. If the concentration has a value of 1 in the respective unit, the loading equals the value of $K_F$.

In principle, the exponent $n$ can take any value (Figure 12.4). In geosorption, mostly $n$ values lower than or equal to 1 are found. With $n = 1$, the isotherm becomes linear.

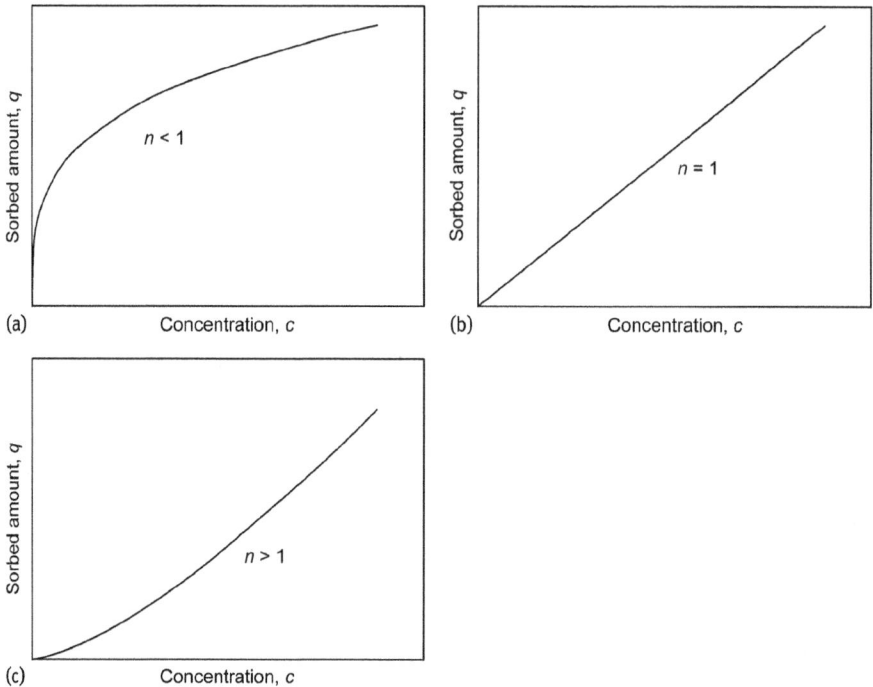

**Figure 12.4:** Different forms of the Freundlich isotherm: (a) favorable ($n < 1$), (b) linear ($n = 1$), (c) unfavorable ($n > 1$).

It has to be noted that the Freundlich isotherm can be considered a composite of Langmuir isotherms with different $K_L$ values representing patches of sorption sites with different sorption energies. Accordingly, summing up a number of Langmuir isotherms for the different sorption sites leads to a Freundlich-type isotherm curve.

The unit of $K_F$ ($= q/c^n$) depends on the units used for $q$ and $c$ and includes the exponent $n$. As mentioned previously, different liquid-phase and solid-phase concentrations can be used in isotherm determination (molar concentrations, mass concentrations) that result in different $K_F$ units. The conversion of these $K_F$ units is not as simple as for other isotherm parameters (Henry or Langmuir isotherm) due to the included exponent $n$. The same problem occurs if the units should be converted to multiples or sub-multiples, for instance conversion of mol to mmol or vice versa.

---

**Example 12.1**
A Freundlich coefficient of $K_F = 5$ (mmol/g)/(mmol/L)$^{0.8}$ was found for a solute with a molecular weight of $M = 96$ g/mol. What is the Freundlich coefficient in (mg/g)/(mg/L)$^{0.8}$?

**Solution:**
The relationship between the molar concentration and the mass concentration is given by:

$$\rho^* = c\,M$$

An analogue relationship can be formulated for the sorbent loading:

$$q^* = q\,M$$

with $q$ in mmol/g and $q^*$ in mg/g. If we denominate the Freundlich coefficient related to mass concentrations as $K_F^*$ and the Freundlich coefficient related to molar concentrations as $K_F$, we can write:

$$K_F^* = \frac{q^*}{(\rho^*)^n} = \frac{q\,M}{(c\,M)^n} = \frac{q\,M}{c^n\,M^n} = \frac{q}{c^n}M^{1-n} = K_F\,M^{1-n}$$

Finally, with $M = 96$ g/mol $= 96$ mg/mmol and $n = 0.8$, we find:

$$K_F^* = [5\,(\text{mmol/g})/(\text{mmol/L})^{0.8}]\,[(96\,\text{mg/mmol})^{1-0.8}] = 12.46\,(\text{mg/g})/(\text{mg/L})^{0.8}$$

---

In order to describe a given set of measured equilibrium concentrations and sorbent loadings mathematically, the applicability of the different isotherm equations has to be tested and the respective isotherm parameters have to be estimated. Whereas the isotherm parameters of the Henry isotherm can be found by simple linear regression, either nonlinear or linear regression can be used to find the isotherm parameters for the two-parameter isotherms. To apply linear regression, the two-parameter isotherm equations have to be linearized.

For the Langmuir isotherm, different types of linearization are possible:

$$\frac{c}{q} = \frac{1}{q_m\,K_L} + \frac{1}{q_m}c \tag{12.9}$$

$$\frac{1}{q} = \frac{1}{q_m} + \frac{1}{q_m\,K_L}\frac{1}{c} \tag{12.10}$$

$$q = q_m - \frac{1}{K_L}\frac{q}{c} \tag{12.11}$$

$$\frac{q}{c} = q_m K_L - q K_L \tag{12.12}$$

The Freundlich isotherm can be linearized by transforming the equation into the logarithmic form:

$$\log q = \log K_F + n \log c \tag{12.13}$$

or

$$\ln q = \ln K_F + n \ln c \tag{12.14}$$

### 12.3.3 Distribution Between Liquid and Solid Phase

Isotherm equations can be used to calculate the distribution of a sorbate between the liquid and the solid phase for a given total concentration. For an arbitrary sorbate C and a solid S we can write a formal reaction equation for the sorption:

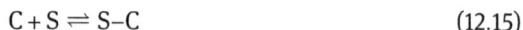

$$C + S \rightleftharpoons S-C \tag{12.15}$$

The related material balance reads:

$$c(C_{total}) = c(C) + q(C) \rho^*(S) \tag{12.16}$$

The sorbent loading has to be multiplied by the mass concentration of the solid in order to make the second term on the right-hand side of the equation volume-related as the other terms in the material balance equation. Now, $q(C)$ can be substituted by the respective isotherm equation, for instance, by the Freundlich isotherm:

$$c(C_{total}) = c(C) + K_F c^n(C) \rho^*(S) \tag{12.17}$$

Since equation (12.17) is nonlinear with respect to $c(C)$, the solution for a given total concentration has to be found by an iterative procedure.

This general speciation approach can be applied, in principle, for all types of sorbate–sorbent interactions. However, it has to be noted that the isotherm parameters are empirical parameters that are found from sorption experiments carried out under defined conditions. Since there are no explicit relationships available that describe the influence of relevant liquid-phase properties such as pH, temperature, or ionic strength, the validity of the isotherm parameters is restricted to the conditions under which they were determined. This is particularly relevant for the sorption of ions onto oxidic sorbents which is strongly influenced by pH and ionic strength.

---

**Example 12.2**

For the sorption of $Cd^{2+}$ at pH = 6 onto solid ferric hydroxide, the Freundlich isotherm parameters $n = 0.36$ and $K_F = 0.88$ $(mmol/g)/(mmol/L)^{0.36}$ were found. What is the fraction of dissolved $Cd^{2+}$ if the total Cd concentration is $1 \times 10^{-8}$ mol/L and the sorbent concentration is 0.1 mg/L?

**Solution:**
The material balance equation reads:

$$c(Cd_{total}) = c(Cd^{2+}) + K_F \, c^n (Cd^{2+}) \, \rho^*(S)$$

Introducing the given data yields:

$$1 \times 10^{-5} \text{ mmol/L} = c(Cd^{2+}) + [0.88 \, (\text{mmol/g})/(\text{mmol/L})^{0.36}][c^{0.36}(Cd^{2+})] \, (1 \times 10^{-4} \text{ g/L})$$

Note that the material balance equation is written here by using the units mmol/L and g/L according to the unit of the isotherm parameter $K_F$. Combining $K_F$ and $\rho^*$ gives:

$$1 \times 10^{-5} \text{ mmol/L} = c(Cd^{2+}) + [8.8 \times 10^{-5} (\text{mmol/L})^{0.64}] \, [c^{0.36}(Cd^{2+})]$$

This equation has to be solved by an iteration procedure where the concentration of $Cd^{2+}$ is varied as long as the right-hand side (RHS) of the equation equals the left-hand side (see the table below).

| Guess | $c(Cd^{2+})$ in mmol/L | RHS in mmol/L |
|---|---|---|
| 1 | $5 \times 10^{-6}$ | $6.09 \times 10^{-6}$ |
| 2 | $8 \times 10^{-6}$ | $9.29 \times 10^{-6}$ |
| 3 | $9 \times 10^{-6}$ | $1.03 \times 10^{-5}$ |
| 4 | $8.8 \times 10^{-6}$ | $1.01 \times 10^{-5}$ |
| 5 | $8.7 \times 10^{-6}$ | $1.00 \times 10^{-5}$ |

The concentration of dissolved $Cd^{2+}$ in the considered system is $8.7 \times 10^{-6}$ mmol/L or $8.7 \times 10^{-9}$ mol/L. The fraction of $Cd^{2+}$ in the aqueous phase is then:

$$f(Cd^{2+}) = \frac{c(Cd^{2+})}{c(Cd_{total})} = \frac{8.7 \times 10^{-9} \text{ mol/L}}{1 \times 10^{-8} \text{ mol/L}} = 0.87$$

Under the given conditions, 87% of the Cd occurs in dissolved form, whereas only 13% is sorbed onto ferric hydroxide.

## 12.4 Sorption Onto Charged Surfaces

### 12.4.1 Introduction

Most of the naturally occurring inorganic sorbents (clay minerals, oxidic solids) carry permanent and/or variable charges. Clay minerals are alumosilicates that have permanent negative charges as a result of the isomorphic substitution of tetravalent silicon atoms by trivalent aluminum atoms in the basic silicate structure. The negative charges are neutralized by positively charged counter ions that can be exchanged by other ions from the aqueous solution. Cation exchange is therefore an important sorption mechanism for clay minerals.

Oxidic materials, such as oxides or hydroxides of silicon, aluminum, manganese or iron, carry variable charges. Clay minerals also show variable charges besides their permanent charges. All these solids are characterized by crystalline structures where positively charged metal or metalloid ions and negatively charged oxygen or hydroxide ions are arranged in such a manner that the different charges compensate for each other. At the surface, this regular structure is disturbed and the charges have to be compensated for by other ions. In aqueous solutions, the negative charges of the surface oxygen ions are neutralized by protons forming OH groups, whereas the positive charges of the surface metal ions are neutralized by hydroxide ions. As a result, the surface of oxidic sorbents is covered with surface OH groups. These groups are subject to protonation or deprotonation depending on the pH value of the solution:

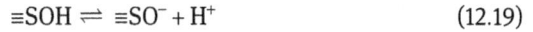

$$\equiv SOH + H^+ \rightleftharpoons \equiv SOH_2^+ \tag{12.18}$$

$$\equiv SOH \rightleftharpoons \equiv SO^- + H^+ \tag{12.19}$$

In these equations, the symbol $\equiv S$ stands for the surface of the solid material.

It follows from equations (12.18) and (12.19) that the surface is positively charged at low pH values and negatively charged at high pH values. Between these regions, a pH value exists at which the sum of negative charges equals the sum of positive charges and the net charge of the surface is zero. This point is referred to as point of zero charge (pzc). The $pH_{pzc}$ is an important sorbent parameter that helps to understand the sorption of charged species and the influence of pH on the sorption process. Generally, the sorption of charged species onto charged surfaces can be expected to be strongly influenced by electrostatic attraction or repulsion forces depending on the charges of the sorbate and the sorbent at the given pH.

The surface charge as a function of pH can be determined by titration of a sorbent suspension with strong acids and bases (e.g., HCl and NaOH) at a specified ionic strength. From such titration curves, the $pH_{pzc}$ can be evaluated.

For each point of the titration curve, the surface charge, $Q_s$, can be calculated from the general mass and charge balance equation:

$$Q_s = q(H^+) - q(OH^-) = \frac{V}{m_{sorbent}}(c_a - c_b - c(H^+) + c(OH^-)) \tag{12.20}$$

where $q(H^+)$ is the surface loading with $H^+$, $q(OH^-)$ is the surface loading with $OH^-$, $c_a$ is the molar concentration of the acid used for titration, $c_b$ is the molar concentration of the base used for titration, $c(H^+)$ is the proton concentration after equilibration (measured as pH), $c(OH^-)$ is the $OH^-$ concentration after equilibration (calculated from the measured pH), $V$ is the volume of the solution, and $m_{sorbent}$ is the sorbent mass. The surface charge is given in mmol/g or mol/kg. The surface charge density, $\sigma_s$ (in $C/m^2$), can be calculated from $Q_s$ by:

$$\sigma_s = \frac{Q_s F}{A_m} \tag{12.21}$$

where $F$ is the Faraday constant (96 485 C/mol) and $A_m$ is the specific (mass-related) surface area ($m^2/kg$). Plotting $Q_s$ or $\sigma_s$ versus pH illustrates the influence of the pH on the surface charge (Figure 12.5). The intersection with the line at $\sigma_s = 0$ $C/m^2$ (or $Q_s = 0$ mol/kg) gives the $pH_{pzc}$. Table 12.2 lists points of zero charge for some solids.

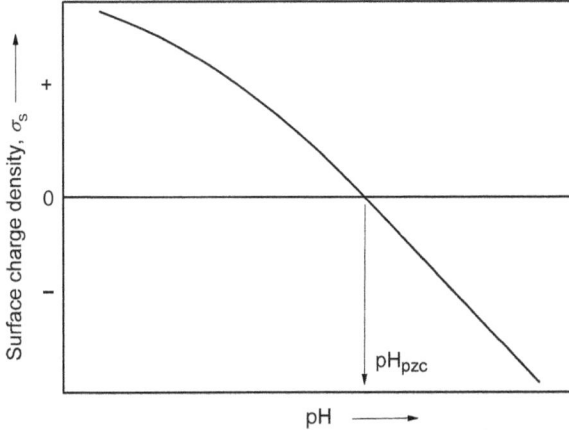

**Figure 12.5:** Schematic representation of a surface titration curve.

**Table 12.2:** Points of zero charge for selected solids (Stumm 1992).

| Solid | $pH_{pzc}$ |
|---|---|
| $\alpha$-$Al_2O_3$ | 9.1 |
| $\alpha$-$Fe_2O_3$ | 8.5 |
| $Fe(OH)_3$ | 8.5 |
| $\gamma$-AlOOH | 8.2 |
| $\alpha$-FeOOH | 7.8 |
| $\beta$-$MnO_2$ | 7.2 |
| $\alpha$-$Al(OH)_3$ | 5.0 |
| Kaolinite | 4.6 |
| $\delta$-$MnO_2$ | 2.8 |
| Montmorillonite | 2.5 |
| $SiO_2$ | 2.0 |

## 12.4.2 Mathematical Description of the Surface Protonation/Deprotonation

The laws of mass action for the protonation and deprotonation of surface OH groups as described by equations (12.18) and (12.19) are:

$$K_{a1} = \frac{c_{ss}(\equiv SOH)\, c(H^+)}{c_{ss}(\equiv SOH_2^+)} \tag{12.22}$$

$$K_{a2} = \frac{c_{ss}(\equiv SO^-)\,c(H^+)}{c_{ss}(\equiv SOH)} \tag{12.23}$$

where $c_{ss}$ is the concentration of the surface sites and $K_{a1}$ and $K_{a2}$ are the acidity constants. Note that $K_{a1}$ is related to the reverse reaction to that shown in equation (12.18), according to the general definition of an acidity constant (Chapter 7, Section 7.4). The surface site concentrations can be expressed in mol/g or mol/m². A conversion of the units can be easily done if the specific surface area, $A_m$ (in m²/g), is known:

$$c_{ss}(mol/m^2) = \frac{c_{ss}(mol/g)}{A_m(m^2/g)} \tag{12.24}$$

For a system with a defined solid concentration, $\rho^*_{solid}$(in g/L), the site concentrations can also be expressed in mol/L:

$$c_{ss}(mol/L) = c_{ss}(mol/g)\,\rho^*_{solid}(g/L) \tag{12.25}$$

As can be derived from equations (12.22) and (12.23), the choice of the unit for $c_{ss}$ has no influence on the unit of the constant, which is mol/L in all cases, the same as we have found for acidity constants of dissolved monoprotic acids (Chapter 7, Section 7.4).

It has to be noted that the acidity constants $K_{a1}$ and $K_{a2}$ are conditional or apparent constants whose values are influenced by the change of the surface charge due to protonation or deprotonation. This effect will be discussed later in more detail. At first, we want to exclude the explicit consideration of this effect and consider $K_{a1}$ and $K_{a2}$ as fixed constants. Under this condition, the site distribution as a function of pH can be found by combining equations (12.22) and (12.23) with the balance equation for the surface sites:

$$c_{ss,\,total} = c_{ss}(\equiv SOH_2^+) + c_{ss}(\equiv SOH) + c_{ss}(\equiv SO^-) \tag{12.26}$$

We start with the calculation of the pH dependence of the fraction $f$ ($\equiv SOH$). Rearranging the laws of mass action:

$$c_{ss}(\equiv SOH_2^+) = \frac{c_{ss}(\equiv SOH)\,c(H^+)}{K_{a1}} \tag{12.27}$$

$$c_{ss}(\equiv SO^-) = \frac{c_{ss}(\equiv SOH)\,K_{a2}}{c(H^+)} \tag{12.28}$$

and introducing these equations into the balance equation gives:

$$c_{ss,\,total} = \frac{c_{ss}(\equiv SOH)\,c(H^+)}{K_{a1}} + c_{ss}(\equiv SOH) + \frac{c_{ss}(\equiv SOH)\,K_{a2}}{c(H^+)} \tag{12.29}$$

$$c_{ss,total} = c_{ss}(\equiv SOH)\left[\frac{c(H^+)}{K_{a1}} + 1 + \frac{K_{a2}}{c(H^+)}\right] \tag{12.30}$$

Finally, we obtain for the fraction $f(\equiv SOH)$:

$$f(\equiv SOH) = \frac{c_{ss}(\equiv SOH)}{c_{ss,total}} = \frac{1}{\dfrac{c(H^+)}{K_{a1}} + 1 + \dfrac{K_{a2}}{c(H^+)}} \tag{12.31}$$

Knowing the fraction of the neutral surface sites, the fractions of the charged surface sites can be found from equations (12.27) and (12.28) after dividing both sides of the equations by the total surface site concentration, $c_{ss,total}$:

$$f(\equiv SOH_2^+) = \frac{c_{ss}(\equiv SOH_2^+)}{c_{ss,total}} = \frac{c_{ss}(\equiv SOH)\,c(H^+)}{c_{ss,total}\,K_{a1}} = f(\equiv SOH)\,\frac{c(H^+)}{K_{a1}} \tag{12.32}$$

$$f(\equiv SO^-) = \frac{c_{ss}(\equiv SO^-)}{c_{ss,total}} = \frac{c_{ss}(\equiv SOH)\,K_{a2}}{c_{ss,total}\,c(H^+)} = f(\equiv SOH)\,\frac{K_{a2}}{c(H^+)} \tag{12.33}$$

As an example, the complete distribution of the different surface sites $\equiv SOH_2^+$, $\equiv SOH$ and $\equiv SO^-$ with the assumed acidity exponents $pK_{a1} = 6.5$ and $pK_{a2} = 9.1$ is shown in Figure 12.6. The site distribution is comparable to the species distribution of a diprotic acid, such as carbonic acid (Section 7.7.2 in Chapter 7).

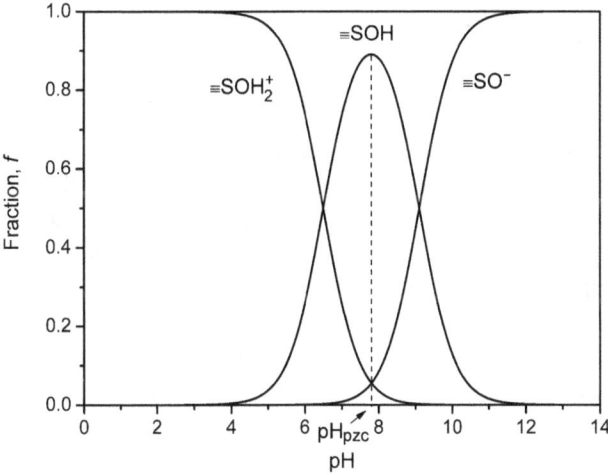

**Figure 12.6:** Site distribution as a function of pH (without charge effect), calculated for $pK_{a1} = 6.5$ and $pK_{a2} = 9.1$.

Based on our previous theoretical considerations, we can also find a relationship between the pH at the point of zero charge ($pH_{pzc}$) and the acidity constants. Given that the net surface charge equals the difference between the positively and negatively charged sites:

$$Q_s = c_{ss}(\equiv SOH_2^+) - c_{ss}(\equiv SO^-) \tag{12.34}$$

we can derive that at the point of zero charge ($Q_s = 0$ mol/kg), the condition:

$$c_{ss}(\equiv SOH_2^+) = c_{ss}(\equiv SO^-) \tag{12.35}$$

must be fulfilled. After introducing equations (12.27) and (12.28) into equation (12.35), we obtain:

$$\frac{c_{pzc}(H^+)}{K_{a1}} = \frac{K_{a2}}{c_{pzc}(H^+)} \tag{12.36}$$

where $c_{pzc}(H^+)$ is the proton concentration at the point of zero charge. In logarithmic form we get:

$$\log c_{pzc}(H^+) - \log K_{a1} = \log K_{a2} - \log c_{pzc}(H^+) \tag{12.37}$$

$$pK_{a1} - pH_{pzc} = pH_{pzc} - pK_{a2} \tag{12.38}$$

which finally leads to the relationship:

$$pH_{pzc} = 0.5(pK_{a1} + pK_{a2}) \tag{12.39}$$

Accordingly, the point of zero charge is $pH_{pzc} = 7.8$ in our example, which also can be seen in Figure 12.6 (intersection of the $\equiv SOH_2^+$ and $\equiv SO^-$ curves and maximum of the $\equiv SOH$ curve).

As already mentioned, the acidity constants are influenced by the surface charge, which changes its value during protonation or deprotonation. Let us first consider the protonation. The more protons added to the surface OH groups, the more the number of positive sites (or the net positive surface charge) increases with the consequence that the tendency to add further protons decreases. This means that the value of the constant $1/K_{a1}$, which describes the extent of proton addition, decreases with decreasing pH (see equations (12.18) and (12.22)). On the other hand, we can expect that $K_{a2}$ for the deprotonation decreases with increasing pH for the same reason (decrease of the tendency to release further protons with increasing negative charge of the surface). From equations (12.32) and (12.33), we can draw some conclusions with respect to the change of the site distribution due to the charge effect in comparison to the simplified case without the charge effect shown in Figure 12.6. We can see from equation (12.32) that a decreased $1/K_{a1}$ (or increased $K_{a1}$) can be compensated for by an equivalent increase of $c(H^+)$ to keep a given ratio of charged and uncharged sites constant. From equation (12.33) we can derive that a decreased $K_{a2}$ can be compensated for by an equivalent decrease of $c(H^+)$ to keep the site ratio constant. Accordingly, we can expect that the curve of $\equiv SOH_2^+$ will be flatter and further extended to lower pH values in comparison to the case where the charge effect is neglected. The $\equiv SO^-$ curve will also be flatter but further extended to higher pH values in comparison to the simplified case shown in Figure 12.6.

In summary, the conditional constants $K_{a1}$ and $K_{a2}$ are not really constant but change with the surface charge and therefore also with pH. This effect can be considered by introducing a correction term. Accordingly, the conditional acidity constant is expressed as the product of a constant that is independent of the electrical charge (intrinsic constant, $K_a^{int}$) and a term describing the influence of the surface potential:

$$K_a = K_a^{int} \exp\left(\frac{-\Delta z_s F \, \Psi_s}{RT}\right) \tag{12.40}$$

or in logarithmic form (with $\ln x = 2.303 \log x$):

$$\log K_a = \log K_a^{int} - \frac{\Delta z_s F \, \Psi_s}{2.303 \, RT} \tag{12.41}$$

where $\Psi_s$ is the surface potential (V), $F$ is the Faraday constant (96 485 C/mol, $R$ is the gas constant (8.3145 J/(mol·K)), $T$ is the absolute temperature (K), and $\Delta z_s$ is the change in the charge due to the surface reaction. Since the acidity constants, $K_{a1}$ and $K_{a2}$, describe the release of one positively charged proton from $\equiv SOH_2^+$ and $\equiv SOH$, respectively, the change in the surface charge is $\Delta z_s = -1$ in both cases.

Given that $1\,C = 1\,A \cdot s$ and $1\,J = 1\,W \cdot s = 1\,V \cdot A \cdot s$, the correction term for 25 °C (298.15 K) is:

$$\frac{\Delta z_s F \, \Psi_s}{2.303 \, RT} = \frac{16.9}{V} \Delta z_s \, \Psi_s \tag{12.42}$$

and with $\Delta z_s = -1$ we have:

$$\log K_a = \log K_a^{int} + \frac{16.9}{V} \Psi_s \tag{12.43}$$

The surface potential, $\Psi_s$ (in V), cannot be determined directly. It has to be calculated from the surface charge density, $\sigma_s$. A simple model approach assumes that there is only one layer of counter ions in the interfacial region and the electric double layer can be considered a parallel plate capacitor. Under this condition, there is a fixed ratio of surface potential and surface charge density, the constant capacitance, $C$:

$$C = \frac{\sigma_s}{\Psi_s}\left[\frac{C/m^2}{V} = \frac{A \cdot s/m^2}{V}\right] \tag{12.44}$$

The surface charge density, $\sigma_s$ (C/m²), is related to the surface charge, $Q_s$ (mol/kg), by equation (12.21) and therefore also to the concentration difference (mol/kg) between the positively and negatively charged surface sites (equation (12.34)).

The speciation of the surface sites at a given pH can be computed if the following data are known: the total site concentration $c_{ss,total}$, the capacitance, $C$, the intrinsic acidity constants, $K_{a1}^{int}$ and $K_{a2}^{int}$, and the specific surface area, $A_m$. Under these conditions, an iterative procedure can be applied to find the site distribution. Starting with a guess of $\Psi_s$, the conditional constants $K_{a1}$ and $K_{a2}$ can be computed by using equations (12.41) or (12.43). Then, the fractions of the different sites and their absolute concentrations can be calculated by means of equations (12.31)–(12.33). With the concentrations of the charged species, the surface charge, $Q_s$, is available (equation (12.34)), which can be converted into $\sigma_s$ by means of equation (12.21). The computed value of $\sigma_s$ has to be compared with that found from equation (12.44) for the assumed $\Psi_s$. If there is no equality, the procedure has to be repeated with a new guess of $\Psi_s$. Such calculations are typically done by means of appropriate speciation software. It has to be noted that the speciation results are often afflicted with relatively high uncertainties mainly due to the limited availability of reliable input data.

Figure 12.7 shows the change in the distribution functions due to the effect of changing surface charge during protonation and deprotonation of the surface OH groups.

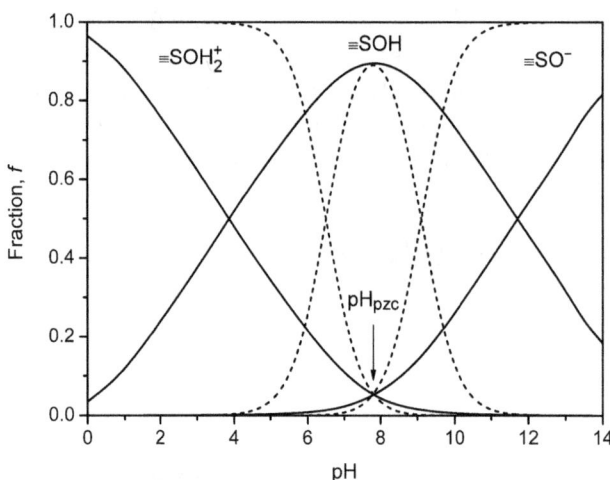

**Figure 12.7:** Site distribution as a function of pH with (solid curves) and without (dashed curves) consideration of the surface charge effect, calculated for $pK_{a1} = 6.5$, $pK_{a2} = 9.1$, $C = 1\,A \cdot s/(V \cdot m^2)$, $A_m = 250\ m^2/g$, and $\rho^*(solid) = 1\ g/L$.

It is noteworthy to state that the relationship between the $pH_{pzc}$ and the acidity constants as given in equation (12.39) remains valid independent of the additional consideration of the charge effect. This follows from the fact that at the point of zero charge the surface potential is zero and the correction term disappears in equations (12.41) and (12.43). Accordingly, $K_a$ equals $K_a^{int}$ and we can write equation (12.39) not only with the conditional but also with the intrinsic constants:

$$pH_{pzc} = 0.5(pK_{a1}^{int} + pK_{a2}^{int})$$ (12.45)

## 12.4.3 Modeling of Ion Sorption

The sorption of dissolved ions onto charged surfaces is frequently described by the surface complexation model (SCM). The formation of surface complexes is strongly related to the structure of the electric double layer which surrounds the charged solid particle. The term "surface complexation" (or surface complex formation) comprises two different types of reactions, the formation of inner-sphere complexes and the formation of outer-sphere complexes (Figure 12.8).

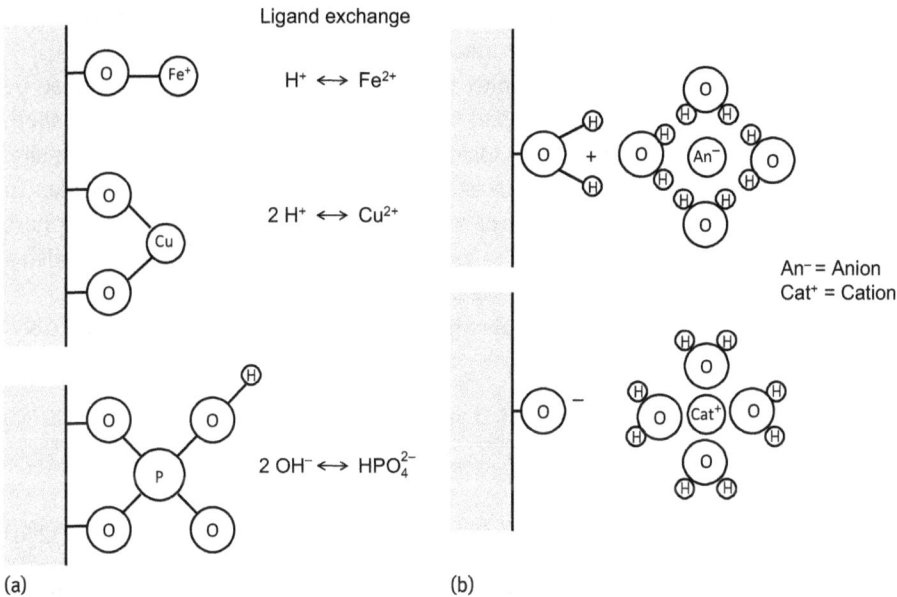

(a)  (b)

**Figure 12.8:** Inner-sphere (a) and outer-sphere (b) complexes (adapted from Sigg and Stumm 2011).

In the case of inner-sphere complexes, the sorbate ions without the water molecules of the hydration shell are directly bound to the surface site by ligand exchange. Cations replace the protons of the surface OH groups as shown in the following reaction equations for a bivalent metal cation ($Me^{2+}$):

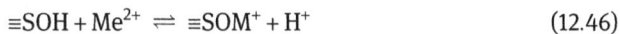

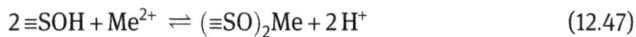

$$\equiv SOH + Me^{2+} \rightleftharpoons \equiv SOM^+ + H^+$$ (12.46)

$$2\equiv SOH + Me^{2+} \rightleftharpoons (\equiv SO)_2Me + 2H^+$$ (12.47)

In the case of anions (here $A^{2-}$), the OH groups are replaced, for instance:

$$\equiv SOH + A^{2-} \rightleftharpoons \equiv SA^- + OH^-$$ (12.48)

$$2 \equiv SOH + A^{2-} \rightleftharpoons (\equiv S)_2 A + 2OH^-$$ (12.49)

According to the equations given above, the sorption of cations increases with increasing pH, whereas the sorption of anions increases with decreasing pH. If the sorbate is an anion of a weak acid or a cation of a weak base, its pH-depending acid/base equilibrium also influences the sorption and has to be considered in the equilibrium calculations in addition to the surface reactions.

The ions sorbed as inner-sphere complexes are strongly bound and located in a compact layer directly attached to the surface. This first part of the electric double layer is also referred to as the surface layer. As can be seen from the given reaction equations, the sorption of ions can lead to neutral or charged surface complexes depending on the ion charge and the number of surface sites that take part in the reaction.

The model of outer-sphere complex formation presumes that ions can also be bound to the surface sites by chemical bonds without losing the hydration shell. This means that a water molecule is located between the ion and the sorption site. Therefore, the distance to the surface is larger and the bond strength is weaker in comparison to the inner-sphere complex formation. The layer where outer-sphere complexation takes place is referred to as the beta ($\beta$) layer. The beta layer is also a part of the compact layer in terms of the double-layer model.

The outer-sphere complexation of cations and anions can be formally described by:

$$\equiv SOH + Me^{2+} + H_2O \rightleftharpoons \equiv SO-H_2O-Me^+ + H^+$$ (12.50)

and

$$\equiv SOH + A^{2-} + H_2O \rightleftharpoons \equiv S-H_2O-A^- + OH^-$$ (12.51)

Beyond the beta layer, a diffuse layer exists where an excess concentration of counter ions (ions charged oppositely to the charge of the surface layer) neutralizes the remaining surface charge. Throughout the diffuse layer, the concentration of the counter ions decreases with increasing distance from the surface until in the bulk liquid the equivalent concentrations of cations equal the equivalent concentrations of the anions. The enrichment of counter ions in the diffuse layer is a result of electrostatic interactions and can be considered a nonspecific sorption.

A number of models were developed to characterize the charge distribution and the accumulation of counter ions. In particular, the constant capacitance model, the diffuse layer model and the triple-layer model are widely used to describe the sorption onto charged surfaces. These models differ mainly in the assumptions concerning the charge distribution and the location of the sorbed species (Figure 12.9).

Other, even more complex models (e.g., the charge distribution-multisite complexa-tion [CD-MUSIC] model) will not be considered here.

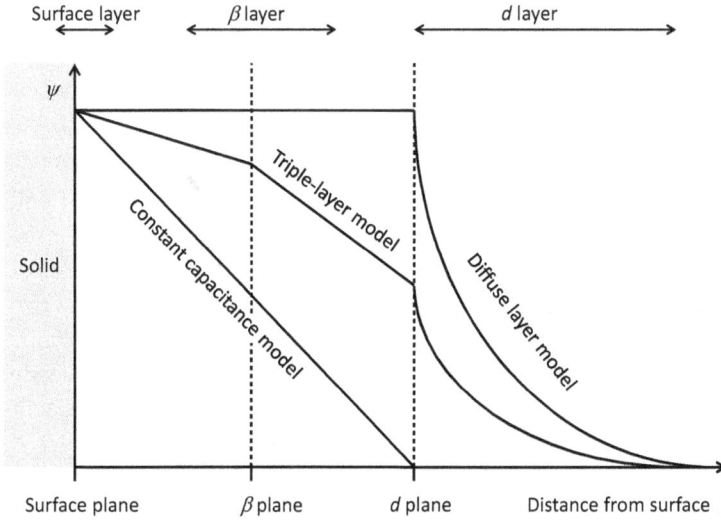

**Figure 12.9:** Surface potential as a function of distance from the surface as assumed in different surface complexation models (adapted from Benjamin 2002).

The triple-layer model assumes specific sorption in the surface plane and in the beta plane as well as charge compensation by nonspecifically sorbed ions distributed in the diffuse layer. The diffuse layer model neglects the beta layer and assumes that the specifically sorbed ions are all located in the surface plane and the charge in the surface plane is neutralized by ions distributed in the diffuse layer. The constant ca-pacitance model also neglects the existence of the beta layer and assumes that all specifically sorbed ions are located in the surface layer and the ions that neutralize the surface charge are all located in the $d$ plane.

The complex formation in the surface or beta plane can be described by laws of mass action with respective equilibrium constants. As already discussed for the pro-tonation/deprotonation (Section 12.4.2), the conditional equilibrium constants of the complex formation depend on the surface charge due to attraction or repulsion caused by the charged surface groups. In analogy to equation (12.40), the condi-tional equilibrium constant, $K$, can be expressed as the product of an intrinsic con-stant, $K^{int}$, and a term describing the influence of the surface potential. For the surface layer, we have to write:

$$K = K^{int} \exp\left(\frac{-\Delta z_s F \Psi_s}{RT}\right)$$

(12.52)

where $\Delta z_s$ is the change of the charge in the surface layer due to sorption and $\Psi_s$ is the potential of the surface plane. If it is assumed that complexation affects the charges in both the surface and beta layers, the apparent equilibrium constant is given by:

$$K = K^{int} \; \exp\left(\frac{-(\Delta z_s \; \Psi_s + \Delta z_\beta \; \Psi_\beta)F}{RT}\right) \tag{12.53}$$

where $\Delta z_\beta$ is the change of the charge in the beta layer and $\Psi_\beta$ is the potential of the beta plane.

To apply a surface complexation model for describing the sorption of ions onto a charged surface, a multitude of equations have to be combined, in particular material balances for all species, laws of mass action for all reactions in all considered layers, charge balances in each layer and charge–potential relationships for all considered layers. To reduce the number of equations, simplifying assumptions can be made, for instance, neglecting the beta layer as in the constant capacitance model or in the diffuse layer model. Commercial speciation software often includes solution algorithms for different surface complex formation approaches.

Relatively strong limitations of such equilibrium models result from uncertainties concerning the model assumptions, the need for simplifications, and the problems in parameter determination as well as from the increasing complexity if a large number of sorbable ions are present in the water. Therefore, for practical purposes, the conventional sorption isotherm equations are frequently used to describe the sorption equilibria under defined conditions instead of applying a complex formation model. Nevertheless, a qualitative characterization of surface chemistry, in particular, the knowledge about the pH-depending charges and the location of $pH_{pzc}$, is helpful for interpreting sorption processes on oxidic or other charged surfaces.

## 12.5 Sorption of Organic Species Onto Organic Material

Neutral organic solutes are mainly sorbed to the organic fractions of the solid sorbents. Relevant mechanisms are hydrophobic interactions, van der Waals forces, and possibly also absorption. Sorption of neutral organic sorbates to mineral surfaces becomes relevant only if the content of organic material in the sorbent is very low (less than about 0.1%). In the case of ionized organic species, the contribution of ionic interactions between charged species and mineral surfaces may be higher and these interactions may dominate even if the organic carbon fraction is higher than the given limit.

In most cases, the sorption of neutral organic solutes onto solid organic material can be described by the linear Henry isotherm (Section 12.3.2). As with all isotherm parameters, the sorption coefficient $K_d$ depends on the properties of both sorbate and sorbent.

Under the assumption that interaction with the organic fraction of the solid is the dominant sorption mechanism, it is reasonable to normalize the sorption coefficient, $K_d$ (in L/kg), to the organic carbon content, $f_{oc}$, of the sorbent:

$$K_{oc} = \frac{K_d}{f_{oc}} \tag{12.54}$$

with

$$f_{oc} = \frac{m_{oc}}{m_{solid}} \tag{12.55}$$

where $m_{oc}$ is the mass of organic carbon in the solid material and $m_{solid}$ is the total mass of the solid material. In the ideal case, this normalization should make the sorption coefficients independent of the kind of sorbent. However, it has to be stated that the following conditions must be fulfilled to make the sorption coefficient really independent of the sorbent type: (i) the sorption onto the solid organic matter is the only sorption mechanism, and (ii) the organic constituents of different sorbents have exactly the same sorption properties. In practice, these conditions are not strictly fulfilled, in particular the second one. Nevertheless, the normalization makes $K_{oc}$ values much more comparable than $K_d$ values.

If we assume that the value of $K_{oc}$ only depends on the sorbate but not on the sorbent, it should be possible to correlate it with other sorbate properties. If we further assume that the sorption of organic solutes is dominated by hydrophobic interactions, it can be expected that the sorption strength increases with increasing hydrophobicity of the sorbate. The hydrophobicity can be characterized by the dimensionless n-octanol-water partition coefficient, $K_{ow}$, that describes the partition of a compound between the organic solvent n-octanol and water. It increases with increasing hydrophobicity. Consequently, $K_{oc}$ should strongly correlate with $K_{ow}$. Indeed, such correlations have been found in numerous studies. The general form of all these correlations is:

$$\log K_{oc} = a \, \log K_{ow} + b \tag{12.56}$$

where $a$ and $b$ are empirical parameters.

Depending on the substances included in the respective studies, two groups of correlations can be distinguished: class-specific correlations and nonspecific correlations. Table 12.3 gives a selection of such correlations. The parameters in the correlations are slightly different. However, the resulting deviations in the predicted $K_{oc}$ values are in most cases smaller than one order of magnitude (see Example 12.3). It has to be noted that the correlations are valid for $K_{oc}$ in L/kg.

**Table 12.3:** Selection of log $K_{oc}$ – log $K_{ow}$ correlations.

| Correlation | Valid for substance class | References |
|---|---|---|
| log $K_{oc}$ = 0.544 log $K_{ow}$ + 1.377 | Not specified | Kenaga and Goring (1980) |
| log $K_{oc}$ = 0.909 log $K_{ow}$ + 0.088 | Not specified | Hassett et al. (1983) |
| log $K_{oc}$ = 0.679 log $K_{ow}$ + 0.663 | Not specified | Gerstl (1990) |
| log $K_{oc}$ = 0.903 log $K_{ow}$ + 0.094 | Not specified | Baker et al. (1997) |
| log $K_{oc}$ = 1.00 log $K_{ow}$ – 0.21 | Benzenes, PAHs | Karickhoff et al. (1979) |
| log $K_{oc}$ = 0.72 log $K_{ow}$ + 0.49 | Chloro and methyl benzenes | Schwarzenbach and Westall (1981) |
| log $K_{oc}$ = 0.89 log $K_{ow}$ – 0.32 | Chlorinated phenols | van Gestel et al. (1991) |
| log $K_{oc}$ = 0.63 log $K_{ow}$ + 0.90<br>log $K_{oc}$ = 0.57 log $K_{ow}$ + 1.08 | Substituted phenols, anilines, nitrobenzenes, chlorinated benzonitriles | Sabljič et al. (1995) |
| log $K_{oc}$ = 1.07 log $K_{ow}$ – 0.98 | Polychlorinated biphenyls | Girvin and Scott (1997) |
| log $K_{oc}$ = 0.42 log $K_{ow}$ + 1.49 | Aromatic amines | Worch et al. (2002) |

**Example 12.3**

Calculate log $K_{oc}$ and $K_{oc}$ for a solute with a log $K_{ow}$ = 3 by using the four nonspecific correlations given in Table 12.3.

**Solution:**

$$\log K_{oc} = 0.544 \log K_{ow} + 1.377 = 3.009 \qquad K_{oc} = 1021\,L/kg$$
$$\log K_{oc} = 0.909 \log K_{ow} + 0.088 = 2.815 \qquad K_{oc} = 653\,L/kg$$
$$\log K_{oc} = 0.679 \log K_{ow} + 0.663 = 2.700 \qquad K_{oc} = 501\,L/kg$$
$$\log K_{oc} = 0.903 \log K_{ow} + 0.094 = 2.803 \qquad K_{oc} = 635\,L/kg$$

The predicted log $K_{oc}$ varies between 2.70 and 3.01 and the respective $K_{oc}$ values are in the range of 501 L/kg to 1021 L/kg.

It has to be stated that the application of such empirical correlations can only give a rough estimate of $K_{oc}$. However, it can be expected that at least the right order of magnitude for $K_{oc}$ can be found from the correlations. The n-octanol-water partition coefficients are available for most solutes from databases or can be estimated by special prediction methods.

Furthermore, it is important to note that the application of the correlations is restricted to neutral species and dominating hydrophobic interactions. For solutes that are subject to protonation or deprotonation, the pH-dependent partition coefficient log $D$ can be used instead of log $K_{ow}$. Here, the concentration in the n-octanol phase ($c_{n\text{-octanol}}$) is related to the sum of the aqueous concentrations of the ionic

$(c_{aq,ionic})$ and neutral species $(c_{aq,neutral})$. This definition includes the assumption that only neutral species occur in the organic $n$-octanol phase:

$$D = \frac{c_{n\text{-octanol}}}{c_{aq,neutral} + c_{aq,ionic}} \tag{12.57}$$

For a dissociating acidic sorbate, HA:

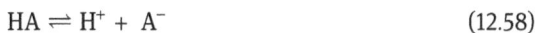

$$HA \rightleftharpoons H^+ + A^- \tag{12.58}$$

we can write:

$$D = \frac{c_{n\text{-octanol}}(HA)}{c_{aq}(HA) + c_{aq}(A^-)} = \frac{c_{n\text{-octanol}}(HA)}{c_{aq}(HA)\left(1 + \dfrac{c_{aq}(A^-)}{c_{aq}(HA)}\right)} \tag{12.59}$$

The ratio of the HA concentrations in the $n$-octanol phase and in the aqueous phase equals $K_{ow}$. Furthermore, as shown in Chapter 7 (Section 7.6.1), the ratio of the anion and the neutral acid can be expressed by means of the acidity exponent, $pK_a$:

$$\frac{c_{aq}(A^-)}{c_{aq}(HA)} = 10^{pH - pK_a} \tag{12.60}$$

Accordingly, we find the following expression for estimating distribution coefficients, $D$, for acidic compounds:

$$D(\text{acid}) = \frac{K_{ow}}{1 + 10^{pH - pK_a}} \tag{12.61}$$

For the dissociation of protonated bases, $BH^+$:

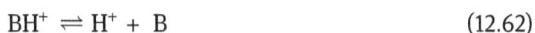

$$BH^+ \rightleftharpoons H^+ + B \tag{12.62}$$

we can derive an analogue expression:

$$D(\text{base}) = \frac{K_{ow}}{1 + 10^{pK_a - pH}} \tag{12.63}$$

where $pK_a$ is here the acidity exponent of the protonated base, $BH^+$ (for the relationship between acidity and basicity constants of conjugate acid/base pairs see Section 7.4 in Chapter 7). Note the different signs of $pK_a$ and pH in comparison to equation (12.61). This difference is in accordance with the fact that in the case of bases the charged species (protonated bases) dominate at $pH < pK_a$, whereas in the case of acids the charged species (acid anions) dominate at $pH > pK_a$.

Equations (12.61) and (12.63) can also be written in logarithmic form as:

$$\log D(\text{acid}) = \log K_{ow} - \log(1 + 10^{pH - pK_a}) \tag{12.64}$$

$$\log D(\text{base}) = \log K_{ow} - \log(1 + 10^{pK_a - pH}) \tag{12.65}$$

---

**Example 12.4**

The log $K_{ow}$ of pentachlorophenol (PCP) is 5.1 and the p$K_a$ is 4.75. What are the log $D$ for pH = 5 and pH = 7 and what log $K_{oc}$ is found if the correlation by van Gestel et al. (Table 12.3) is applied?

**Solution:**

log $D$ can be found by equation (12.64):

$$\log D(\text{PCP, pH} = 5) = \log K_{ow} - \log(1 + 10^{\text{pH} - \text{p}K_a}) = 5.1 - \log(1 + 10^{5 - 4.75}) = 5.1 - 0.44 = 4.66$$

$$\log D(\text{PCP, pH} = 7) = \log K_{ow} - \log(1 + 10^{\text{pH} - \text{p}K_a}) = 5.1 - \log(1 + 10^{7 - 4.75}) = 5.1 - 2.25 = 2.85$$

If we introduce the log $D$ into the correlation proposed by van Gestel et al., we find:

$$\log K_{oc}(\text{PCP, pH} = 5) = 0.89 \log D(\text{PCP, pH} = 5) - 0.32 = (0.89)(4.66) - 0.32 = 3.83$$

$$\log K_{oc}(\text{PCP, pH} = 7) = 0.89 \log D(\text{PCP, pH} = 7) - 0.32 = (0.89)(2.85) - 0.32 = 2.22$$

---

It has to be noted that the introduction of log $D$ into the correlations used to estimate log $K_{oc}$ considers only the decrease of hydrophobic interactions with increasing fraction of ionic (more hydrophilic) species. However, sorption of organic ions based on other binding mechanisms, such as ion exchange, cannot be predicted in this way.

An alternative approach to describe the pH-dependent sorption of organic acids is to measure $K_d$ (or $K_{oc}$) values at different pH values and to apply the following equation to fit the measured data:

$$K_d(\text{pH}) = (K_{d,n} - K_{d,i})(1 - \alpha) + K_{d,i} \tag{12.66}$$

where $K_{d,n}$ and $K_{d,i}$ are the sorption coefficients of the neutral and ionized species, respectively. $\alpha$ is the degree of protolysis (degree of dissociation):

$$\alpha = \frac{1}{1 + 10^{\text{p}K_a - \text{pH}}} \tag{12.67}$$

The term $(1 - \alpha)$ in equation (12.66) represents the fraction of the undissociated acid (Chapter 7, Section 7.6) and is given by:

$$1 - \alpha = \frac{1}{1 + 10^{\text{pH} - \text{p}K_a}} \tag{12.68}$$

Plotting the measured $K_d$ values over $(1 - \alpha)$ will allow to calculate $K_{d,n}$ and $K_{d,i}$. If these coefficients are once known, $K_d$ can be predicted for each pH of interest by means of equation (12.66).

Frequently, the ionized species does not contribute significantly to the overall sorption because of its strong hydrophilicity. Under this condition, equation (12.66) can be simplified to:

$$K_d(\text{pH}) = K_{d,n} (1 - \alpha) \tag{12.69}$$

Dividing equations (12.66) and (12.69) by $f_{oc}$, we find the corresponding relationships for $K_{oc}$:

$$K_{oc}(pH) = (K_{oc,n} - K_{oc,i})(1 - \alpha) + K_{oc,i} \tag{12.70}$$

$$K_{oc}(pH) = K_{oc,n}(1 - \alpha) \tag{12.71}$$

---

**Example 12.5**

The sorption coefficients of undissociated pentachlorophenol and its anion were found to be $K_{oc,n} = 6\,800$ L/kg and $K_{oc,i} = 320$ L/kg (Amiri et al. 2004). Calculate the sorption coefficients $K_{oc}$ (pH = 5) and $K_{oc}$ (pH = 7) and the respective logarithmic values. The $pK_a$ of pentachlorophenol is 4.75.

**Solution:**

The fractions of the undissociated acid, $(1 - \alpha)$, for pH = 5 and pH = 7 are:

$$1 - \alpha = \frac{1}{1 + 10^{pH - pK_a}} = \frac{1}{1 + 10^{5 - 4.75}} = 0.360 \quad (pH = 5)$$

$$1 - \alpha = \frac{1}{1 + 10^{pH - pK_a}} = \frac{1}{1 + 10^{7 - 4.75}} = 0.0056 \quad (pH = 7)$$

In the case of pentachlorophenol, the sorption coefficient of the anion is relatively high. Therefore, the simplified equation (12.71) cannot be used here and equation (12.70) has to be applied:

$$K_{oc}(pH) = (K_{oc,n} - K_{oc,i})(1 - \alpha) + K_{oc,i}$$

$$K_{oc}(pH = 5) = (6\,800 \text{ L/kg} - 320 \text{ L/kg})(0.36) + 320 \text{ L/kg} = 2\,653 \text{ L/kg}$$

$$\log K_{oc}(pH = 5) = 3.42$$

$$K_{oc}(pH = 7) = (6\,800 \text{ L/kg} - 320 \text{ L/kg})(0.0056) + 320 \text{ L/kg} = 356 \text{ L/kg}$$

$$\log K_{oc}(pH = 7) = 2.55$$

As expected, the sorption coefficient decreases with increasing degree of protolysis (or dissociation) and approaches the value of the ionized species. The $\log K_{oc}$ calculated here on the basis of experimental data compare well with those predicted from $\log D$ by means of the empirical correlation (see Example 12.4).

---

# 12.6 Sorption-Influenced Subsurface Transport of Dissolved Substances

Sorption plays an important role for the transport of dissolved substances with groundwater or during riverbank filtration where water flows in the underground through porous rock and sediment, which can act as sorbents. The porous material with water in between is referred to as aquifer. Sorption onto the aquifer material reduces the transport velocity of sorbing solutes in comparison to the flow velocity of water or non-sorbing substances. This effect is referred to as retardation. The strength of this effect can be expressed by the retardation factor. The retardation factor, $R_d$, describes the ratio of the mean pore water velocity in the aquifer (interstitial water velocity), $v_w$, and the velocity of the concentration front of the considered solute, $v_c$, (more exactly the velocity of the center of mass of the concentration front):

$$R_d = \frac{v_w}{v_c} \tag{12.72}$$

If no sorption takes place, the velocity of the concentration front is the same as the pore water velocity and $R_d$ is 1. If sorption takes place, $v_c$ is lower than $v_w$ and $R_d$ becomes greater than 1. The stronger the sorption is, the higher the value of $R_d$ and the lower the velocity of the concentration front.

It can be shown by solute transport modeling that a relationship between the retardation factor, $R_d$, and the sorption isotherm exists. The general relationship reads:

$$R_d = 1 + \frac{\rho_B}{\varepsilon_B} \frac{dq}{dc} \tag{12.73}$$

where $\rho_B$ is the bulk density of the aquifer material and $\varepsilon_B$ is the effective porosity. If we express $q$ by the isotherm equation and differentiate the resulting expression with respect to $c$, we find for the linear isotherm (equation (12.4)):

$$R_d = 1 + \frac{\rho_B}{\varepsilon_B} K_d \tag{12.74}$$

In the case of the nonlinear Freundlich isotherm (equation (12.8)), $R_d$ depends on the concentration:

$$R_d = 1 + \frac{\rho_B}{\varepsilon_B} K_F \, n \, c^{n-1} \tag{12.75}$$

By using these equations, the subsurface transport velocity of solutes can be estimated if the sorption isotherm parameter(s) and the aquifer material properties density and porosity are known. Conversely, the equations can also be used to find the sorption parameters ($R_d$ or $K_d$) for the sorbent material of interest from column experiments (breakthrough curve measurements to determine $v_w$ and $v_c$).

---

**Example 12.6**
What is the travel velocity of chlorobenzene in an aquifer that is characterized by the following properties: $\rho_B = 1.6$ g/cm³, $\varepsilon_B = 0.40$, $f_{oc} = 0.01$. The mean pore water velocity is assumed to be 1 m/day. The n-octanol-water partition coefficient of chlorobenzene is $K_{ow} = 2.58$. To find the sorption coefficient, the correlation given by Schwarzenbach and Westall (Table 12.3) can be used.

**Solution:**
At first, $K_{oc}$ is calculated by the Schwarzenbach-Westall correlation:

$$\log K_{oc} = 0.72 \log K_{ow} + 0.49 = (0.72)\,(2.58) + 0.49 = 2.35$$
$$K_{oc} = 10^{2.35} \, \text{L/kg} = 223.9 \, \text{L/kg}$$

$K_d$ can be found from $K_{oc}$ and $f_{oc}$ (see equation (12.54)):

$$K_d = K_{oc}\, f_{oc} = (223.9\ L/kg)(0.01) = 2.239\ L/kg \approx 2.24\ L/kg$$

Note that the standard unit of $K_{oc}$ and $K_d$ is L/kg.

With $K_d$, the retardation factor, $R_d$, can be calculated:

$$R_d = 1 + \frac{\rho_B}{\varepsilon_B}\, K_d = 1 + \frac{1.6\ kg/L}{0.40}\, 2.24\ L/kg = 9.96$$

The retardation factor is 9.96. This means that the travel velocity of chlorobenzene is by a factor of about 10 lower than the pore water velocity:

$$v_c = \frac{v_w}{R_d} = \frac{1\ m/day}{9.96} = 0.1\ m/day$$

## 12.7 Problems

**12.1.** 30 mg granular ferric hydroxide was added to 1 L of a phosphate solution containing 1 mg/L P. The measured equilibrium concentration was 0.45 mg/L P. What is the sorbed amount $q^*_{eq}(P)$ in mg/g?

**12.2.** For a sorbent/sorbate system, the Freundlich isotherm parameters $n = 0.9$ and $K_F = 0.2$ (mmol/g)/(mmol/L)$^{0.9}$ were found. What is the Freundlich coefficient in (µmol/g)/(µmol/L)$^{0.9}$?

**12.3.** Take the data for the $Cd^{2+}$ sorption onto ferric hydroxide from Example 12.2 (Section 12.3.3) and discuss the change in the speciation if: a) the total cadmium concentration is increased from $1 \times 10^{-8}$ mol/L to $5 \times 10^{-8}$ mol/L (sorbent concentration: $\rho^* = 0.1$ mg/L) and b) the sorbent mass concentration is increased from 0.1 to 0.5 mg/L (total Cd concentration: $c\ (Cd_{total}) = 1 \times 10^{-8}$ mol/L).

**12.4.** For the sorption of dibenzothiophene on a sandy aquifer material with an organic carbon content of 0.21 mg/g, the linear sorption coefficient was found to be $K_d = 2.3$ L/kg. Calculate the organic carbon normalized sorption coefficient $K_{oc}$ and compare $K_{oc}$ and log $K_{oc}$ with the predictions made by using the correlation proposed by Baker et al. (Table 12.3). The n-octanol-water partition coefficient of dibenzothiophene is given as log $K_{ow} = 4.5$.

**12.5.** The anti-inflammatory drug naproxen is an acidic drug with a $pK_a$ of 4.8. The log $K_{ow}$ of the neutral acid is 3. What is the log $K_{oc}$ for pH = 6 that will be found by using the correlation proposed by Baker et al. (Table 12.3)?

**12.6.** The sorption coefficient of neutral 2-methyl-4,6-dinitrophenol (2-M-4,6-DNP) was found to be $K_{oc} = 2020$ L/kg, whereas its anion shows no significant sorption. The $pK_a$ of 2-M-4,6-DNP is 4.31. What is the $K_{oc}$ at pH = 6?

# 13 Solutions to the Problems

## Chapter 2

### Problem 2.1

At first, the volume ($V = 413\ 000\ \text{km}^3 = 4.13 \times 10^{14}\ \text{m}^3$) is converted into the mass by means of the density ($\rho = 1\ \text{g/cm}^3 = 1 \times 10^3\ \text{kg/m}^3$):

$$m = V\rho = (4.13 \times 10^{14}\ \text{m}^3)\,(1 \times 10^3\ \text{kg/m}^3) = 4.13 \times 10^{17}\ \text{kg}$$

Then, the mass is converted to the substance amount, $n$ (in mol), by means of the molecular weight ($M = 18\ \text{g/mol} = 1.8 \times 10^{-2}\ \text{kg/mol}$):

$$n = \frac{m}{M} = \frac{4.13 \times 10^{17}\ \text{kg}}{1.8 \times 10^{-2}\ \text{kg/mol}} = 2.29 \times 10^{19}\ \text{mol}$$

Finally, the total enthalpy can be computed from the molar enthalpy:

$$\Delta H_{\text{vap, total}} = n\,\Delta H_{\text{vap}} = (2.29 \times 10^{19}\ \text{mol})\,(44.3\ \text{kJ/mol}) = 1.01 \times 10^{21}\ \text{kJ}$$

### Problem 2.2

The enthalpy of freezing has the same value as the enthalpy of fusion. The only difference consists of the negative sign (exothermic process):

$$\Delta H_{\text{freez}} = -\Delta H_{\text{fus}} = -6.01\ \text{kJ/mol}$$

To calculate the total enthalpy that is released during freezing of 100 L water, we need the number of moles. The mass of 100 L ($= 1 \times 10^5\ \text{cm}^3$) water can be found by means of the density:

$$m = \rho\,V = (1\ \text{g/cm}^3)(1 \times 10^5\ \text{cm}^3) = 1 \times 10^5\ \text{g}$$

With the molecular weight of water ($M = 18\ \text{g/mol}$), we get for the number of moles:

$$n = \frac{m}{M} = \frac{1 \times 10^5\ \text{g}}{18\ \text{g/mol}} = 5.56 \times 10^3\ \text{mol}$$

The total enthalpy released is then:

$$\Delta H = n\,\Delta H_{\text{freez}} = (5.56 \times 10^3\ \text{mol})\,(-6.01\ \text{kJ/mol}) = -3.34 \times 10^4\ \text{kJ}$$

https://doi.org/10.1515/9783110758788-013

## Problem 2.3

The change of the molar enthalpy is given by:

$$\Delta H = \bar{c}_p(T_2 - T_1)$$

The difference of the absolute temperatures equals the difference of the Celsius temperatures:

$$\vartheta_2 - \vartheta_1 = T_2 - T_1 = 50\,\text{K}$$

Therefore, we can write:

$$\Delta H = (75.5\,\text{J}/(\text{K}\cdot\text{mol}))\,(50\,\text{K}) = 3\,775\,\text{J}/\text{mol} = 3.775\,\text{kJ}/\text{mol}$$

This is the enthalpy needed to heat 1 mol water. The total enthalpy for 1 L can be found by multiplying the molar enthalpy with the number of moles, which is in our case:

$$n = \frac{m}{M} = \frac{\rho V}{M} = \frac{(1\,\text{g/cm}^3)\,(1\,000\,\text{cm}^3)}{18\,\text{g/mol}} = 55.56\,\text{mol}$$

$$\Delta H_{total} = n\,\Delta H = (55.56\,\text{mol})\,(3.775\,\text{kJ}/\text{mol}) = 209.74\,\text{kJ}$$

## Problem 2.4

The relationship between kinematic and dynamic viscosity is given by:

$$\nu = \frac{\eta}{\rho}$$

With $\eta = 1.308 \times 10^{-3}$ Pa·s $= 1.308 \times 10^{-3}$ kg/(m·s) and $\rho = 1$ g/cm$^3 = 1 \times 10^3$ kg/m$^3$, we receive:

$$\nu = \frac{\eta}{\rho} = \frac{1.308 \times 10^{-3}\,\text{kg/(m}\cdot\text{s})}{1 \times 10^3\,\text{kg/m}^3} = 1.308 \times 10^{-6}\,\text{m}^2/\text{s}$$

## Problem 2.5

The capillary rise in the case of $\theta = 0°$ (cos $\theta = 1$) is given by:

$$h = \frac{2\,\sigma\cos\theta}{\rho\,g\,r} = \frac{2\,\sigma}{\rho\,g\,r}$$

For 10 °C, we obtain with 1 g/cm$^3 = 1 \times 10^3$ kg/m$^3$:

$$h = \frac{2\,\sigma}{\rho\,g\,r} = \frac{2\,(74.23 \times 10^{-3}\,\text{kg/s}^2)}{(1 \times 10^3\,\text{kg/m}^3)\,(9.81\,\text{m/s}^2)\,(6 \times 10^{-5}\,\text{m})} = 0.252\,\text{m}$$

and for 30 °C:

$$h = \frac{2\,\sigma}{\rho\,g\,r} = \frac{2\,(71.2 \times 10^{-3}\,\text{kg/s}^2)}{(1 \times 10^3\,\text{kg/m}^3)\,(9.81\,\text{m/s}^2)\,(6 \times 10^{-5}\,\text{m})} = 0.242\,\text{m}$$

The temperature increase from 10 °C to 30 °C leads to a decrease of 1 cm (or approximately 4%) in the capillary rise.

### Problem 2.6

To solve the problem, we have to apply the equation:

$$\frac{p(\text{droplets})}{p_0} = \exp\left(\frac{2\,\sigma\,V_m}{r\,R\,T}\right)$$

After rearranging, we receive an expression for the radius:

$$r = \frac{2\,\sigma\,V_m}{R\,T} \cdot \frac{1}{\ln \dfrac{p(\text{droplets})}{p_0}}$$

Then, the searched radius can be computed after inserting the given data:

$$r = \frac{2\,(7.423 \times 10^{-2}\,\text{N/m})\,(1.8 \times 10^{-5}\,\text{m}^3/\text{mol})}{(8.3145\,\text{N} \cdot \text{m}/(\text{mol} \cdot \text{K}))\,(283.15\,\text{K})} \cdot \frac{1}{\ln 1.5} = 2.8 \times 10^{-9}\,\text{m} = 2.8\,\text{nm}$$

## Chapter 3

### Problem 3.1

The molar concentrations can be found by dividing the mass concentrations by the molecular weights. In the case of sulfate, at first the molecular weight of $SO_4^{2-}$ has to be calculated from the molecular weights of the atoms. To find the equivalent concentrations, the molar concentrations have to be multiplied by the charge number (as an absolute value):

$$c(Ca^{2+}) = \frac{\rho^*(Ca^{2+})}{M(Ca^{2+})} = \frac{50\,\text{mg/L}}{40.1\,\text{mg/mmol}} = 1.25\,\text{mmol/L}$$

$$c\left(\frac{1}{2}Ca^{2+}\right) = 2\,c(Ca^{2+}) = 2.5\,\text{mmol/L}$$

$$M(SO_4^{2-}) = M(S) + 4\,M(O) = 32.1\,\text{mg/mmol} + 64\,\text{mg/mmol} = 96.1\,\text{mg/mmol}$$

$$c(SO_4^{2-}) = \frac{\rho^*(SO_4^{2-})}{M(SO_4^{2-})} = \frac{20\,\text{mg/L}}{96.1\,\text{mg/mmol}} = 0.21\,\text{mmol/L}$$

$$c\left(\frac{1}{2}SO_4^{2-}\right) = 2\,c(SO_4^{2-}) = 0.42\,\text{mmol/L}$$

## Problem 3.2

Under consideration of the conversions $1\,\text{kg} = 1 \times 10^3\,\text{g} = 1 \times 10^6\,\text{mg}$ and $60\,\text{mg} = 0.06\,\text{g}$, we can calculate the concentrations in the different units:

a.

$$w(Na^+) = \frac{0.06\,\text{g}}{1 \times 10^3\,\text{g}} = 6 \times 10^{-5}\,\text{g/g}$$

Alternatively,

$$w(Na^+) = \frac{60\,\text{mg}}{1 \times 10^6\,\text{mg}} = 6 \times 10^{-5}\,\text{mg/mg}\ (= \text{g/g})$$

b.

$$\text{Mass percent } (Na^+) = w(Na^+)\,(100\%) = 6 \times 10^{-3}\%$$

c.

$$1\,\text{ppm} = \frac{1\,\text{mg}}{1 \times 10^6\,\text{mg}} = 1\,\text{mg/kg} \rightarrow 60\,\text{mg/kg}\,Na^+ = 60\,\text{ppm}\,Na^+$$

## Problem 3.3

The mole fraction is given by:

$$x(Ca^{2+}) = \frac{n(Ca^{2+})}{\sum n}$$

The term in the denominator includes the moles of solvent and solutes. Since the solution is assumed to be a dilute solution (and since no other concentrations are known), $\sum n$ is set equal to $n(H_2O)$:

$$x(Ca^{2+}) = \frac{n(Ca^{2+})}{n(H_2O)} = \frac{n(Ca^{2+})\,V_{\text{solution}}}{n(H_2O)\,V_{\text{solution}}} = \frac{c(Ca^{2+})}{c(H_2O)}$$

$c(H_2O)$ can be found by dividing the density (1 000 g/L) by the molecular weight (18 g/mol):

$$c(H_2O) = \frac{1\,000 \text{ g/L}}{18 \text{ g/mol}} = 55.56 \text{ mol/L}$$

Now, the mole fraction can be calculated:

$$x(Ca^{2+}) = \frac{c(Ca^{2+})}{c(H_2O)} = \frac{2 \text{ mmol/L}}{55.56 \text{ mol/L}} = \frac{2 \times 10^{-3} \text{ mol/L}}{55.56 \text{ mol/L}} = 3.6 \times 10^{-5}$$

## Problem 3.4

At first, we have to calculate the partial pressure of nitrogen and then we can use the ideal gas law to convert the partial pressure into the molar concentration. The partial pressure can be found from:

$$p(N_2) = y(N_2)\, p_{total}$$

Since for ideal gases the mole fraction equals the volume fraction and the volume fraction is 0.78 (vol% divided by 100%), we find for the partial pressure at $p_{total} = 1$ bar:

$$p(N_2) = (0.78)\,(1\,\text{bar}) = 0.78 \text{ bar}$$

The ideal gas law reads:

$$p_i\, V = n_i\, R\, T$$

Accordingly, the relationship between the partial pressure and the molar concentration is:

$$\frac{n_i}{V} = c_i = \frac{p_i}{R\,T}$$

With $R = 0.083145$ L·bar/(mol·K) and $T = 298.15$ K (25 °C), we find:

$$c(N_2) = \frac{p(N_2)}{R\,T} = \frac{0.78 \text{ bar}}{(0.083145\,\text{L}\cdot\text{bar}/(\text{mol}\cdot\text{K}))\,(298.15\,\text{K})} = 3.15 \times 10^{-2} \text{ mol/L}$$

## Problem 3.5

To establish an ion balance and to calculate the balance error, the equivalent concentrations of all ions have to be known. Therefore, at first, the molar concentrations have to be calculated according to:

$$c = \frac{\rho^*}{M}$$

where the molecular weights of sulfate and hydrogencarbonate have to be calculated by summing up the molecular weights of the atoms:

$$M(SO_4^{2-}) = (32.1 + 4 \times 16) \text{ mg/mmol} = 96.1 \text{ mg/mmol}$$

$$M(HCO_3^-) = (1 + 12 + 3 \times 16) \text{ mg/mmol} = 61 \text{ mg/mmol}$$

The equivalent concentrations are found by multiplying the molar concentrations with the absolute value of the ion charge. For the cations, we find:

$$c(Na^+) = \frac{118 \text{ mg/L}}{23 \text{ mg/mmol}} = 5.13 \text{ mmol/L} \qquad c\left(\frac{1}{1} Na^+\right) = 5.13 \text{ mmol/L}$$

$$c(K^+) = \frac{11 \text{ mg/L}}{39.1 \text{ mg/mmol}} = 0.28 \text{ mmol/L} \qquad c\left(\frac{1}{1} K^+\right) = 0.28 \text{ mmol/L}$$

$$c(Ca^{2+}) = \frac{348 \text{ mg/L}}{40.1 \text{ mg/mmol}} = 8.68 \text{ mmol/L} \qquad c\left(\frac{1}{2} Ca^{2+}\right) = 17.36 \text{ mmol/L}$$

$$c(Mg^{2+}) = \frac{108 \text{ mg/L}}{24.3 \text{ mg/mmol}} = 4.44 \text{ mmol/L} \qquad c\left(\frac{1}{2} Mg^{2+}\right) = 8.88 \text{ mmol/L}$$

Summing up gives:

$$\sum \text{cation equivalent concentrations} = (5.13 + 0.28 + 17.36 + 8.88) \text{ mmol/L}$$

$$= 31.65 \text{ mmol/L}$$

For the anions, we obtain:

$$c(Cl^-) = \frac{40 \text{ mg/L}}{35.5 \text{ mg/mmol}} = 1.13 \text{ mmol/L}, \qquad c\left(\frac{1}{1} Cl^-\right) = 1.13 \text{ mmol/L}$$

$$c(HCO_3^-) = \frac{1\,816 \text{ mg/L}}{61 \text{ mg/mmol}} = 29.77 \text{ mmol/L}, \qquad c\left(\frac{1}{1} HCO_3^-\right) = 29.77 \text{ mmol/L}$$

$$c(SO_4^{2-}) = \frac{38 \text{ mg/L}}{96.1 \text{ mg/mmol}} = 0.40 \text{ mmol/L}, \qquad c\left(\frac{1}{2} SO_4^{2-}\right) = 0.80 \text{ mmol/L}$$

and the sum is:

$$\sum \text{anion equivalent concentrations} = (1.13 + 29.77 + 0.80) \text{ mmol/L} = 31.70 \text{ mmol/L}$$

With the equivalent concentrations of the cations and anions, we can calculate the balance error as follows:

$$\text{Balance error} = \frac{\left| \sum\limits_{\text{cations}} c_i z_i - \sum\limits_{\text{anions}} c_i z_i \right|}{0.5 \left( \sum\limits_{\text{cations}} c_i z_i + \sum\limits_{\text{anions}} c_i z_i \right)} \times 100\%$$

$$\text{Balance error} = \frac{|31.65 \text{ mmol/L} - 31.70 \text{ mmol/L}|}{(0.5)\,(31.65 \text{ mmol/L} + 31.70 \text{ mmol/L})} \times 100\%$$

$$\text{Balance error} = \frac{0.05}{31.675} \times 100\% = 0.16\%$$

The balance error is much lower than the guide value of 5%. Hence, the analysis is plausible.

## Problem 3.6

The total hardness is given as the sum of the molar concentrations of $Ca^{2+}$ and $Mg^{2+}$:

$$TH = c(Ca^{2+}) + c(Mg^{2+}) = 1.25 \text{ mmol/L} + 0.35 \text{ mmol/L} = 1.60 \text{ mmol/L}$$
$$= 1.6 \times 10^{-3} \text{ mol/L}$$

The carbonate hardness can be found from the hydrogencarbonate concentration according to:

$$CH = 0.5\,c(HCO_3^-) = 1.23 \text{ mmol/L} = 1.23 \times 10^{-3} \text{ mol/L}$$

The latter relationship is only valid if $CH \leq TH$. This condition is fulfilled. Therefore, the non-carbonate hardness can be calculated according to:

$$NCH = TH - CH = 1.60 \text{ mmol/L} - 1.23 \text{ mmol/L} = 0.37 \text{ mmol/L} = 0.37 \times 10^{-3} \text{ mol/L}$$

To express the hardness in mg/L $CaCO_3$, the molar ion concentrations have to be converted into mass concentrations and then multiplied with the specific stoichiometric factors for the conversion of the ion concentration into the $CaCO_3$ concentration:

$$\text{Hardness [mg/L } CaCO_3] = 2.5 \; \rho^*(Ca^{2+})[\text{mg/L}] + 4.12 \; \rho^*(Mg^{2+})[\text{mg/L}]$$
$$\rho^*(Ca^{2+}) = c(Ca^{2+})\,M(Ca^{2+}) = (1.25 \text{ mmol/L})\,(40.1 \text{ mg/mmol})$$
$$= 50.13 \text{ mg/L}$$
$$\rho^*(Mg^{2+}) = c(Mg^{2+})\,M(Mg^{2+}) = (0.35 \text{ mmol/L})\,(24.3 \text{ mg/mmol}) = 8.51 \text{ mg/L}$$
$$\text{Hardness [mg/L } CaCO_3] = (2.5)(50.13 \text{ mg/L}) + (4.12)(8.51 \text{ mg/L})$$
$$\text{Hardness [mg/L } CaCO_3] = (125.33 + 35.06) \text{ mg/L} = 160.39 \text{ mg/L}$$

Alternatively, we can multiply the molar concentration of the total hardness by the molecular weight of $CaCO_3$:

$$\text{Hardness [mg/L CaCO}_3] = (1.6\,\text{mmol/L})\,(100.1\,\text{mg/L}) = 160.16\,\text{mg/L}$$

The small difference in the results is due to round-off errors.

## Problem 3.7

Since the concentrations of all major ions are given, the ionic strength can be calculated by using the definition equation:

$$I = 0.5 \sum_i c_i z_i^2$$

Introducing the concentrations gives:

$$I = 0.5 \sum_i c_i z_i^2 = (0.5)[(0.58 + 0.05 + 1.71 \times 4 + 0.51 \times 4 + 0.65 + 2.51 + 0.94 \times 4)\,\text{mmol/L}]$$

$$I = (0.5)[(0.58 + 0.05 + 6.84 + 2.04 + 0.65 + 2.51 + 3.76)\,\text{mmol/L}]$$

$$I = 8.215\,\text{mmol/L} = 8.215 \times 10^{-3}\,\text{mol/L}$$

The Güntelberg equation reads:

$$\log \gamma_1 = -0.5\,\frac{\sqrt{\dfrac{I}{\text{mol/L}}}}{1 + 1.4\sqrt{\dfrac{I}{\text{mol/L}}}}$$

With the calculated ionic strength (in mol/L), we find:

$$\sqrt{\frac{I}{\text{mol/L}}} = \sqrt{8.215 \times 10^{-3}} = 0.0906$$

For $\log \gamma_1$, we obtain:

$$\log \gamma_1 = -0.5\,\frac{\sqrt{\dfrac{I}{\text{mol/L}}}}{1 + 1.4\sqrt{\dfrac{I}{\text{mol/L}}}} = -0.5\,\frac{0.0906}{1 + (1.4)(0.0906)} = -0.0402$$

and for $\gamma_1$, we find:

$$\gamma_1 = 0.912$$

For multivalent ions, we have to use the general relationship:

$$\gamma_z = \gamma_1^{z^2}$$

and with $z = 2$ for bivalent ions, we find:

$$\gamma_2 = \gamma_1^{z^2} = 0.912^4 = 0.692$$

## Chapter 4

### Problem 4.1

The vapor pressure lowering is given by:

$$\frac{p_{01} - p_1}{p_{01}} = x_2$$

where $x_2$ is the mole fraction of the solute. Since NaCl dissociates completely into two ions ($Na^+$, $Cl^-$), the van't Hoff factor is $i = 2$. Therefore, the mole fraction $x_2$ has to be calculated according to:

$$x_2 = \frac{2n(\text{NaCl})}{n(H_2O) + 2n(\text{NaCl})}$$

To find the mole fraction, at first, the substance amounts of NaCl and water have to be derived from the given molality. According to the molality $b = 1$ mol/kg, the substance amount of NaCl in 1 kg water is $n = 1$ mol. The substance amount of water that is equivalent to 1 kg can be calculated by means of the molecular weight:

$$n(\text{water}) = \frac{m(\text{water})}{M(\text{water})} = \frac{1\,000 \text{ g}}{18 \text{ g/mol}} = 55.56 \text{ mol}$$

With these data, the relative vapor pressure lowering can be computed:

$$\frac{p_{01} - p_1}{p_{01}} = x_2 = \frac{2\,(1 \text{ mol})}{55.56 \text{ mol} + 2\,(1 \text{ mol})} = 0.0347$$

Finally, the vapor pressure of the solution, $p_1$, can be calculated by:

$$p_1 = p_{01} - 0.0347\,p_{01} = 23.4 \text{ mbar} - (0.0347)\,(23.4 \text{ mbar}) = 22.59 \text{ mbar}$$

### Problem 4.2

The freezing point depression is −4 K. After rearranging the equation for the freezing point depression caused by a dissociating salt, we obtain:

$$ib = \frac{\Delta T^{SL}}{K_F} = -\frac{-4\,K}{-1.86\,K \cdot kg/mol} = 2.15\,mol/kg$$

If we assume that the numerical values of molality and molarity are the same, we find:

$$ic = 2.15\,mol/L$$

and the osmotic pressure is:

$$\pi = icRT = (2.15\,mol/L)\,(0.083145\ bar \cdot L/(mol \cdot K))\,(293.15\,K) = 52.4\,bar$$

**Problem 4.3**

The freezing point depression of a multicomponent ideal solution can be calculated by:

$$\Delta T^{SL} = \sum_i b_i\,K_F$$

At first, the sum of the molalities has to be computed. Since the concentrations are given as ppm (mg per kg solution), the masses of the solutes have to be divided by the molecular weights to find the concentrations in mol per kg solution. The results are: $Cl^-$: 0.545 mol/kg, $Na^+$: 0.468 mol/kg, $SO_4^{2-}$: 0.028 mol/kg, $Mg^{2+}$: 0.053 mol/kg. Furthermore, it has to be considered that the concentrations are given here as mol per kg solution but the molality is related to kg solvent. As can be derived from the total salinity (3.5%), there are 35 g salts and 965 g water in 1 kg solution. Therefore, the concentrations in mol per kg solution have to be divided by 0.965 (= 965 g/1 000 g) to find the moles per kg water. The molalities are: $b(Cl^-) = 0.565$ mol/kg, $b(Na^+) = 0.485$ mol/kg, $b(SO_4^{2-}) = 0.029$ mol/kg, and $b(Mg^{2+}) = 0.055$ mol/kg. The sum of the molalities is then:

$$\sum_i b_i = 0.565\,mol/kg + 0.485\,mol/kg + 0.029\,mol/kg + 0.055\,mol/kg = 1.134\,mol/kg$$

Finally, we find for the freezing point depression:

$$\Delta T^{SL} = \sum_i b_i\,K_F = (1.134\,mol/kg)\,(-1.86\,K \cdot kg/mol) = -2.11\,K$$

The freezing temperature of seawater is 2.11 K (= 2.11 °C) lower than that of pure water (0 °C). Accordingly, the freezing point is −2.11 °C.

**Problem 4.4**

The molality of ethylene glycol that is needed to reduce the freezing point from 0 to −20 °C ($\Delta T^{SL} = -20$ K) is:

$$b = \frac{\Delta T^{SL}}{K_F} = \frac{-20\,\text{K}}{-1.86\,\text{K}\cdot\text{kg/mol}} = 10.75\,\text{mol/kg}$$

Thus, we have to add 10.75 mol of ethylene glycol to 1 kg (= 1 L) water. The respective mass is:

$$m = n\,M = (10.75\,\text{mol})\,(62\,\text{g/mol}) = 666.5\,\text{g}$$

and the volume that has to be added to 1 L water is:

$$V = \frac{m}{\rho} = \frac{666.5\,\text{g}}{1.11\,\text{g/cm}^3} = 600.5\,\text{cm}^3$$

**Problem 4.5**

The equation for the osmotic pressure is:

$$\pi = \sum_i c_i R T$$

At first, the sum of the concentrations has to be computed:

$$\sum_i c_i = (0.50 + 0.03 + 1.25 + 0.40 + 3.00 + 0.23 + 0.30)\,\text{mmol/L} = 5.71\,\text{mmol/L}$$

According to the unit of the gas constant, the sum of the concentrations has to be expressed in mol/L:

$$\sum_i c_i = 5.71\times 10^{-3}\,\text{mol/L}$$

With the gas constant ($0.083145$ bar · L/(mol · K)) and the temperature (10 °C = 283.15 K), we can find the osmotic pressure:

$$\pi = \sum_i c_i R T = (5.71\times 10^{-3}\,\text{mol/L})\,(0.083145\,\text{bar}\cdot\text{L/(mol}\cdot\text{K}))\,(283.15\,\text{K}) = 0.134\,\text{bar}$$

**Problem 4.6**

During the reverse osmosis process, water passes through the membrane from the solution side to the side of pure water. Accordingly, the concentration and the osmotic pressure increase on the solution side. The process stops when the osmotic pressure reaches the applied external pressure, in this case 80 bar. The related concentration can be found after rearranging the van't Hoff equation:

$$\sum_i c_i = \frac{\pi}{RT} = \frac{80\,\text{bar}}{(0.083145\,\text{bar}\cdot\text{L}/(\text{mol}\cdot\text{K}))\,(298.15\,\text{K})} = 3.23\,\text{mol/L}$$

It follows from the law of conservation of mass that the number of moles before ($n_1$) and after ($n_2$) the concentration process must be the same. Furthermore, the substance amount can be expressed as product of molar concentration and volume:

$$n_1 = n_2$$
$$c_1 V_1 = c_2 V_2$$

Accordingly, the volume of the concentrated solution after the reverse osmosis process is:

$$V_2 = \frac{c_1 V_1}{c_2} = \frac{(1.12\,\text{mol/L})\,(1\,\text{m}^3)}{3.23\,\text{mol/L}} = 0.347\,\text{m}^3$$

Therefore, the volume of pure water that can be produced from seawater until the osmotic pressure reaches the applied external pressure of 80 bar amounts to:

$$V_{prod} = V_1 - V_2 = 1\,\text{m}^3 - 0.347\,\text{m}^3 = 0.653\,\text{m}^3$$

# Chapter 5

## Problem 5.1

The standard Gibbs energy of reaction can be found from the Gibbs energies of formation by:

$$\Delta_R G^0 = \sum_i \nu_i\, G^0_{f,i}(\text{products}) - \sum_i \nu_i\, G^0_{f,i}(\text{reactants})$$

$$\Delta_R G^0 = -552.8\,\text{kJ/mol} - 528.0\,\text{kJ/mol} - (-1\,129.2\,\text{kJ/mol}) = 48.4\,\text{kJ/mol}$$

Since the equilibrium constant is related to the standard Gibbs energy by:

$$\Delta_R G^0 = -2.303\,R\,T \log K^*$$

we can find the constant after rearranging this equation:

$$\log K^* = -\frac{\Delta_R G^0}{2.303\,R\,T} = -\frac{48.4\,\text{kJ/mol}}{(2.303)\,(8.3145\times 10^{-3}\,\text{kJ}/(\text{mol}\,\text{K}))\,(298.15\,\text{K})} = -8.48$$

$$K^* = 10^{-8.48} \text{ mol}^2/\text{L}^2 = 3.31 \times 10^{-9} \text{ mol}^2/\text{L}^2$$

The unit of the constant follows from the law of mass action.

## Problem 5.2

To solve the problem, we have to compare the reaction quotient with the equilibrium constant. By analogy to the law of mass action, the reaction quotient is given as:

$$Q = \frac{c^2(\text{H}^+)}{c(\text{Pb}^{2+})\, c^{0.5}(\text{O}_2)} = \frac{1 \times 10^{-14} \text{ mol}^2/\text{L}^2}{(1 \times 10^{-8} \text{ mol}/\text{L}) \, (0.0162 \text{ mol}^{0.5}/\text{L}^{0.5})} = 6.17 \times 10^{-5} \text{ mol}^{0.5}/\text{L}^{0.5}$$

Since $Q > K$, the reaction does not proceed spontaneously in the written direction under the given conditions. In contrast, the reverse reaction will proceed spontaneously.

## Problem 5.3

The temperature dependence of the equilibrium constant for a temperature-independent molar standard enthalpy of reaction is given by:

$$\ln\frac{K^*_{T_2}}{K^*_{T_1}} = \frac{\Delta_R H^0}{R}\left(\frac{1}{T_1} - \frac{1}{T_2}\right) = \frac{\Delta_R H^0}{R}\left(\frac{T_2 - T_1}{T_1 T_2}\right)$$

Rearranging gives:

$$\Delta_R H^0 = R \ln\frac{K^*_{T_2}}{K^*_{T_1}}\left(\frac{T_1 T_2}{T_2 - T_1}\right)$$

With the given data $T_1 = 283.15$ K, $K^*_{T_1} = 3.93 \times 10^{-9}$ mol$^2$/L$^2$, $T_2 = 298.15$ K, and $K^*_{T_2} = 3.30 \times 10^{-9}$ mol$^2$/L$^2$, we get:

$$\Delta_R H^0 = 8.3145\frac{\text{J}}{\text{mol}\cdot\text{K}}\ln\frac{3.30 \times 10^{-9} \text{ mol}^2/\text{L}^2}{3.93 \times 10^{-9} \text{ mol}^2/\text{L}^2}\left(\frac{(283.15\text{ K})(298.15\text{ K})}{298.15\text{ K} - 283.15\text{ K}}\right)$$

$$\Delta_R H^0 = (8.3145)(-0.175)(5628.08) \text{ J/mol} = -8189.07 \text{ J/mol} = -8.19 \text{ kJ/mol}$$

The reaction enthalpy is negative. This means that the dissolution of calcium carbonate is an exothermic reaction.

## Problem 5.4

The reaction $CO_{2(aq)} + H_2O \rightleftharpoons 2H^+ + CO_3^{2-}$ is the overall reaction that results from the addition of the elementary reactions according to:

$$CO_{2(aq)} + H_2O \rightleftharpoons H^+ + HCO_3^-$$

$$HCO_3^- \rightleftharpoons H^+ + CO_3^{2-}$$

---

$$CO_{2(aq)} + H_2O \rightleftharpoons 2H^+ + CO_3^{2-}$$

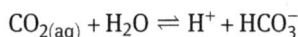

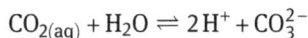

Therefore, the equilibrium constants of the elementary reactions have to be multiplied to find the equilibrium constant of the overall reaction:

$$K^*_{overall} = K_1^* \, K_2^* = (4.4 \times 10^{-7} \text{ mol/L}) \, (4.7 \times 10^{-11} \text{ mol/L}) = 2.1 \times 10^{-17} \text{ mol}^2/\text{L}^2$$

## Chapter 6

### Problem 6.1

The unit of the given Henry constant is $L \cdot \text{bar/mol}$. This unit corresponds to Henry's law written in the form:

$$p = H_{inv} \, c$$

which is the inverse form of the basic equation used here in the book. The Henry constant $H$ for the basic form is therefore:

$$H = \frac{1}{H_{inv}} = \frac{1}{5.495 \text{ L} \cdot \text{bar/mol}} = 0.182 \text{ mol}/(\text{L} \cdot \text{bar})$$

$K_c \, (= c_{aq}/c_g)$ can be found from:

$$K_c = H \, R \, T = (0.182 \text{ mol}/(\text{L} \cdot \text{bar})) \, (0.083145 \text{ bar} \cdot \text{L}/(\text{mol} \cdot \text{K})) \, (298.15 \text{ K}) = 4.51$$

Alternatively, we can find $K_c$ directly from $H_{inv}$ by:

$$K_c = \frac{R \, T}{H_{inv}} = \frac{(0.083145 \text{ bar} \cdot \text{L}/(\text{mol} \cdot \text{K})) \, (298.15 \text{ K})}{5.495 \text{ L} \cdot \text{bar/mol}} = 4.51$$

### Problem 6.2

Since the gas phase consists of pure oxygen (gas-phase mole fraction $y = 1$), the partial pressure equals the total pressure and amounts to 10 bar:

$$p(O_2) = y(O_2) \, p_{total} = 1 \, (10 \text{ bar}) = 10 \text{ bar}$$

The aqueous-phase concentration resulting from Henry's law is:

$c(O_2) = H(O_2) p(O_2) = (1.247 \text{ mol}/(m^3 \cdot \text{bar})) (10 \text{ bar}) = 12.47 \text{ mol}/m^3 = 0.01247 \text{ mol}/L$

Conversion of the molar concentration into the mass concentration gives:

$\rho^*(O_2) = c(O_2) M(O_2) = (0.01247 \text{ mol}/L) (32 \text{ g}/\text{mol}) = 0.399 \text{ g}/L = 399 \text{ mg}/L$

## Problem 6.3

Rearranging equation (6.4) for $\ln H$ gives:

$$\ln H = \ln H_0 + \frac{\Delta H_{sol}}{R} \left( \frac{1}{T_0} - \frac{1}{T_1} \right)$$

By introducing $T = 293.15$ K, $T_0 = 298.15$ K, $H_0 = 3.342 \times 10^{-2}$ mol/(L · bar), $\Delta H_{sol} = -20.79$ kJ/mol, and $R = 8.3145$ J/(mol · K), we get:

$$\ln H = -3.3986 + \frac{-20\,790 \text{ J}/\text{mol}}{8.3145 \text{ J}/(\text{mol} \cdot \text{K})} \left( \frac{1}{298.15 \text{ K}} - \frac{1}{293.15 \text{ K}} \right)$$

$\ln H = -3.3986 - 2500.5 (0.003354 - 0.003411) = -3.3986 + 0.1425 = -3.2561$

The Henry constant at 20 °C is then:

$$H = e^{\ln H} = e^{-3.2561} \text{ mol}/(L \cdot \text{bar}) = 3.854 \times 10^{-2} \text{ mol}/(L \cdot \text{bar})$$

## Problem 6.4

If the $CO_2$ content in the soil atmosphere is 50 times higher compared to the normal atmosphere, the respective partial pressure is:

$$p(CO_2) = 50 (0.000415 \text{ bar}) = 0.02075 \text{ bar}$$

Thus, the concentration in the aqueous phase, given by Henry's law, is:

$c(CO_2) = H(CO_2) p(CO_2)$

$c(CO_2) = (0.0525 \text{ mol}/(L \cdot \text{bar})) (0.02075 \text{ bar}) = 1.09 \times 10^{-3} \text{ mol}/L = 1.09 \text{ mmol}/L$

## Problem 6.5

At first, the distribution constant, $K_c$, of trichloroethylene (TCE) has to be calculated according to:

$K_c(TCE) = H(TCE) R\ T$

$K_c(TCE) = (0.085 \text{ mol}/(L \cdot \text{bar})) (0.083145 \text{ bar} \cdot L/(\text{mol} \cdot K)) (298.15 \text{ K}) = 2.107$

The total masses of trichloroethylene in the flasks can be found by multiplying the initial concentration with the respective volume of the aqueous phase. The volume of the aqueous phase is 0.8 L in flask 1 and 0.2 L in flask 2 and the concentration is 100 mg/L in both cases. Consequently, the total mass, $m_{total}$, is 80 mg in flask 1 and 20 mg in flask 2.

With these values and under consideration of the respective gas volumes (0.2 L in flask 1 and 0.8 L in flask 2), the aqueous-phase concentrations in both flasks can be calculated by means of equation (6.27).

Flask 1:

$$p_{aq}^*(TCE) = \frac{m_{total}(TCE)}{\left(\dfrac{V_g}{K_c(TCE)} + V_{aq}\right)} = \frac{80\,mg}{\dfrac{0.2\,L}{2.107} + 0.8\,L} = 89.39\,mg/L$$

Flask 2:

$$p_{aq}^*(TCE) = \frac{m_{total}(TCE)}{\left(\dfrac{V_g}{K_c(TCE)} + V_{aq}\right)} = \frac{20\,mg}{\dfrac{0.8\,L}{2.107} + 0.2\,L} = 34.50\,mg/L$$

**Problem 6.6**

The minimum $CO_2$ concentration in the water, which can be reached by stripping with air, is the concentration that is in equilibrium with the $CO_2$ partial pressure in the air. At first, the partial pressure of $CO_2$ has to be calculated. The $CO_2$ content is given here as 415 ppm$_v$. This means 415 parts per million parts (volume per volume). To find the percentage (= parts per 100 parts), the given value has to be divided by 10 000. Consequently, the $CO_2$ content is 0.0415 vol% and the mole fraction in the gas phase is $y(CO_2) = 0.000415$ (volume percent divided by 100, for ideal gases). With the total pressure of $p_{total} = 1$ bar, we can calculate the partial pressure:

$$p(CO_2) = y(CO_2)\,p_{total} = (0.000415)\,(1\,bar) = 0.000415\,bar$$

The lowest molar $CO_2$ concentration, which can be reached by stripping with air, follows from Henry's law with the previously calculated partial pressure:

$$c(CO_2) = H(CO_2)\,p(CO_2) = (0.0525\,mol/(L\cdot bar))\,(0.000415\,bar) = 2.18 \times 10^{-5}\,mol/L$$

$$c(CO_2) = 0.0218\,mmol/L$$

Finally, the molar concentration has to be converted into the mass concentration according to:

$$p^*(CO_2) = c(CO_2)\,M(CO_2) = (0.0218\,mmol/L)\,(44\,mg/mmol) = 0.959\,mg/L \approx 0.96\,mg/L$$

## Chapter 7

### Problem 7.1

The proton concentration is given by:

$$c(H^+) = 10^{-pH} \, mol/L = 10^{-8.5} \, mol/L = 3.16 \times 10^{-9} \, mol/L$$

The pOH value can be found as follows:

$$pOH = pK_w - pH = 14 - 8.5 = 5.5$$

and the resulting hydroxide ion concentration is:

$$c(OH^-) = 10^{-pOH} \, mol/L = 10^{-5.5} \, mol/L = 3.16 \times 10^{-6} \, mol/L$$

### Problem 7.2

First, the initial molar concentrations have to be calculated:

$$c_0 = \frac{\rho_0^*}{M} = \frac{5 \, g/L}{128.6 \, g/mol} = 3.89 \times 10^{-2} \, mol/L$$

and

$$c_0 = \frac{\rho_0^*}{M} = \frac{0.5 \, g/L}{128.6 \, g/mol} = 3.89 \times 10^{-3} \, mol/L$$

The $pK_a$ of 9.4 characterizes 4-chlorophenol as a weak acid. Therefore, the pH can be calculated in a simplified manner according to:

$$pH = 0.5 \, (pK_a - \log c_0(4\text{–}CP))$$

The results for the different concentrations are:

$$pH = (0.5) \, (9.4 + 1.41) = 5.405 \approx 5.4 \quad \text{for} \quad \rho_0^* = 5 \, g/L$$

and

$$pH = (0.5) \, (9.4 + 2.41) = 5.905 \approx 5.9 \quad \text{for} \quad \rho_0^* = 0.5 \, g/L$$

### Problem 7.3

Given that phenol is a weak acid, we can try to calculate the pH without consideration of the autoprotolysis of water by using the simplified equation (7.64) with $c_0 = 1 \times 10^{-5} \, mol/L$:

$$\text{pH} = 0.5\,(\text{p}K_a - \log c_0(\text{phenol})) = (0.5)\,(9.9 + 5) = 7.45$$

The result is not plausible. Since phenol is an acid, the resulting pH of the solution cannot be higher than pH = 7 or, in other words, dissolution of an acid cannot lead to an increase of the pH in comparison to the pH of pure water. Obviously, the simplified equation fails in this case and the exact cubic equation:

$$c^3(\text{H}^+) + K_a\,c^2(\text{H}^+) - (K_a\,c_0(\text{phenol}) + K_w)\,c(\text{H}^+) - K_w\,K_a = 0$$

has to be used to find the pH. In principle, this equation has to be solved by an iteration procedure. However, we can simplify this equation if we consider the boundary condition that the pH can be at maximum 7 (= minimum proton concentration $c(\text{H}^+) = 1 \times 10^{-7}$ mol/L). Furthermore, we expect a pH not much lower than this maximum pH (weak acid, low concentration). Introducing the minimum proton concentration, the acidity constant ($K_a = 10^{-9.9}$ mol/L = $1.259 \times 10^{-10}$ mol/L), the ion product of water ($K_w = 1 \times 10^{-14}$ mol$^2$/L$^2$), and the initial concentration of phenol ($1 \times 10^{-5}$ mol/L) into the cubic equation, we find:

$$1 \times 10^{-21} \text{ mol}^3/\text{L}^3 + 1.259 \times 10^{-24} \text{ mol}^3/\text{L}^3 - 1.126 \times 10^{-21} \text{ mol}^3/\text{L}^3$$
$$- 1.259 \times 10^{-24} \text{ mol}^3/\text{L}^3 = 0$$

Obviously, only the first and the third term are relevant, because the values of the second and the fourth are much lower and, furthermore, compensate for each other due to the opposite signs. Therefore, we can simplify the cubic equation to:

$$c^3(\text{H}^+) = (K_a\,c_0(\text{phenol}) + K_w)\,c(\text{H}^+)$$

After dividing both sides of the equation by $c(\text{H}^+)$, we find:

$$c^2(\text{H}^+) = K_a\,c_0(\text{phenol}) + K_w$$

and

$$c(\text{H}^+) = \sqrt{K_a\,c_0(\text{phenol}) + K_w}$$

Thus, the proton concentration is:

$$c(\text{H}^+) = \sqrt{(1.259 \times 10^{-10} \text{ mol/L})\,(1 \times 10^{-5} \text{ mol/L}) + 1 \times 10^{-14} \text{ mol}^2/\text{L}^2}$$
$$c(\text{H}^+) = 1.061 \times 10^{-7} \text{ mol/L}$$

and finally, the pH is:

$$\text{pH} = -\log c(\text{H}^+) = -\log(1.061 \times 10^{-7}) = 6.97$$

This result is plausible, since the pH is lower than 7 (see previous discussion), but it also shows that addition of a low concentration of a very weak acid causes only a marginal shift in pH compared to the pH of pure water.

**Problem 7.4**

a. The $pK_b$ of the conjugate base can be found according to:

$$pK_b = pK_w - pK_a = 14 - 4.58 = 9.42$$

The related basicity constant is:

$$K_b = 10^{-pK_b}\,\text{mol/L} = 10^{-9.42}\,\text{mol/L} = 3.8 \times 10^{-10}\,\text{mol/L}$$

b. The ratio of protonated and neutral (unprotonated) aniline (abbreviations $AnH^+$ and $An$) can be found from the law of mass action for the protonated form:

$$K_a(AnH^+) = \frac{c(H^+)\,c(An)}{c(AnH^+)}$$

With:

$$c(H^+) = 10^{-pH}\,\text{mol/L} = 10^{-5.5}\,\text{mol/L} = 3.16 \times 10^{-6}\,\text{mol/L}$$

and

$$K_a = 10^{-pK_a}\,\text{mol/L} = 10^{-4.58}\,\text{mol/L} = 2.63 \times 10^{-5}\,\text{mol/L}$$

we obtain:

$$\frac{c(AnH^+)}{c(An)} = \frac{c(H^+)}{K_a(AnH^+)} = \frac{3.16 \times 10^{-6}\,\text{mol/L}}{2.63 \times 10^{-5}\,\text{mol/L}} = 0.12$$

Thus, the ratio of protonated and neutral aniline at pH = 5.5 is 0.12.

c. The given reaction equation describes a reaction that is inverse to the dissociation of the protonated form for which the acidity constant is defined. Since the constant of a reverse reaction is the reciprocal of the constant of a forward reaction, it holds:

$$K_{reverse} = \frac{1}{K_a(AnH^+)} = \frac{1}{2.63 \times 10^{-5}\,\text{mol/L}} = 3.8 \times 10^4\,\text{L/mol}$$

**Problem 7.5**

Sodium carbonate $Na_2CO_3$ is a salt consisting of the cation of the very strong base NaOH and the anion of the weak acid $CO_{2(aq)}$. Therefore, it can be expected that the solution is alkaline. The $OH^-$ producing reaction is:

$$CO_3^{2-} + H_2O \rightleftharpoons HCO_3^- + OH^-$$

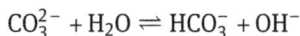

Under the expected pH conditions, the further reaction of $HCO_3^-$ to $CO_2$ can be neglected.

The $pK_b$ for the carbonate ion can be estimated from the $pK_a$ of the conjugate acid hydrogencarbonate by:

$$pK_b = pK_w - pK_a = 14 - 10.33 = 3.67$$

The corresponding basicity constant is:

$$K_b = 10^{-pK_b} \text{ mol/L} = 10^{-3.67} \text{ mol/L} = 2.14 \times 10^{-4} \text{ mol/L}$$

Since $pK_b = 3.67$ indicates a relatively strong base, first the equation:

$$c(OH^-) = -\frac{K_b}{2} \pm \sqrt{\frac{K_b^2}{4} + K_b\,c_0(\text{base})}$$

will be used. Later, the simplified version:

$$pOH = 0.5\,(pK_b - \log c_0)$$

will be proved.

With the initial concentration $c_0(CO_3^{2-}) = c_0\ (Na_2CO_3) = 2$ mol/L, we have:

$$c(OH^-) = -\frac{2.14 \times 10^{-4} \text{ mol/L}}{2}$$

$$\pm \sqrt{\frac{4.58 \times 10^{-8} \text{ mol}^2/L^2}{4} + (2.14 \times 10^{-4} \text{ mol/L})\,(2 \text{ mol/L})}$$

$$c(OH^-) = -1.07 \times 10^{-4} \text{ mol/L} \pm \sqrt{1.145 \times 10^{-8} \text{ mol}^2/L^2 + 4.28 \times 10^{-4} \text{ mol}^2/L^2}$$

$$c(OH^-) = -1.07 \times 10^{-4} \text{ mol/L} + 2.07 \times 10^{-2} \text{ mol/L} = 2.06 \times 10^{-2} \text{ mol/L}$$

$$\text{(negative result canceled)}$$

The corresponding pOH is:

$$pOH = -\log c(OH^-) = 1.69$$

and the pH is:

$$pH = pK_w - pOH = 14 - 1.69 = 12.31$$

By using the simplified equation, we receive:

$$pOH = 0.5\,(pK_b - \log c_0) = 0.5(3.67 - 0.30) = 1.69$$

and the related pH is:

$$pH = pK_w - pOH = 14 - 1.69 = 12.31$$

We can see that in this case there is no difference between the results of the different solution methods.

## Problem 7.6

During the dissolution of iron(III) chloride, hydrated iron(III) ions are formed that act as an acid according to:

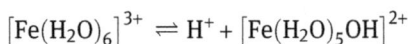

$$[Fe(H_2O)_6]^{3+} \rightleftharpoons H^+ + [Fe(H_2O)_5OH]^{2+}$$

The $pK_a$ of 2.2 indicates that the acid is relatively strong and therefore the quadratic equation:

$$c(H^+) = -\frac{K_a}{2} \pm \sqrt{\frac{K_a^2}{4} + K_a\, c_0(\text{acid})}$$

has to be used to find the pH. The initial concentration is:

$$c_0(FeCl_3) = c_0(Fe^{3+}) = c_0([Fe(H_2O)_6]^{3+}) = 0.05\,\text{mol/L}$$

and the constant $K_a$ is given by:

$$K_a = 10^{-pK_a}\,\text{mol/L} = 10^{-2.2}\,\text{mol/L} = 6.31 \times 10^{-3}\,\text{mol/L}$$

For the concentration of protons, we find:

$$c(H^+) = -\frac{6.31 \times 10^{-3}\,\text{mol/L}}{2}$$

$$\pm \sqrt{\frac{3.98 \times 10^{-5}\,\text{mol}^2/\text{L}^2}{4} + (6.31 \times 10^{-3}\,\text{mol/L})(0.05\,\text{mol/L})}$$

$$c(H^+) = -3.155 \times 10^{-3}\,\text{mol/L} \pm \sqrt{9.95 \times 10^{-6}\,\text{mol}^2/\text{L}^2 + 3.155 \times 10^{-4}\,\text{mol}^2/\text{L}^2}$$

$$c(H^+) = -3.155 \times 10^{-3}\,\text{mol/L} + 1.804 \times 10^{-2}\,\text{mol/L} = 0.0149$$

$$\text{(negative result canceled)}$$

and the corresponding pH is:

$$pH = -\log c(H^+) = -\log(0.0149) = 1.83$$

## Problem 7.7

The degree of protolysis is given by:

$$\alpha = \frac{1}{10^{pK_a - pH} + 1}$$

For $pH = pK_a + 1$, we find:

$$\alpha = \frac{1}{10^{pK_a - (pK_a + 1)} + 1} = \frac{1}{10^{-1} + 1} = \frac{1}{1.1} = 0.909 \approx 0.91$$

The degree of protolysis at $pH = pK_a + 1$ is $\alpha = 0.91$ or 91%.

## Problem 7.8

a.  The pH of the buffer system can be calculated by:

$$pH = pK_a + \log \frac{c(\text{buffer base})}{c(\text{buffer acid})}$$

Therefore, the pH of the buffer system is:

$$pH = 7.12 + \log \frac{0.1 \, \text{mol/L}}{0.1 \, \text{mol/L}} = 7.12$$

b.  Since during dilution, the ratio of buffer base and buffer acid remains constant, the pH will not change:

$$pH = 7.12 + \log \frac{0.025 \, \text{mol/L}}{0.025 \, \text{mol/L}} = 7.12$$

c.  The buffer effect is here based on the consumption of the added protons by the buffer base under formation of an equivalent amount of buffer acid. Accordingly, the change of pH in the original buffer solution after addition of 0.005 mol/L of the very strong acid HCl (complete dissociation, acid concentration = proton concentration) can be calculated as follows:

$$pH = pK_a + \log \frac{c(\text{buffer base}) - c(\text{H}^+)}{c(\text{buffer acid}) + c(\text{H}^+)}$$

$$pH = 7.12 + \log \frac{0.1 \, \text{mol/L} - 0.005 \, \text{mol/L}}{0.1 \, \text{mol/L} + 0.005 \, \text{mol/L}} = 7.12 + \log \frac{0.095}{0.105} = 7.12 - 0.04 = 7.08$$

There is only a slight change in pH from 7.12 to 7.08.
In the case of the diluted solution, we find:

$$pH = 7.12 + \log \frac{0.025\,\text{mol/L} - 0.005\,\text{mol/L}}{0.025\,\text{mol/L} + 0.005\,\text{mol/L}} = 7.12 + \log \frac{0.02}{0.03} = 7.12 - 0.18 = 6.94$$

The pH change is here a bit higher, because the relative impact of HCl on the lower buffer base and buffer acid concentrations is higher.

d.  The pH of the 0.005 M solution of the very strong (completely dissociated) acid HCl alone would be:

$$pH = -\log c_0(\text{acid}) = -\log 0.005 = 2.3$$

Comparison of the results given in b. or c. with that in d. clearly demonstrates the strong buffer effect of the given buffer system.

## Problem 7.9

The constants necessary for the calculations are:

$$K_{a1} = 10^{-pK_{a1}}\,\text{mol/L} = 10^{-6.36}\,\text{mol/L} = 4.37 \times 10^{-7}\,\text{mol/L}$$

$$K_{a2} = 10^{-pK_{a2}}\,\text{mol/L} = 10^{-10.33}\,\text{mol/L} = 4.68 \times 10^{-11}\,\text{mol/L}$$

$$K_{a1}\,K_{a2} = (4.37 \times 10^{-7}\,\text{mol/L})\,(4.68 \times 10^{-11}\,\text{mol/L}) = 2.05 \times 10^{-17}\,\text{mol}^2/\text{L}^2$$

The fractions can be calculated according to:

$$f(CO_2) = \frac{c(CO_2)}{c(DIC)} = \frac{1}{1 + \dfrac{K_{a1}}{c(H^+)} + \dfrac{K_{a1}\,K_{a2}}{c^2(H^+)}}$$

$$f(CO_2) = \frac{1}{1 + \dfrac{4.37 \times 10^{-7}\,\text{mol/L}}{1 \times 10^{-7}\,\text{mol/L}} + \dfrac{2.05 \times 10^{-17}\,\text{mol}^2/\text{L}^2}{1 \times 10^{-14}\,\text{mol}^2/\text{L}^2}}$$

$$f(CO_2) = \frac{1}{1 + 4.37 + 2.05 \times 10^{-3}} = 0.186$$

$$f(HCO_3^-) = \frac{c(HCO_3^-)}{c(DIC)} = \frac{1}{\dfrac{c(H^+)}{K_{a1}} + 1 + \dfrac{K_{a2}}{c(H^+)}}$$

$$f(HCO_3^-) = \frac{1}{\dfrac{1 \times 10^{-7}\,\text{mol/L}}{4.37 \times 10^{-7}\,\text{mol/L}} + 1 + \dfrac{4.68 \times 10^{-11}\,\text{mol/L}}{1 \times 10^{-7}\,\text{mol/L}}}$$

$$f(HCO_3^-) = \frac{1}{0.229 + 1 + 4.68 \times 10^{-4}} = 0.813$$

$$f(CO_3^{2-}) = \frac{c(CO_3^{2-})}{c(DIC)} = \frac{1}{\frac{c^2(H^+)}{K_{a1}K_{a2}} + \frac{c(H^+)}{K_{a2}} + 1}$$

$$f(CO_3^{2-}) = \frac{1}{\frac{1 \times 10^{-14} \text{ mol}^2/L^2}{2.05 \times 10^{-17} \text{ mol}^2/L^2} + \frac{1 \times 10^{-7} \text{ mol}/L}{4.68 \times 10^{-11} \text{ mol}/L} + 1}$$

$$f(CO_3^{2-}) = \frac{1}{487.8 + 2136.8 + 1} = 0.00038$$

At pH = 7, $CO_2$ and $HCO_3^-$ mainly contribute to the total inorganic carbon content of the water, whereas the fraction of $CO_3^{2-}$ is negligibly small.

**Problem 7.10**

At pH = 7.7, the $BNC_{8.2}$ can be set equal to the $CO_2$ concentration and the $ANC_{4.3}$ equals the $HCO_3^-$ concentration, whereas the $CO_3^{2-}$ concentration is very small and can be neglected in the DIC balance (see Figure 7.4). Accordingly, the concentrations are:

$$c(CO_2) = BNC_{8.2} = 0.08 \text{ mmol/L}$$
$$c(HCO_3^-) = ANC_{4.3} = 2.1 \text{ mmol/L}$$
$$c(CO_3^{2-}) \approx 0 \text{ mmol/L}$$
$$c(DIC) = c(CO_2) + c(HCO_3^-) + c(CO_3^{2-})$$
$$c(DIC) = 0.08 \text{ mmol/L} + 2.1 \text{ mmol/L} + 0 \text{ mmol/L} = 2.18 \text{ mmol/L}$$

Alternatively, we can calculate $c(DIC)$ from the equation:

$$c(DIC) = m - p$$

Under consideration of the relationships:

$$ANC_{4.3} = m$$

and

$$BNC_{8.2} = -p$$

we obtain:

$$c(DIC) = m - p = 2.1 \text{ mmol/L} - (-0.08 \text{ mmol/L}) = 2.18 \text{ mmol/L}$$

## Problem 7.11

A $CO_2$ concentration of 600 ppm equals 0.06 vol%. The corresponding mole fraction is $y = 0.0006$ and the partial pressure can be calculated by:

$$p(CO_2) = y(CO_2) \, p_{total} = 0.0006 \, (1 \text{ bar}) = 0.0006 \text{ bar}$$

With equation (7.172), we find:

$$c(H^+) = \sqrt{H \, K_{a1} \, p(CO_2)} = \sqrt{(0.033 \text{ mol}/(L \cdot bar)) \, (4.4 \times 10^{-7} \text{ mol}/L)(0.0006 \text{ bar})}$$

$$c(H^+) = 2.95 \times 10^{-6} \text{ mol}/L$$

$$pH = 5.53$$

As expected, a further increase of the $CO_2$ concentration in the atmosphere would decrease the pH of rain water.

## Problem 7.12

The species distribution for $pH = 8.2$ and $m = 1.05 \times 10^{-3}$ mol/L can be found from the relationship between $m$ and $c(DIC)$. Rearranging equation (7.176) for $c(DIC)$ gives:

$$c(DIC) = \frac{m - c(OH^-) + c(H^+)}{f_1 + 2f_2}$$

$f_1$ and $f_2$ can be calculated from the equilibrium constants ($K_{a1} = 4.11 \times 10^{-7}$ mol/L, $K_{a2} = 4.21 \times 10^{-11}$ mol/L) and the proton concentration ($c(H^+) = 10^{-pH}$ mol/L $= 6.31 \times 10^{-9}$ mol/L):

$$f_1 = f(HCO_3^-) = \frac{1}{\frac{c(H^+)}{K_{a1}} + 1 + \frac{K_{a2}}{c(H^+)}} = \frac{1}{\frac{6.31 \times 10^{-9} \text{ mol}/L}{4.11 \times 10^{-7} \text{ mol}/L} + 1 + \frac{4.21 \times 10^{-11} \text{ mol}/L}{6.31 \times 10^{-9} \text{ mol}/L}} = 0.978$$

$$f_2 = f(CO_3^{2-}) = \frac{1}{\frac{c^2(H^+)}{K_{a1} K_{a2}} + \frac{c(H^+)}{K_{a2}} + 1} = \frac{1}{\frac{3.98 \times 10^{-17} \text{ mol}^2/L^2}{1.73 \times 10^{-17} \text{ mol}^2/L^2} + \frac{6.31 \times 10^{-9} \text{ mol}/L}{4.21 \times 10^{-11} \text{ mol}/L} + 1} = 0.0065$$

The $OH^-$ concentration can be found from the dissociation constant of water:

$$c(OH^-) = \frac{K_w}{c(H^+)} = \frac{6.82 \times 10^{-15} \text{ mol}^2/L^2}{6.31 \times 10^{-9} \text{ mol}/L} = 1.08 \times 10^{-6} \text{ mol}/L$$

The DIC concentration is then:

$$c(\text{DIC}) = \frac{m - c(\text{OH}^-) + c(\text{H}^+)}{f_1 + 2f_2}$$

$$= \frac{1.05 \times 10^{-3} \text{ mol/L} - 1.08 \times 10^{-6} \text{ mol/L} + 6.31 \times 10^{-9} \text{ mol/L}}{0.978 + 2(0.0065)}$$

$$c(\text{DIC}) = 1.06 \times 10^{-3} \text{ mol/L}$$

With $c(\text{DIC})$, the species concentrations can be calculated:

$$c(\text{HCO}_3^-) = f_1\, c(\text{DIC}) = (0.978)\,(1.06 \times 10^{-3} \text{ mol/L}) = 1.04 \times 10^{-3} \text{ mol/L}$$

$$c(\text{CO}_3^{2-}) = f_2\, c(\text{DIC}) = (0.0065)\,(1.06 \times 10^{-3} \text{ mol/L}) = 6.89 \times 10^{-6} \text{ mol/L}$$

$$c(\text{CO}_2) = f(\text{CO}_2)\, c(\text{DIC}) = \frac{1}{1 + \dfrac{K_{a1}}{c(\text{H}^+)} + \dfrac{K_{a1} K_{a2}}{c^2(\text{H}^+)}} c(\text{DIC})$$

$$c(\text{CO}_2) = \frac{1}{1 + \dfrac{4.11 \times 10^{-7} \text{ mol/L}}{6.31 \times 10^{-9} \text{ mol/L}} + \dfrac{1.73 \times 10^{-17} \text{ mol}^2/\text{L}^2}{3.98 \times 10^{-17} \text{ mol}^2/\text{L}^2}} 1.06 \times 10^{-3} \text{ mol/L}$$

$$= 1.59 \times 10^{-5} \text{ mol/L}$$

Verification of the results:

$$c(\text{DIC}) = c(\text{CO}_2) + c(\text{HCO}_3^-) + c(\text{CO}_3^{2-})$$

$$c(\text{DIC}) = 1.59 \times 10^{-5} \text{ mol/L} + 1.04 \times 10^{-3} \text{ mol/L} + 6.89 \times 10^{-6} \text{ mol/L} = 1.06 \times 10^{-3} \text{ mol/L}$$

The same calculation for pH = 10.5 ($c(\text{H}^+) = 3.16 \times 10^{-11}$ mol/L) gives:

$$f_1 = \frac{1}{\dfrac{c(\text{H}^+)}{K_{a1}} + 1 + \dfrac{K_{a2}}{c(\text{H}^+)}} = \frac{1}{\dfrac{3.16 \times 10^{-11} \text{ mol/L}}{4.11 \times 10^{-7} \text{ mol/L}} + 1 + \dfrac{4.21 \times 10^{-11} \text{ mol/L}}{3.16 \times 10^{-11} \text{ mol/L}}} = 0.429$$

$$f_2 = \frac{1}{\dfrac{c^2(\text{H}^+)}{K_{a1} K_{a2}} + \dfrac{c(\text{H}^+)}{K_{a2}} + 1} = \frac{1}{\dfrac{9.99 \times 10^{-22} \text{ mol}^2/\text{L}^2}{1.73 \times 10^{-17} \text{ mol}^2/\text{L}^2} + \dfrac{3.16 \times 10^{-11} \text{ mol/L}}{4.21 \times 10^{-11} \text{ mol/L}} + 1} = 0.571$$

$$c(\text{OH}^-) = \frac{K_w}{c(\text{H}^+)} = \frac{6.82 \times 10^{-15} \text{ mol}^2/\text{L}^2}{3.16 \times 10^{-11} \text{ mol/L}} = 2.16 \times 10^{-4} \text{ mol/L}$$

$$c(\text{DIC}) = \frac{m - c(\text{OH}^-) + c(\text{H}^+)}{f_1 + 2f_2}$$

$$= \frac{1.05 \times 10^{-3} \text{ mol/L} - 2.16 \times 10^{-4} \text{ mol/L} + 3.16 \times 10^{-11} \text{ mol/L}}{0.429 + 2(0.571)}$$

$$c(\text{DIC}) = 5.31 \times 10^{-4} \text{ mol/L}$$

$$c(\text{HCO}_3^-) = f_1\, c(\text{DIC}) = (0.429)\,(5.31 \times 10^{-4} \text{ mol/L}) = 2.28 \times 10^{-4} \text{ mol/L}$$

$$c(\text{CO}_3^{2-}) = f_2\, c(\text{DIC}) = (0.571)\,(5.31 \times 10^{-4} \text{ mol/L}) = 3.03 \times 10^{-4} \text{ mol/L}$$

$$c(\text{CO}_2) = f(\text{CO}_2)\, c(\text{DIC}) = \frac{1}{1 + \dfrac{K_{a1}}{c(\text{H}^+)} + \dfrac{K_{a1}\,K_{a2}}{c^2(\text{H}^+)}} c(\text{DIC})$$

$$c(\text{CO}_2) = \frac{1}{1 + \dfrac{4.11 \times 10^{-7} \text{ mol/L}}{3.16 \times 10^{-11} \text{ mol/L}} + \dfrac{1.73 \times 10^{-17} \text{ mol}^2/\text{L}^2}{9.99 \times 10^{-22} \text{ mol}^2/\text{L}^2} + 1} 5.31 \times 10^{-4} \text{ mol/L}$$

$$= 1.75 \times 10^{-8} \text{ mol/L}$$

Verification of the results:

$$c(\text{DIC}) = c(\text{CO}_2) + c(\text{HCO}_3^-) + c(\text{CO}_3^{2-})$$

$$c(\text{DIC}) = 1.75 \times 10^{-8} \text{ mol/L} + 2.28 \times 10^{-4} \text{ mol/L} + 3.03 \times 10^{-4} \text{ mol/L} = 5.31 \times 10^{-4} \text{ mol/L}$$

## Chapter 8

### Problem 8.1

The reaction equation for the dissolution of calcium hydroxide is:

$$\text{Ca(OH)}_2 \rightleftharpoons \text{Ca}^{2+} + 2\,\text{OH}^-$$

and the law of mass action reads:

$$K_{sp} = c(\text{Ca}^{2+})\, c^2(\text{OH}^-)$$

It follows from the stoichiometry of the reaction that:

$$c_S = c(\text{Ca}^{2+}) = 0.5\, c(\text{OH}^-)$$

Accordingly, the law of mass action can be rewritten as:

$$K_{sp} = 4\, c_S^3$$

and the saturation concentration can be found by:

$$c_S = \sqrt[3]{\frac{K_{sp}}{4}}$$

Considering the relationship between $c_s$ and $c(OH^-)$, we get:

$$c(OH^-) = 2\sqrt[3]{\frac{K_{sp}}{4}} = 2\sqrt[3]{\frac{5 \times 10^{-6}\,\text{mol}^3/\text{L}^3}{4}} = 2\sqrt[3]{1.25 \times 10^{-6}\,\text{mol}^3/\text{L}^3}$$

$$= 2.15 \times 10^{-2}\,\text{mol/L}$$

The corresponding proton concentration results from the ion product of water:

$$c(H^+) = \frac{K_w}{c(OH^-)} = \frac{1 \times 10^{-14}\,\text{mol}^2/\text{L}^2}{2.15 \times 10^{-2}\,\text{mol/L}} = 4.65 \times 10^{-13}\,\text{mol/L}$$

Finally, the pH is:

$$pH = -\log c(H^+) = -\log(4.65 \times 10^{-13}) = 12.33$$

## Problem 8.2

The reaction equation for the dissolution of nickel hydroxide reads:

$$Ni(OH)_2 \rightleftharpoons Ni^{2+} + 2OH^-$$

and the value of the solubility product can be calculated from the solubility exponent according to:

$$K_{sp} = 10^{-pK_{sp}}\,\text{mol}^3/\text{L}^3 = 10^{-13.8}\,\text{mol}^3/\text{L}^3 = 1.58 \times 10^{-14}\,\text{mol}^3/\text{L}^3$$

The molar concentration of the limiting value for $Ni^{2+}$ is:

$$c(Ni^{2+}) = \frac{\rho^*(Ni^{2+})}{M(Ni^{2+})} = \frac{0.5\,\text{mg/L}}{58.7\,\text{mg/mmol}} = 8.52 \times 10^{-3}\,\text{mmol/L} = 8.52 \times 10^{-6}\,\text{mol/L}$$

This concentration has to be inserted into the law of mass action to find the related $OH^-$ concentration:

$$K_{sp} = c(Ni^{2+})\,c^2(OH^-)$$

$$c(OH^-) = \sqrt{\frac{K_{sp}}{c(Ni^{2+})}} = \sqrt{\frac{1.58 \times 10^{-14}\,\text{mol}^3/\text{L}^3}{8.52 \times 10^{-6}\,\text{mol/L}}} = 4.31 \times 10^{-5}\,\text{mol/L}$$

The corresponding $H^+$ concentration can be derived from the ion product of water:

$$c(H^+) = \frac{K_w}{c(OH^-)} = \frac{1 \times 10^{-14}\,\text{mol}^2/\text{L}^2}{4.31 \times 10^{-5}\,\text{mol/L}} = 2.32 \times 10^{-10}\,\text{mol/L}$$

and the pH is:

$$pH = -\log c(H^+) = -\log(2.32 \times 10^{-10}) = 9.63$$

This is the minimum pH that is necessary to receive a nickel concentration not higher than $8.52 \times 10^{-6}$ mol/L (= 0.5 mg/L). Since the solubility product is constant, a higher pH value would lead to a lower nickel concentration.

Alternatively, we can also compute the pH from the pOH according to:

$$pOH = -\log c(OH^-) = -\log(4.31 \times 10^{-5}) = 4.37$$

$$pH = pK_w - pOH = 14 - 4.37 = 9.63$$

## Problem 8.3

The reaction equation for the cadmium hydroxide dissolution reads:

$$Cd(OH)_2 \rightleftharpoons Cd^{2+} + 2\,OH^-$$

The allowed molar residual (equilibrium) concentration is:

$$c(Cd^{2+}) = \frac{\rho^*(Cd^{2+})}{M(Cd^{2+})} = \frac{0.1\,mg/L}{112.4\,mg/mmol} = 8.90 \times 10^{-4}\,mmol/L = 8.90 \times 10^{-7}\,mol/L$$

The minimum $OH^-$ concentration that is necessary to meet the residual $Cd^{2+}$ concentration can be found from the solubility product according to:

$$K_{sp} = c(Cd^{2+})\,c^2(OH^-)$$

$$c(OH^-) = \sqrt{\frac{K_{sp}}{c(Cd^{2+})}} = \sqrt{\frac{1.6 \times 10^{-14}\,mol^3/L^3}{8.9 \times 10^{-7}\,mol/L}} = 1.34 \times 10^{-4}\,mol/L$$

Finally, the respective proton concentration and the pH are:

$$c(H^+) = \frac{K_w}{c(OH^-)} = \frac{1 \times 10^{-14}\,mol^2/L^2}{1.34 \times 10^{-4}\,mol/L} = 7.46 \times 10^{-11}\,mol/L$$

$$pH = -\log c(H^+) = -\log(7.46 \times 10^{-11}) = 10.13$$

The same value is found by the following calculation:

$$pOH = -\log c(OH^-) = -\log(1.34 \times 10^{-4}) = 3.87$$

$$pH = pK_w - pOH = 14 - 3.87 = 10.13$$

It follows from the result that pH = 9 is not high enough to attain a $Cd^{2+}$ concentration not higher than 0.1 mg/L. A pH of at least 10.13 is necessary.

**Problem 8.4**

To solve the problem, the products of the actually measured concentrations have to be compared with the respective solubility products. To do that, at first the solubility products and the concentration of $OH^-$ have to be calculated. The solubility products can be found from the solubility exponents by:

$$K_{sp} = 10^{-pK_{sp}}$$

The respective solubility products are:

$$K_{sp}(MgCO_3) = c(Mg^{2+})\,c(CO_3^{2-}) = 10^{-5.2}\,mol^2/L^2 = 6.31 \times 10^{-6}\,mol^2/L^2$$

$$K_{sp}(CaCO_3) = c(Ca^{2+})\,c(CO_3^{2-}) = 10^{-8.5}\,mol^2/L^2 = 3.16 \times 10^{-9}\,mol^2/L^2$$

$$K_{sp}(Mg(OH)_2) = c(Mg^{2+})\,c^2(OH^-) = 10^{-11.3}\,mol^3/L^3 = 5.01 \times 10^{-12}\,mol^3/L^3$$

$$K_{sp}(Ca(OH)_2) = c(Ca^{2+})\,c^2(OH^-) = 10^{-5.3}\,mol^3/L^3 = 5.01 \times 10^{-6}\,mol^3/L^3$$

The concentration of the hydroxide ions can be calculated from the pH according to:

$$pOH = pK_w - pH = 14 - 11 = 3$$

$$c(OH^-) = 10^{-pOH}\,mol/L = 1 \times 10^{-3}\,mol/L$$

Now, all data necessary to compare $Q$ and $K_{sp}$ are available.
For $MgCO_3$ and $Mg(OH)_2$, we find with the measured concentrations:

$$Q(MgCO_3) = c(Mg^{2+})\,c(CO_3^{2-}) = (0.5 \times 10^{-3}\,mol/L)\,(2 \times 10^{-3}\,mol/L)$$

$$= 1 \times 10^{-6}\,mol^2/L^2$$

$$Q(MgCO_3) < K_{sp}(MgCO_3)$$

$$Q(Mg(OH)_2) = c(Mg^{2+})\,c^2(OH^-) = (0.5 \times 10^{-3}\,mol/L)\,(1 \times 10^{-6}\,mol/L)$$

$$= 5 \times 10^{-10}\,mol^3/L^3$$

$$Q(Mg(OH)_2) > K_{sp}(Mg(OH)_2)$$

It follows from the results that $Mg(OH)_2$ is expected to precipitate, whereas $MgCO_3$ will not precipitate.

Under the same conditions ($c(Ca^{2+}) = 0.5 \times 10^{-3}\,mol/L$, $c(CO_3^{2-}) = 2 \times 10^{-3}\,mol/L$, pH = 11), we find with $Q(CaCO_3) = 1 \times 10^{-6}\,mol^2/L^2$ and $Q(Ca(OH)_2) = 5 \times 10^{-10}\,mol^3/L^3$ that calcium carbonate will precipitate whereas calcium hydroxide will not, because

$$Q(CaCO_3) > K_{sp}(CaCO_3) \text{ and } Q(Ca(OH)_2) < K_{sp}(Ca(OH)_2).$$

Thus, the situation is contrary to that which we have found for magnesium.

## Chapter 9

### Problem 9.1

The $CO_2$ concentration can be calculated by means of the Tillmans equation:

$$c(CO_2) = \frac{K_T}{f_T} c^2(HCO_3^-) \, c(Ca^{2+})$$

$K_T$ is given by:

$$K_T = 10^{\log K_T} \, L^2/mol^2 = 10^{4.383} \, L^2/mol^2 = 2.415 \times 10^4 \, L^2/mol^2$$

The activity coefficient $f_T$ can be found from the ionic strength by means of the Güntelberg equation:

$$\log \gamma_1 = -0.5 \frac{\sqrt{\frac{I}{mol/L}}}{1+1.4\sqrt{\frac{I}{mol/L}}} = -0.5 \frac{\sqrt{0.01}}{1+1.4\sqrt{0.01}} = -0.5 \frac{0.1}{1+0.14} = -0.0439$$

$$\gamma_1 = 10^{\log \gamma_1} = 10^{-0.0439} = 0.904$$

$$f_T = \frac{1}{\gamma_1^6} = \frac{1}{0.904^6} = 1.832$$

Introducing the data into the Tillmans equation gives:

$$c(CO_2) = \frac{2.415 \times 10^4 \, L^2/mol^2}{1.832} (3.8 \times 10^{-3} mol/L)^2 (1.4 \times 10^{-3} \, mol/L)$$

$$= 2.66 \times 10^{-4} \, mol/L$$

This $CO_2$ concentration and the related $HCO_3^-$ concentration represent one point of the Tillmans curve with the curve parameter $c(HCO_3^-) - 2\,c(Ca^{2+}) = 0.001$ mol/L. The corresponding pH can be found from the law of mass action of the first dissociation step of the carbonic acid with $f_{a1} = \gamma_1$ and $K_{a1}^* = 3.428 \times 10^{-7}$ mol/L:

$$a(H^+) = \frac{K_{a1}^* \, c(CO_2)}{f_{a1} \, c(HCO_3^-)} = \frac{(3.428 \times 10^{-7} \, mol/L)\,(2.66 \times 10^{-4} \, mol/L)}{(0.904)\,(3.8 \times 10^{-3} \, mol/L)}$$

$$a(H^+) = 2.654 \times 10^{-8} \, mol/L$$

$$pH = -\log a(H^+) = -\log(2.654 \times 10^{-8}) = 7.58$$

**Problem 9.2**

The assessment of the calcite saturation state can be carried out by means of the Langelier equation:

$$pH_{eq} = -\log K_{La} - \log f_{La} - \log c(HCO_3^-) - \log c(Ca^{2+})$$

The logarithm of the Langelier constant is given as:

$$\log K_{La} = -2.082$$

and the logarithms of the concentrations of $HCO_3^-$ and $Ca^{2+}$ (in mol/L!) can easily be calculated according to:

$$\log c(HCO_3^-) = \log(1.7 \times 10^{-3}) = -2.770$$

$$\log c(Ca^{2+}) = \log(1.2 \times 10^{-3}) = -2.921$$

To find the logarithm of the activity coefficient $f_{La}$, we need the ionic strength, $I$. Since we do not know the concentrations of all major ions, we cannot apply the definition equation of the ionic strength. Instead, we have to apply the empirical correlation that allows us to find the ionic strength from the given electrical conductivity, $\kappa$, at 25 °C:

$$I(\text{mol/L}) = \frac{\kappa_{25}(\text{mS/m})}{6\,200} = \frac{45}{6\,200} = 7.26 \times 10^{-3}$$

Now, we can apply the Güntelberg equation to find the activity coefficient for univalent ions:

$$\log \gamma_1 = -0.5 \frac{\sqrt{\dfrac{I}{\text{mol/L}}}}{1 + 1.4\sqrt{\dfrac{I}{\text{mol/L}}}}$$

With

$$\sqrt{\frac{I}{\text{mol/L}}} = 0.085$$

we get:

$$\log \gamma_1 = -0.5 \frac{0.085}{1 + (1.4)\,(0.085)} = -0.038$$

Finally, $\log f_{La}$ can be calculated by:

$$\log f_{La} = 5 \log \gamma_1 = -0.19$$

Introducing all data into the Langelier equation gives:

$$pH_{eq} = -\log K_{La} - \log f_{La} - \log c(HCO_3^-) - \log c(Ca^{2+})$$

$$pH_{eq} = 2.082 + 0.190 + 2.770 + 2.921 = 7.963$$

Finally, we can calculate the saturation index:

$$SI = pH_{meas} - pH_{eq} = 8.2 - 7.963 = 0.237$$

Since $SI$ is positive, the water under consideration is a calcite-precipitating water.

## Problem 9.3

The $CO_2$ concentration can be calculated from the pH and the given hydrogencar-bonate concentration by using the rearranged law of mass action for the first disso-ciation step of the carbonic acid:

$$c(CO_2) = \frac{a(H^+) f_{a1} \, c(HCO_3^-)}{K^*_{a1}}$$

$f_{a1}$ equals $\gamma_1$, which can be calculated as shown in the solution to Problem 9.2:

$$\log \gamma_1 = -0.038$$

$$f_{a1} = \gamma_1 = 0.916$$

With the other given data:

$$a(H^+) = 10^{-pH} = 10^{-8.2} \, mol/L = 6.31 \times 10^{-9} \, mol/L$$

$$c(HCO_3^-) = 1.7 \, mmol/L = 1.7 \times 10^{-3} \, mol/L$$

$$K^*_{a1} = 3.428 \times 10^{-7} \, mol/L$$

we can calculate the $CO_2$ concentration:

$$c(CO_2) = \frac{(6.31 \times 10^{-9} \, mol/L) \, (0.916) \, (1.7 \times 10^{-3} \, mol/L)}{3.428 \times 10^{-7} \, mol/L} = 2.87 \times 10^{-5} \, mol/L$$

In the state of the calco-carbonic equilibrium, the Tillmans equation:

$$c(CO_2) = \frac{K_T}{f_T} c^2(HCO_3^-) c(Ca^{2+})$$

has to be fulfilled. The required Tillmans constant, $K_T$ can be calculated from the single equilibrium constants according to:

$$K_T = \frac{K_{a2}^*}{K_{a1}^* K_{sp}^*} = \frac{3.251 \times 10^{-11} \, mol/L}{(3.428 \times 10^{-7} \, mol/L)(3.926 \times 10^{-9} \, mol^2/L^2)} = 2.416 \times 10^4 \, L^2/mol^2$$

whereas $f_T$ is given by:

$$f_T = \frac{1}{\gamma_1^6} = \frac{1}{0.916^6} = 1.693$$

Introducing all data into the Tillmans equation gives:

$$c(CO_2) = \frac{2.416 \times 10^4 \, L^2/mol^2}{1.693}(2.89 \times 10^{-6} \, mol^2/L^2)(1.2 \times 10^{-3} \, mol/L)$$

$$c(CO_2) = 4.949 \times 10^{-5} \, mol/L$$

If we compare the calculated $CO_2$ concentrations, we see that the water contains less $CO_2$ ($2.87 \times 10^{-5}$ mol/L) than in the state of equilibrium ($4.95 \times 10^{-5}$ mol/L) and is therefore calcite-precipitating. As expected, the assessment is the same as in Problem 9.2.

## Problem 9.4

At first, the law of mass action for the overall reaction is formulated under consideration of the material balance $c(Ca^{2+}) = 0.5 \, c(HCO_3^-)$:

$$K_{overall} = \frac{c(Ca^{2+}) c^2(HCO_3^-)}{p(CO_2)} = \frac{0.5 \, c(HCO_3^-) c^2(HCO_3^-)}{p(CO_2)} = \frac{0.5 \, c^3(HCO_3^-)}{p(CO_2)}$$

With $K_{overall} = 2.172 \times 10^{-6}$ mol³/(L³·bar) and $p(CO_2) = (100)(0.000415$ bar$) = 0.0415$ bar and after rearranging the equilibrium relationship, we find:

$$c(HCO_3^-) = \sqrt[3]{2p(CO_2) K_{overall}} = \sqrt[3]{2(0.0415 \, bar)(2.172 \times 10^{-6} \, mol^3/(L^3 \cdot bar))}$$

$$c(HCO_3^-) = \sqrt[3]{1.803 \times 10^{-7} mol^3/L^3} = 5.649 \times 10^{-3} \, mol/L$$

The $CO_2$ concentration can be calculated by means of Henry's law:

$$c(CO_2) = H\,p(CO_2) = (5.247 \times 10^{-2}\ mol/(L \cdot bar))\,(0.0415\ bar) = 2.178 \times 10^{-3}\ mol/L$$

Now, we can calculate the pH and the carbonate concentration by using the equilibrium relationships of the carbonic acid system:

$$c(H^+) = \frac{c(CO_2)\,K_{a1}}{c(HCO_3^-)} = \frac{(2.178 \times 10^{-3}\ mol/L)\,(3.428 \times 10^{-7}\ mol/L)}{5.649 \times 10^{-3}\ mol/L} = 1.322 \times 10^{-7}\ mol/L$$

$$pH = -\log c(H^+) = -\log (1.322 \times 10^{-7}) = 6.88$$

$$c(CO_3^{2-}) = \frac{K_{a2}\,c(HCO_3^-)}{c(H^+)} = \frac{(3.251 \times 10^{-11}\ mol/L)\,(5.649 \times 10^{-3}\ mol/L)}{1.322 \times 10^{-7}\ mol/L}$$

$$c(CO_3^{2-}) = 1.389 \times 10^{-6}\ mol/L$$

Finally, the calcium concentration can be computed from the solubility product:

$$c(Ca^{2+}) = \frac{K_{sp}}{c(CO_3^{2-})} = \frac{3.926 \times 10^{-9}\ mol^2/L^2}{1.389 \times 10^{-6}\ mol/L} = 2.826 \times 10^{-3}\ mol/L$$

## Chapter 10

### Problem 10.1

The general equation for the redox intensity reads:

$$pe = pe^0 + \frac{1}{n_e}\log \frac{\Pi\,a^\nu(Ox)}{\Pi\,a^\nu(Red)}$$

For the given half-reaction of the $Fe(OH)_{3(s)}/Fe^{2+}$ redox couple, we obtain:

$$pe = pe^0 + \log \frac{c^3(H^+)}{c(Fe^{2+})} = 16.3 - 3\,pH - \log c(Fe^{2+}) = 16.3 - 18 + 4.7 = 3.0$$

### Problem 10.2

The general relationship between the standard redox intensity and the standard redox potential is:

$$pe^0 = \frac{F}{2.303\,R\,T}\,E_H^0$$

After rearranging this relationship and introducing the given data, we find:

$$E_H^0 = \frac{2.303\,R\,T}{F}\,\text{pe}^0 = \frac{(2.303)(8.3145\,\text{V}\cdot\text{A}\cdot\text{s}/(\text{mol}\cdot\text{K}))\,(298.15\,\text{K})}{96\,485\,\text{A}\cdot\text{s}/\text{mol}}\,13 = 0.77\,\text{V}$$

Note that $1\,\text{J} = 1\,\text{W}\cdot\text{s} = 1\,\text{V}\cdot\text{A}\cdot\text{s}$ and that $1\,\text{C} = 1\,\text{A}\cdot\text{s}$.
For 25 °C, we can also use the simple relationship:

$$\text{pe}^0 = \frac{1}{0.059\,\text{V}}\,E_H^0$$

Rearranging the relationship and introducing $\text{pe}^0$ gives:

$$E_H^0 = (0.059\,\text{V})\,\text{pe}^0 = (0.059\,\text{V})\,13 = 0.77\,\text{V}$$

## Problem 10.3

The Henry constant describes the distribution of $H_2S$ between the gas and the aqueous phase:

$$H_2S_{(g)} \rightleftharpoons H_2S_{(aq)}$$

and the $pK_a$ is related to the $H_2S$ dissociation according to:

$$H_2S_{(aq)} \rightleftharpoons H^+ + HS^-$$

The second redox equation can be expressed as an overall reaction equation that results from the addition of the first redox reaction equation, the reaction equation for the gas/water distribution, and the reaction equation for the $H_2S$ dissociation:

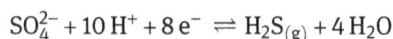

$$SO_4^{2-} + 10\,H^+ + 8\,e^- \rightleftharpoons H_2S_{(g)} + 4\,H_2O$$
$$H_2S_{(g)} \rightleftharpoons H_2S_{(aq)}$$
$$H_2S_{(aq)} \rightleftharpoons H^+ + HS^-$$

$$\overline{\phantom{SO_4^{2-} + 9\,H^+ + 8\,e^- \rightleftharpoons HS^- + 4\,H_2O}}$$

$$SO_4^{2-} + 9\,H^+ + 8\,e^- \rightleftharpoons HS^- + 4\,H_2O$$

Now, we can write all given constants as logarithms and then add the logarithms to find the logarithm of the equilibrium constant of the overall reaction. In the case of the Henry constant, we have to convert the unit $m^3$ to L.
First reaction:

$$\log K = n_e\,\text{pe}^0 = 8(5.25) = 42$$

Second reaction (with $H = 0.1022$ mol/(L $\cdot$ bar)):

$$\log H = \log (0.1022) = -0.99$$

Third reaction:

$$\log K_a = -pK_a = -7$$

Overall reaction:

$$\log K_{overall} = \log K + \log H + \log K_a = 42 - 0.99 - 7 = 34.01$$

Accordingly, the related standard redox intensity for the redox couple $SO_4^{2-}/HS^-$ is:

$$pe^0 = \frac{1}{n_e}\log K_{overall} = \frac{1}{8}\log K_{overall} = \frac{34.01}{8} = 4.25$$

## Problem 10.4

The redox intensity equation for the $Mn^{2+}/MnO_2$ redox couple reads:

$$pe = pe^0 + \frac{1}{2}\log\frac{c^4(H^+)}{c(Mn^{2+})} = pe^0 - 2\,pH - 0.5\log c(Mn^{2+})$$

After rearranging the equation and considering that $\log K = 2\,pe^0$, we obtain for $\log c(Mn^{2+})$:

$$\log c(Mn^{2+}) = 2\,pe^0 - 2\,pe - 4\,pH = 41.6 - 20 - 24 = -2.4$$

and for $c(Mn^{2+})$:

$$c(Mn^{2+}) = 10^{-2.4}\ mol/L = 3.98 \times 10^{-3}\ mol/L$$

## Problem 10.5

The complete redox reaction can be found by adding the half-reactions after writing the second reaction in the reverse direction and multiplying it by a factor of 4:

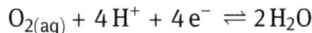

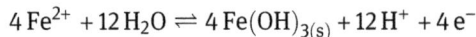

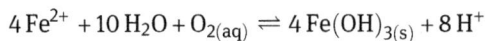

$$O_{2(aq)} + 4\,H^+ + 4\,e^- \rightleftharpoons 2\,H_2O$$

$$4\,Fe^{2+} + 12\,H_2O \rightleftharpoons 4\,Fe(OH)_{3(s)} + 12\,H^+ + 4\,e^-$$

---

$$4\,Fe^{2+} + 10\,H_2O + O_{2(aq)} \rightleftharpoons 4\,Fe(OH)_{3(s)} + 8\,H^+$$

If we want to calculate the difference of the standard redox intensities and the related equilibrium constant, we have to bear in mind that the subtrahend in the equation:

$$\log K = n_e \, \Delta pe^0 = n_e (pe_1^0 - pe_2^0)$$

is the $pe^0$ that belongs to the half-reaction, which has to be written in reverse direction. This is, in our case, the half-reaction of the $Fe^{2+}/Fe(OH)_{3(s)}$ redox couple. Consequently, we can calculate:

$$\Delta pe^0 = pe_1^0 - pe_2^0 = 21.48 - 16.3 = 5.18$$

$$\log K = n_e \, \Delta pe^0 = n_e (pe_1^0 - pe_2^0) = 4(5.18) = 20.72$$

and for the standard Gibbs energy of the complete reaction, we have:

$$\Delta_R G^0 = -2.303 \, R \, T \, \log K = -2.303 \, n_e \, R \, T \, \Delta pe^0$$

$$\Delta_R G^0 = (-2.303) \, (8.3145 \, J/(mol \cdot K)) \, (298.15 \, K) \, (20.72)$$

$$\Delta_R G^0 = -118 \; 292 \; J/mol = -118.292 \, kJ/mol$$

The standard redox intensities at pH = 7 can be calculated by using the equation:

$$pe^0 (pH = 7) = pe^0 - \frac{n_p}{n_e} 7$$

For the redox couple $O_{2(aq)}/H_2O$, the numbers of protons and electrons are $n_p = 4$ and $n_e = 4$, respectively:

$$pe^0 (pH = 7) = 21.48 - 7 = 14.48$$

The analogue calculation for $Fe^{2+}/Fe(OH)_{3(s)}$ with $n_p = 3$ and $n_e = 1$ gives:

$$pe^0 (pH = 7) = 16.3 - 21 = -4.7$$

and the difference of the standard redox intensities at pH = 7 is:

$$\Delta pe^0 (pH = 7) = pe_1^0 (pH = 7) - pe_2^0 (pH = 7) = 14.48 + 4.7 = 19.18$$

The standard Gibbs energy of reaction for pH = 7 is then:

$$\Delta_R G^0 (pH = 7) = -2.303 \, n_e \, R \, T \, \Delta pe^0 (pH = 7)$$

$$\Delta_R G^0 (pH = 7) = (-2.303) \, (4) \, (8.3145 \, J/(K \cdot mol)) \, (298.15 \, K) \, (19.18)$$

$$\Delta_R G^0 (pH = 7) = -437 \; 999 \; J/mol \approx -438 \, kJ/mol$$

## Problem 10.6

The complete redox reaction can be found by adding the half-reactions after multiplying the first equation by two and writing the second in the reverse direction:

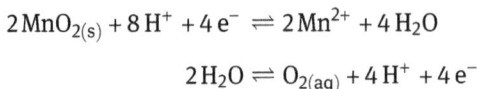

$$2\,MnO_{2(s)} + 8\,H^+ + 4\,e^- \rightleftharpoons 2\,Mn^{2+} + 4\,H_2O$$

$$2\,H_2O \rightleftharpoons O_{2(aq)} + 4\,H^+ + 4\,e^-$$

---

$$2\,MnO_{2(s)} + 4\,H^+ \rightleftharpoons 2\,Mn^{2+} + O_{2(aq)} + 2\,H_2O$$

The law of mass action of the overall reaction is:

$$K = \frac{c^2(Mn^{2+})\,c(O_2)}{c^4(H^+)}$$

The logarithm of the constant of the overall reaction can be found from the standard redox intensities that are related to the equilibrium constants of the half-reactions:

$$pe_1^0 = \frac{1}{n_e}\log K_1 = \frac{41.6}{2} = 20.8$$

$$pe_2^0 = \frac{1}{n_e}\log K_2 = \frac{85.9}{4} = 21.48$$

$$\log K = n_e(pe_1^0 - pe_2^0) = 4(20.8 - 21.48) = -2.72$$

Accordingly, the equilibrium constant $K$ is:

$$K = 10^{\log K}\,L/mol = 10^{-2.72}\,L/mol = 1.91 \times 10^{-3}\,L/mol$$

Furthermore, the concentrations of oxygen and of the protons are:

$$c(O_2) = \frac{0.261\,mmol/L}{100} = 0.00261\,mmol/L = 2.61 \times 10^{-6}\,mol/L$$

and

$$c(H^+) = 10^{-pH}\,mol/L = 1 \times 10^{-6}\,mol/L$$

The equilibrium concentration of $Mn^{2+}$ can be found after rearranging the law of mass action:

$$c^2(\text{Mn}^{2+}) = \frac{K\,c^4(\text{H}^+)}{c(\text{O}_2)} = \frac{(1.91 \times 10^{-3}\,\text{L/mol})\,(1 \times 10^{-24}\,\text{mol}^4/\text{L}^4)}{2.61 \times 10^{-6}\,\text{mol/L}}$$

$$= 7.32 \times 10^{-22}\,\text{mol}^2/\text{L}^2$$

$$c(\text{Mn}^{2+}) = \sqrt{7.32 \times 10^{-22}\,\text{mol}^2/\text{L}^2} = 2.71 \times 10^{-11}\,\text{mol/L}$$

Finally, we have to multiply the molar concentration with the molecular weight to find the mass concentration:

$$\rho^*(\text{Mn}^{2+}) = (2.71 \times 10^{-11}\,\text{mol/L})\,(55\,\text{g/mol}) = 1.49 \times 10^{-9}\,\text{g/L} = 1.49 \times 10^{-6}\,\text{mg/L}$$

## Problem 10.7

The overall reaction equation can be found by multiplying the $\text{PbO}_2/\text{Pb}^{2+}$ half-reaction by two, writing this reaction equation in reverse direction, and adding both half-reactions:

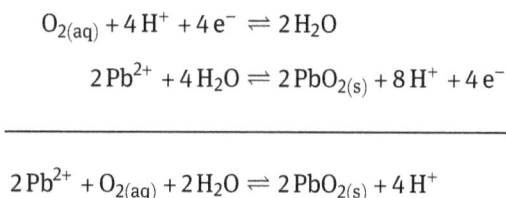

$$\text{O}_{2(\text{aq})} + 4\,\text{H}^+ + 4\,\text{e}^- \rightleftharpoons 2\,\text{H}_2\text{O}$$

$$2\,\text{Pb}^{2+} + 4\,\text{H}_2\text{O} \rightleftharpoons 2\,\text{PbO}_{2(\text{s})} + 8\,\text{H}^+ + 4\,\text{e}^-$$

_____

$$2\,\text{Pb}^{2+} + \text{O}_{2(\text{aq})} + 2\,\text{H}_2\text{O} \rightleftharpoons 2\,\text{PbO}_{2(\text{s})} + 4\,\text{H}^+$$

The law of mass action reads:

$$K = \frac{c^4(\text{H}^+)}{c^2(\text{Pb}^{2+})\,c(\text{O}_2)}$$

The value of $K$ can be found from:

$$\log K = n_e(\text{pe}_1^0 - \text{pe}_2^0) = 4(21.48 - 24.7) = -12.88$$

$$K = 10^{-12.88}\,\text{mol/L} = 1.32 \times 10^{-13}\,\text{mol/L}$$

Note that $\text{pe}_2^0$ is the standard redox intensity of the half-reaction that has to be written in the reverse direction (here the $\text{PbO}_2/\text{Pb}^{2+}$ half-reaction). The reaction quotient, $Q$, can be calculated with the actual concentrations $c(\text{O}_2) = 2.61 \times 10^{-4}\,\text{mol/L}$, $c(\text{H}^+) = 1 \times 10^{-7}\,\text{mol/L}$, and

$$c(\text{Pb}^{2+}) = \frac{\rho^*(\text{Pb}^{2+})}{M(\text{Pb}^{2+})} = \frac{5 \times 10^{-5}\,\text{g/L}}{207\,\text{g/mol}} = 2.42 \times 10^{-7}\,\text{mol/L}$$

$$Q = \frac{c^4(H^+)}{c^2(Pb^{2+})\, c(O_2)} = \frac{1 \times 10^{-28} \ mol^4/L^4}{(5.86 \times 10^{-14} \ mol^2/L^2)\,(2.61 \times 10^{-4} \ mol/L)}$$

$$= 6.54 \times 10^{-12} \ mol/L$$

Since $Q$ is larger than $K$, the reaction will not proceed in the written but in the reverse direction.

### Problem 10.8

The complete redox reaction equation can be established after multiplying the $O_2/H_2O$ half-reaction by two and writing the $SO_4^{2-}/HS^-$ half-reaction in the reverse direction:

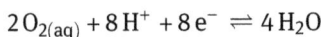

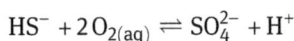

$$2\,O_{2(aq)} + 8\,H^+ + 8\,e^- \rightleftharpoons 4\,H_2O$$

$$HS^- + 4\,H_2O \rightleftharpoons SO_4^{2-} + 9\,H^+ + 8\,e^-$$

$$\overline{\phantom{xxxxxxxxxxxxxxxxxxxxxxxx}}$$

$$HS^- + 2\,O_{2(aq)} \rightleftharpoons SO_4^{2-} + H^+$$

with

$$K = \frac{c(SO_4^{2-})\, c(H^+)}{c^2(O_2)\, c(HS^-)}$$

The equilibrium constant is given by:

$$\log K = n_e\,(pe_1^0 - pe_2^0) = 8\,(21.48 - 4.25) = 137.84$$

Rearranging the law of mass action and considering the condition $c(SO_4^{2-}) = c(HS^-)$ gives:

$$c(O_2) = \sqrt{\frac{c(H^+)}{K}}$$

Due to the large value of $K$ ($\approx 10^{138}$ L/mol) it is more convenient to use the logarithms. Thus, we receive:

$$\log c(O_2) = 0.5\,(\log c(H^+) - \log K) = 0.5\,(-pH - \log K)$$

$$\log c(O_2) = 0.5\,(-7.5 - 137.84) = -72.67 \approx -73$$

and

$$c(O_2) = 10^{-73} \ mol/L$$

This means that HS⁻ occurs in a significant portion only in waters that are practically free of oxygen.

## Problem 10.9

To solve the problem, it is useful to calculate the concentration ratio $NO_3^-/NH_4^+$ in the state of equilibrium. The total redox reaction for the oxidation of ammonium results from the addition of the half-reactions after writing the second equation in reverse direction and multiplying the first reaction by two to have the same numbers of electrons ($n_e = 8$):

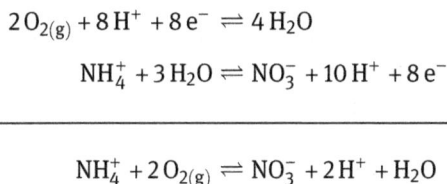

$$2\,O_{2(g)} + 8\,H^+ + 8\,e^- \rightleftharpoons 4\,H_2O$$

$$NH_4^+ + 3\,H_2O \rightleftharpoons NO_3^- + 10\,H^+ + 8\,e^-$$

$$NH_4^+ + 2\,O_{2(g)} \rightleftharpoons NO_3^- + 2\,H^+ + H_2O$$

The law of mass action reads:

$$K = \frac{c(NO_3^-)\,c^2(H^+)}{c(NH_4^+)\,p^2(O_2)}$$

and the equilibrium constant results from:

$$\log K = n_e(pe_1^0 - pe_2^0) = 8\,(20.75 - 14.9) = 46.8$$

$$K = 10^{46.8}\,mol^2/(L^2 \cdot bar^2) = 6.31 \times 10^{46}\,mol^2/(L^2 \cdot bar^2)$$

The concentration ratio of $NO_3^-$ and $NH_4^+$ can be found by rearranging the law of mass action:

$$\frac{c(NO_3^-)}{c(NH_4^+)} = \frac{p^2(O_2)\,K}{c^2(H^+)} = \frac{(0.0437\,bar^2)\,(6.31 \times 10^{46}\,mol^2/(bar^2 \cdot L^2))}{1 \times 10^{-14}\,mol^2/L^2} = 2.76 \times 10^{59}$$

The large value of the ratio indicates that in an oxygen-saturated water ammonium will be nearly completely oxidized to nitrate.

## Chapter 11

### Problem 11.1

The individual complex formation constants are valid for the reactions:

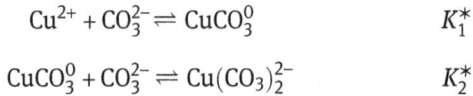

$$Cu^{2+} + CO_3^{2-} \rightleftharpoons CuCO_3^0 \qquad\qquad K_1^*$$

$$CuCO_3^0 + CO_3^{2-} \rightleftharpoons Cu(CO_3)_2^{2-} \qquad\qquad K_2^*$$

whereas the overall complex formation constant is related to the reaction:

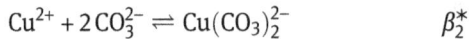

$$Cu^{2+} + 2CO_3^{2-} \rightleftharpoons Cu(CO_3)_2^{2-} \qquad\qquad \beta_2^*$$

If reaction equations are added to an overall equation, the equilibrium constants have to be multiplied. Therefore, the overall complex formation constant is the product of the individual complex formation constants:

$$\beta_2^* = K_1^* K_2^* = (5.37 \times 10^6 \, \text{L/mol})\,(1.26 \times 10^3 \, \text{L/mol}) = 6.77 \times 10^9 \, \text{L}^2/\text{mol}^2$$

The overall constant for the complex dissociation:

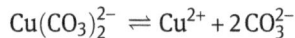

$$Cu(CO_3)_2^{2-} \rightleftharpoons Cu^{2+} + 2CO_3^{2-}$$

is the reciprocal of the overall complex formation constant:

$$\beta_{2,\,\text{diss}}^* = \frac{1}{\beta_2^*} = \frac{1}{6.77 \times 10^9 \, \text{L}^2/\text{mol}^2} = 1.48 \times 10^{-10} \, \text{mol}^2/\text{L}^2$$

### Problem 11.2

At first, the mass concentration of the limiting value (2 mg/L zinc) has to be converted into the molar concentration. Here, it has to be taken into account that 1 mol Zn is equivalent to 1 mol $Zn(OH)_4^{2-}$:

$$c(Zn(OH)_4^{2-}) = c(Zn_{\text{total}}) = \frac{\rho^*(Zn_{\text{total}})}{M(Zn)} = \frac{2 \times 10^{-3} \, \text{g/L}}{65.4 \, \text{g/mol}} = 3.06 \times 10^{-5} \, \text{mol/L}$$

To find the $OH^-$ concentration that is related to this complex concentration, we have to apply the law of mass action for the complex formation together with the solubility product, because both equilibrium conditions have to be fulfilled simultaneously. The law of mass action for the complex formation reads:

$$\beta_4 = \frac{c(Zn(OH)_4^{2-})}{c(Zn^{2+}) \, c^4(OH^-)}$$

The unknown $Zn^{2+}$ concentration can be substituted by means of the solubility product:

$$K_{sp} = c(Zn^{2+}) \, c^2(OH^-)$$

$$c(Zn^{2+}) = \frac{K_{sp}}{c^2(OH^-)}$$

$$\beta_4 = \frac{c(Zn(OH)_4^{2-})}{c(Zn^{2+}) \, c^4(OH^-)} = \frac{c(Zn(OH)_4^{2-}) \, c^2(OH^-)}{K_{sp} \, c^4(OH^-)} = \frac{c(Zn(OH)_4^{2-})}{K_{sp} \, c^2(OH^-)}$$

After introducing the given data:

$$c(OH^-) = \sqrt{\frac{c(Zn(OH)_4^{2-})}{K_{sp} \, \beta_4}} = \sqrt{\frac{3.06 \times 10^{-5} \, \text{mol/L}}{(1 \times 10^{-17} \, \text{mol}^3/\text{L}^3) \, (2.5 \times 10^{15} \, \text{L}^4/\text{mol}^4)}}$$

we find the concentration of $OH^-$ and finally the pH:

$$c(OH^-) = \sqrt{1.22 \times 10^{-3} \, \text{mol}^2/\text{L}^2}$$

$$c(OH^-) = 0.035 \, \text{mol/L} \rightarrow pOH = 1.46$$

$$pH = pK_w - pOH = 14 - 1.46 = 12.54$$

At pH = 12.54, the zinc concentration is exactly 2 mg/L. From the equation derived for the simultaneous equilibrium, we can also see that the zinc complex concentration increases with increasing $OH^-$ concentration (increasing pH). This means that at pH > 12.54 the zinc concentration in the liquid phase exceeds the mass concentration of 2 mg/L.

## Problem 11.3

We start with the Cd balance:

$$c(Cd_{total}) = c(Cd^{2+}) + c(CdCO_3^0) + c(CdOH^+) + c(Cd(OH)_2^0)$$

Substituting the complex species by means of the equilibrium relationships gives:

$$c(\text{Cd}_{\text{total}}) = c(\text{Cd}^{2+}) + K_{\text{carb},1}\, c(\text{Cd}^{2+})\, c(\text{CO}_3^{2-}) + K_{\text{hydr},1}\, c(\text{Cd}^{2+})\, c(\text{OH}^-)$$
$$+ K_{\text{hydr},1}\, K_{\text{hydr},2}\, c(\text{Cd}^{2+})\, c^2(\text{OH}^-)$$

After factoring out the $\text{Cd}^{2+}$ concentration:

$$c(\text{Cd}_{\text{total}}) = c(\text{Cd}^{2+})\left[1 + K_{\text{carb},1}\, c(\text{CO}_3^{2-}) + K_{\text{hydr},1}\, c(\text{OH}^-) + K_{\text{hydr},1}\, K_{\text{hydr},2}\, c^2(\text{OH}^-)\right]$$

we find an expression for the desired fraction of $\text{Cd}^{2+}$:

$$\frac{c(\text{Cd}^{2+})}{c(\text{Cd}_{\text{total}})} = \frac{1}{1 + K_{\text{carb},1}\, c(\text{CO}_3^{2-}) + K_{\text{hydr},1}\, c(\text{OH}^-) + K_{\text{hydr},1}\, K_{\text{hydr},2}\, c^2(\text{OH}^-)}$$

With

$$c(\text{OH}^-) = 10^{-(\text{pKw} - \text{pH})}\ \text{mol}/\text{L} = 10^{-(14-8)}\ \text{mol}/\text{L} = 1 \times 10^{-6}\ \text{mol}/\text{L}$$

and

$$K_{\text{carb},1}\, c(\text{CO}_3^{2-}) = (2.5 \times 10^5\ \text{L}/\text{mol})\ (5 \times 10^{-5}\ \text{mol}/\text{L}) = 12.5$$

$$K_{\text{hydr},1}\, c(\text{OH}^-) = (8.3 \times 10^3\ \text{L}/\text{mol})\ (1 \times 10^{-6}\ \text{mol}/\text{L}) = 8.3 \times 10^{-3}$$

$$K_{\text{hydr},1}\, K_{\text{hydr},2}\, c^2(\text{OH}^-) = (8.3 \times 10^3\ \text{L}/\text{mol})\ (5.42 \times 10^3\ \text{L}/\text{mol})\ (1 \times 10^{-12}\ \text{mol}^2/\text{L}^2)$$
$$= 4.5 \times 10^{-5}$$

we finally find:

$$\frac{c(\text{Cd}^{2+})}{c(\text{Cd}_{\text{total}})} = \frac{1}{1 + 12.5 + 8.3 \times 10^{-3} + 4.5 \times 10^{-5}} = 0.074$$

At pH $= 8$, only 7.4% of the total Cd concentration occurs as free (more exactly hydrated) $\text{Cd}^{2+}$.

## Problem 11.4

The iron(III) balance reads:

$$c(\text{Fe(III)}_{\text{total}}) = c(\text{Fe}^{3+}) + c(\text{FeOH}^{2+}) + c(\text{Fe(OH)}_2^+) + c(\text{Fe(OH)}_3^0) + c(\text{Fe(OH)}_4^-)$$

Replacing the complex species by means of the equilibrium relationships gives:

$$c(\text{Fe(III)}_{\text{total}}) = c(\text{Fe}^{3+}) + \beta_1\, c(\text{Fe}^{3+})\, c(\text{OH}^-) + \beta_2\, c(\text{Fe}^{3+})\, c^2(\text{OH}^-)$$
$$+ \beta_3\, c(\text{Fe}^{3+})\, c^3(\text{OH}^-) + \beta_4\, c(\text{Fe}^{3+})\, c^4(\text{OH}^-)$$

Then, the $Fe^{3+}$ concentration can be factored out. After rearranging the equation, we obtain:

$$c(Fe^{3+}) = \frac{c(Fe(III)_{total})}{1+\beta_1 c(OH^-)+\beta_2 c^2(OH^-)+\beta_3 c^3(OH^-)+\beta_4 c^4(OH^-)}$$

Introducing the values of the constants and the concentration of the hydroxide ions, $c(OH^-) = 1 \times 10^{-7}$ mol/L (pH = pOH = 7), we can calculate the terms in the denominator:

$$\beta_1 c(OH^-) = (6.3\times 10^{11}\,L/mol)\,(1\times 10^{-7}\,mol/L) = 6.3\times 10^4$$

$$\beta_2 c^2(OH^-) = (1.6\times 10^{22}\,L^2/mol^2)\,(1\times 10^{-14}\,mol^2/L^2) = 1.6\times 10^8$$

$$\beta_3 c^3(OH^-) = (2.5\times 10^{28}\,L^3/mol^3)\,(1\times 10^{-21}\,mol^3/L^3) = 2.5\times 10^7$$

$$\beta_4 c^4(OH^-) = (2.5\times 10^{34}\,L^4/mol^4)\,(1\times 10^{-28}\,mol^4/L^4) = 2.5\times 10^6$$

Thus, the $Fe^{3+}$ concentration is:

$$c(Fe^{3+}) = \frac{1\times 10^{-5}\,mol/L}{1+6.3\times 10^4 +1.6\times 10^8 +2.5\times 10^7 +2.5\times 10^6} = \frac{1\times 10^{-5}\,mol/L}{1.88\times 10^8}$$

$$c(Fe^{3+}) = 5.32\times 10^{-14}\,mol/L$$

The concentrations of the complex species are found by rearranging the equilibrium relationships and introducing the $Fe^{3+}$ concentration:

$$c(FeOH^{2+}) = c(Fe^{3+})\beta_1 c(OH^-)$$

$$c(FeOH^{2+}) = (5.32\times 10^{-14}\,mol/L)\,(6.3\times 10^{11}\,L/mol)\,(1\times 10^{-7}\,mol/L)$$

$$c(FeOH^{2+}) = 3.35\times 10^{-9}\,mol/L$$

$$c(Fe(OH)_2^+) = c(Fe^{3+})\beta_2 c^2(OH^-)$$

$$c(Fe(OH)_2^+) = (5.32\times 10^{-14}\,mol/L)\,(1.6\times 10^{22}\,L^2/mol^2)\,(1\times 10^{-14}\,mol^2/L^2)$$

$$c(Fe(OH)_2^+) = 8.51\times 10^{-6}\,mol/L$$

$$c(Fe(OH)_3^0) = c(Fe^{3+})\beta_3 c^3(OH^-)$$

$$c(Fe(OH)_3^0) = (5.32\times 10^{-14}\,mol/L)\,(2.5\times 10^{28}\,L^3/mol^3)\,(1\times 10^{-21}\,mol^3/L^3)$$

$$c(Fe(OH)_3^0) = 1.33\times 10^{-6}\,mol/L$$

$$c(Fe(OH)_4^-) = c(Fe^{3+})\beta_4 c^4(OH^-)$$

$$c(Fe(OH)_4^-) = (5.32\times 10^{-14}\,mol/L)\,(2.5\times 10^{34}\,L^4/mol^4)\,(1\times 10^{-28}\,mol^4/L^4)$$

$$c(Fe(OH)_4^-) = 1.33 \times 10^{-7}\,mol/L$$

To verify the result, we compare the sum of the species concentrations with the given total concentration of Fe(III):

$$c(Fe(III)_{total}) = c(Fe^{3+}) + c(FeOH^{2+}) + c(Fe(OH)_2^+) + c(Fe(OH)_3^0) + c(Fe(OH)_4^-)$$

$$= 5.32 \times 10^{-14}\,mol/L + 3.35 \times 10^{-9}\,mol/L + 8.51 \times 10^{-6}\,mol/L$$

$$+ 1.33 \times 10^{-6}\,mol/L + 1.33 \times 10^{-7}\,mol/L = 9.98 \times 10^{-6}\,mol/L$$

The small difference to $1 \times 10^{-5}\,mol/L$ is due to round-off errors and can be neglected.

**Problem 11.5**

The aqueous solubility (saturation concentration, $c_s$) of AgCl can be calculated from the solubility product as shown in Chapter 8:

$$c_s = c(Ag^+) = c(Cl^-)$$
$$K_{sp} = c(Ag^+)\,c(Cl^-) = c_s^2$$
$$c_s = \sqrt{K_{sp}} = \sqrt{1.8 \times 10^{-10}\,mol^2/L^2} = 1.34 \times 10^{-5}\,mol/L$$

If complex formation takes place, the initially dissolved $Ag^+$ is partly converted into the complex species. The saturation concentration after dissolution and complex formation is therefore given by the material balance equation:

$$c_s = c(Ag^+) + c(AgNH_3^+) + c(Ag(NH_3)_2^+)$$

Substituting the complex concentrations by means of the equilibrium relationships and factoring out the $Ag^+$ concentration gives:

$$c(AgNH_3^+) = \beta_1\,c(Ag^+)\,c(NH_3)$$
$$c(Ag(NH_3)_2^+) = \beta_2\,c(Ag^+)\,c^2(NH_3)$$
$$c_s = c(Ag^+)\left[1 + \beta_1\,c(NH_3) + \beta_2\,c^2(NH_3)\right]$$

Rearranging the equation gives:

$$c(Ag^+) = \frac{c_s}{1 + \beta_1\,c(NH_3) + \beta_2\,c^2(NH_3)}$$

Now, this equation can be introduced into the solubility product. Note that $c_s$ also equals $c(Cl^-)$:

$$K_{sp} = c(Ag^+)\,c(Cl^-) = \frac{c_s}{1 + \beta_1\,c(NH_3) + \beta_2\,c^2(NH_3)}\quad c_s = \frac{c_s^2}{1 + \beta_1\,c(NH_3) + \beta_2\,c^2(NH_3)}$$

After rearranging this equation, we can calculate $c_s$:

$$c_s = \sqrt{K_{sp}[1 + \beta_1\,c(NH_3) + \beta_2\,c^2(NH_3)]}$$

$$\beta_1\,c(NH_3) = (2.5 \times 10^3\,L/mol)\,(0.5\,mol/L) = 1.25 \times 10^3$$

$$\beta_2\,c^2(NH_3) = (2.5 \times 10^7\,L^2/mol^2)\,(0.25\,mol^2/L^2) = 6.25 \times 10^6$$

$$c_s = \sqrt{(1.8 \times 10^{-10}\,mol^2/L^2)(1 + 1.25 \times 10^3 + 6.25 \times 10^6)}$$

$$c_s = \sqrt{1.125 \times 10^{-3}\,mol^2/L^2} = 3.35 \times 10^{-2}\,mol/L$$

If we compare the solubility in the presence of 0.5 mol/L $NH_3$ with the solubility in pure water, we see an increase of the solubility by a factor of 2 500 due to the complex formation.

## Problem 11.6

At first, the explanations for the simplifying assumptions will be given.

$c(CdNTA^-) \ll c(NTA_{total})$: According to the composition of the complex, the $CdNTA^-$ concentration cannot be higher than the total Cd concentration ($1 \times 10^{-9}$ mol/L), which is much lower than the NTA concentration ($1 \times 10^{-7}$ mol/L).

$c(CaNTA^-) \ll c(Ca_{total})$: According to the composition of the complex, the $CaNTA^-$ concentration cannot be higher than the total NTA concentration ($1 \times 10^{-7}$ mol/L), which is much lower than the total Ca concentration ($1 \times 10^{-3}$ mol/L).

$c(CdCO_3^0) \ll c(DIC)$: According to composition of the complex, the $CdCO_3^0$ concentration cannot be higher than the total Cd concentration ($1 \times 10^{-9}$ mol/L), which is much lower than the DIC concentration ($1 \times 10^{-3}$ mol/L).

Further simplifications can be derived from a comparison of the $pK_a$ of the DIC and NTA species with the pH (pH = 8 in our case). Due to their low $pK_a$ in comparison to the pH, $CO_2$ ($pK_a = 6.4$), $H_3NTA$ ($pK_a = 1.7$), and $H_2NTA^-$ ($pK_a = 2.9$) can be neglected in the respective balance equations.

Now, the material balance equations can be formulated under consideration of the simplifying assumptions:

$$c(DIC) = c(CO_2) + c(HCO_3^-) + c(CO_3^{2-}) + c(CdCO_3^0) \approx c(HCO_3^-) + c(CO_3^{2-})$$

$$c(Ca_{total}) = c(Ca^{2+}) + c(CaNTA^-) \approx c(Ca^{2+})$$

$$c(Cd_{total}) = c(Cd^{2+}) + c(CdOH^+) + c(Cd(OH)_2^0) + c(CdCO_3^0) + c(CdNTA^-)$$

$$c(NTA_{total}) = c(H_3NTA) + c(H_2NTA^-) + c(HNTA^{2-}) + c(NTA^{3-}) + c(CaNTA^-)$$

$$+ c(CdNTA^-) \approx c(HNTA^{2-}) + c(NTA^{3-}) + c(CaNTA^-)$$

From the given pH we can derive that $c(H^+)$ amounts to $1 \times 10^{-8}$ mol/L and $c(OH^-)$ amounts to $1 \times 10^{-6}$ mol/L.

According to the above-mentioned simplifying assumption, the $Ca^{2+}$ concentration equals the total calcium concentration:

$$c(Ca^{2+}) = c(Ca_{total}) = 1 \times 10^{-3} \text{ mol/L}$$

Next, the concentrations of $CO_3^{2-}$ and $NTA^{3-}$ are computed by combining the material balance equations with the equilibrium relationships:

$$c(DIC) = c(HCO_3^-) + c(CO_3^{2-}) = c(CO_3^{2-})\left(1 + \frac{c(H^+)}{K_a(HCO_3^-)}\right)$$

$$c(CO_3^{2-}) = \frac{c(DIC)}{1 + \dfrac{c(H^+)}{K_a(HCO_3^-)}} = \frac{1 \times 10^{-3} \text{ mol/L}}{1 + \dfrac{1 \times 10^{-8} \text{ mol/L}}{4.7 \times 10^{-11} \text{ mol/L}}} = 4.68 \times 10^{-6} \text{ mol/L}$$

$$c(NTA_{total}) = c(HNTA^{2-}) + c(NTA^{3-}) + c(CaNTA^-)$$

$$c(NTA_{total}) = c(NTA^{3-})\left(\frac{c(H^+)}{K_a(HNTA^{2-})} + 1 + K_1(CaNTA^-)\,c(Ca^{2+})\right)$$

$$c(NTA^{3-}) = \frac{c(NTA_{total})}{\dfrac{c(H^+)}{K_a(HNTA^{2-})} + 1 + K_1(CaNTA^-)\,c(Ca^{2+})}$$

$$= \frac{1 \times 10^{-7} \text{ mol/L}}{\dfrac{1 \times 10^{-8} \text{ mol/L}}{5 \times 10^{-11} \text{ mol/L}} + 1 + (4 \times 10^7 \text{ L/mol})\,(1 \times 10^{-3} \text{ mol/L})} = 2.49 \times 10^{-12} \text{ mol/L}$$

After introducing the equilibrium relationships into the Cd balance, we can calculate the $Cd^{2+}$ concentration:

$$c(Cd_{total}) = c(Cd^{2+})[1 + \beta_1(CdOH^+)\,c(OH^-) + \beta_2(Cd(OH)_2^0)\,c^2(OH^-)$$

$$+ K_1(CdCO_3^0)\,c(CO_3^{2-}) + K_1(CdNTA^-)\,c(NTA^{3-})]$$

The summands within the squared brackets are:

$$\beta_1(CdOH^+)\,c(OH^-) = (8.3 \times 10^3 \text{ L/mol})\,(1 \times 10^{-6} \text{ mol/L}) = 8.3 \times 10^{-3}$$

$$\beta_2(Cd(OH)_2^0)\,c^2(OH^-) = (4.5 \times 10^7 \text{ L}^2/\text{mol}^2)\,(1 \times 10^{-12} \text{ mol}^2/\text{L}^2) = 4.5 \times 10^{-5}$$

$$K_1(CdCO_3^0)\,c(CO_3^{2-}) = (2.5 \times 10^5 \text{ L/mol})\,(4.68 \times 10^{-6} \text{ mol/L}) = 1.17$$

$$K_1(CdNTA^-)\,c(NTA^{3-}) = (1 \times 10^{10} \text{ L/mol})\,(2.49 \times 10^{-12} \text{ mol/L}) = 2.49 \times 10^{-2}$$

Introducing these values into the rearranged Cd balance equation gives:

$$c(Cd^{2+}) = \frac{c(Cd_{total})}{1 + 8.3 \times 10^{-3} + 4.5 \times 10^{-5} + 1.17 + 2.49 \times 10^{-2}}$$

$$c(Cd^{2+}) = \frac{1 \times 10^{-9} \text{ mol/L}}{2.203} = 4.54 \times 10^{-10} \text{ mol/L}$$

With the $Cd^{2+}$ concentration, all cadmium complex concentrations can be calculated:

$$c(CdOH^+) = \beta_1(CdOH^+)\,c(Cd^{2+})\,c(OH^-)$$
$$= (8.3 \times 10^3 \text{ L/mol})\,(4.54 \times 10^{-10} \text{ mol/L})\,(1 \times 10^{-6} \text{ mol/L})$$
$$= 3.77 \times 10^{-12} \text{ mol/L}$$

$$c(Cd(OH)_2^0) = \beta_2(Cd(OH)_2^0)\,c(Cd^{2+})\,c^2(OH^-)$$
$$= (4.5 \times 10^7 \text{ L}^2/\text{mol}^2)\,(4.54 \times 10^{-10} \text{ mol/L})\,(1 \times 10^{-12} \text{ mol}^2/\text{L}^2)$$
$$= 2.04 \times 10^{-14} \text{ mol/L}$$

$$c(CdCO_3^0) = K_1(CdCO_3^0)\,c(Cd^{2+})\,c(CO_3^{2-})$$
$$= (2.5 \times 10^5 \text{ L/mol})\,(4.54 \times 10^{-10} \text{ mol/L})\,(4.68 \times 10^{-6} \text{ mol/L})$$
$$= 5.31 \times 10^{-10} \text{ mol/L}$$

$$c(CdNTA^-) = K_1(CdNTA^-)\,c(Cd^{2+})\,c(NTA^{3-})$$
$$= (1 \times 10^{10} \text{ L/mol})\,(4.54 \times 10^{-10} \text{ mol/L})\,(2.49 \times 10^{-12} \text{ mol/L})$$
$$= 1.13 \times 10^{-11} \text{ mol/L}$$

To check the results for the cadmium complexes, we can add the calculated concentrations and compare the sum with the given total concentration:

$$c(Cd_{total}) = c(Cd^{2+}) + c(CdOH^+) + c(Cd(OH)_2^0) + c(CdCO_3^0) + c(CdNTA^-)$$
$$= 4.54 \times 10^{-10} \text{ mol/L} + 3.77 \times 10^{-12} \text{ mol/L} + 2.04 \times 10^{-14} \text{ mol/L}$$
$$+ 5.31 \times 10^{-10} \text{ mol/L} + 1.13 \times 10^{-11} \text{ mol/L} = 1.0 \times 10^{-9} \text{ mol/L}$$

Finally, we have to calculate the concentration of the $CaNTA^-$ complex:

$$c(CaNTA^-) = K_1(CaNTA^-)\, c(Ca^{2+})\, c(NTA^{3-})$$
$$= (4 \times 10^7 \text{ L/mol})\, (1 \times 10^{-3} \text{ mol/L})\, (2.49 \times 10^{-12} \text{ mol/L})$$
$$= 9.96 \times 10^{-8} \text{ mol/L}$$

To check the NTA balance, we need at first the $HNTA^{2-}$ concentration, which can be found from the acidity constant:

$$c(HNTA^{2-}) = \frac{c(H^+)\, c(NTA^{3-})}{K_a(HNTA^{2-})} = \frac{(1 \times 10^{-8} \text{ mol/L})\, (2.49 \times 10^{-12} \text{ mol/L})}{5 \times 10^{-11} \text{ mol/L}}$$
$$= 4.98 \times 10^{-10} \text{ mol/L}$$

Now, we can compare the sum of the calculated NTA species concentrations with the given total concentration:

$$c(NTA_{total}) = c(HNTA^{2-}) + c(NTA^{3-}) + c(CaNTA^-) + c(CdNTA^-)$$
$$= 4.98 \times 10^{-10} \text{ mol/L} + 2.49 \times 10^{-12} \text{ mol/L} + 9.96 \times 10^{-8} \text{ mol/L}$$
$$+ 1.13 \times 10^{-11} \text{ mol/L}$$
$$= 1.0 \times 10^{-7} \text{ mol/L}$$

It can be seen from the results that the concentration of the $CaNTA^-$ complex $(9.96 \times 10^{-8} \text{ mol/L})$ is much higher than that of the $CdNTA^-$ complex $(1.13 \times 10^{-11} \text{ mol/L})$ although the calcium complex is weaker $(K_1 = 4 \times 10^7 \text{ L/mol})$ than the cadmium complex $(K_1 = 1 \times 10^{10} \text{ L/mol})$. This is due to the much higher calcium concentration in comparison to the cadmium concentration. However, the fraction, $f$, of the NTA complex related to the total metal concentration is higher for Cd (Ca: $f \approx 10^{-4}$, Cd: $f \approx 10^{-2}$).

## Chapter 12

### Problem 12.1

To find the sorbed amount, the balance equation:

$$q_{eq}^* = \frac{V}{m_{sorbent}} (\rho_0^* - \rho_{eq}^*)$$

has to be used. After introducing the given data, we find:

$$q_{eq}^* = \frac{1\,L}{0.030\,g}\,(1\ mg/L - 0.45\ mg/L) = 18.33\ mg/g$$

## Problem 12.2

With the conversion factor:

$$1\ mmol = 1\,000\ \mu mol$$

we can write:

$$K_F = 0.2\frac{mmol/g}{(mmol/L)^{0.9}} = 0.2\frac{1\,000\ \mu mol/g}{(1\,000\ \mu mol/L)^{0.9}} = 0.2(1\,000^{1-0.9})\frac{\mu mol/g}{(\mu mol/L)^{0.9}}$$

Accordingly, the Freundlich coefficient is:

$$K_F = 0.2(1\,000^{0.1})\frac{\mu mol/g}{(\mu mol/L)^{0.9}} = 0.399\frac{\mu mol/g}{(\mu mol/L)^{0.9}}$$

## Problem 12.3

a. The balance equation reads:

$$c(Cd_{total}) = c(Cd^{2+}) + q(Cd^{2+})\,\rho^*\,(sorbent) = c(Cd^{2+}) + K_F\,c^n(Cd^{2+})\,\rho^*(sorbent)$$

Introducing the total concentration $c(Cd_{total}) = 5 \times 10^{-8}$ mol/L $= 5 \times 10^{-5}$ mmol/L, the sorbent concentration $\rho^* = 0.1$ mg/L $= 1 \times 10^{-4}$ g/L, and the isotherm parameters $K_F = 0.88$ (mmol/g)/(mmol/L)$^{0.36}$ and $n = 0.36$ gives:

$$5 \times 10^{-5}\ mmol/L = c(Cd^{2+}) + [0.88(mmol/g)/(mmol/L)^{0.36}]\,c^{0.36}(Cd^{2+})$$
$$\cdot (1 \times 10^{-4}\ g/L)$$

$$5 \times 10^{-5}\ mmol/L = c(Cd^{2+}) + [8.8 \times 10^{-5}(mmol/L)^{0.64}]\,c^{0.36}(Cd^{2+})$$

By applying an iteration procedure under variation of $c(Cd^{2+})$, we find $c(Cd^{2+}) = 4.76 \times 10^{-5}$ mmol/L:

$$5 \times 10^{-5}\ mmol/L = 4.76 \times 10^{-5}\ mmol/L + [8.8 \times 10^{-5}(mmol/L)^{0.64}]$$
$$\cdot (4.76 \times 10^{-5}\ mmol/L)^{0.36}$$

$$5 \times 10^{-5}\ mmol/L \equiv 5 \times 10^{-5}\ mmol/L$$

Accordingly, the fraction of $Cd^{2+}$ is:

$$\frac{c(Cd^{2+})}{c(Cd_{total})} = \frac{4.76 \times 10^{-5}\ \text{mmol/L}}{5 \times 10^{-5}\ \text{mmol/L}} = 0.952$$

If we compare this result with that of Example 12.2 in Section 12.3.3, we can conclude that an increase of the total Cd concentration at constant sorbent dose leads to an increase of the fraction of dissolved (i.e., nonsorbed) $Cd^{2+}$.

b.  The same procedure has to be carried out for the increased sorbent concentration. Introducing the total concentration $c(Cd_{total}) = 1 \times 10^{-8}$ mol/L $= 1 \times 10^{-5}$ mmol/L, the sorbent concentration $\rho^* = 0.5$ mg/L $= 5 \times 10^{-4}$ g/L, and the isotherm parameters $K_F = 0.88$ (mmol/g)/(mmol/L)$^{0.36}$ and $n = 0.36$ gives:

$$1 \times 10^{-5}\ \text{mmol/L} = c(Cd^{2+}) + [0.88(\text{mmol/g})/(\text{mmol/L})^{0.36}]\ c^{0.36}(Cd^{2+})$$
$$\cdot (5 \times 10^{-4}\ \text{g/L})$$

$$1 \times 10^{-5}\ \text{mmol/L} = c(Cd^{2+}) + [4.4 \times 10^{-4}(\text{mmol/L})^{0.64}]\ c^{0.36}(Cd^{2+})$$

From that, we find by iteration $c(Cd^{2+}) = 4.7 \times 10^{-6}$ mmol/L:

$$1 \times 10^{-5}\ \text{mmol/L} = 4.7 \times 10^{-6}\ \text{mmol/L} + [4.4 \times 10^{-4}(\text{mmol/L})^{0.64}]$$
$$\cdot (4.7 \times 10^{-6}\ \text{mmol/L})^{0.36}$$

$$1 \times 10^{-5}\ \text{mmol/L} \equiv 1 \times 10^{-5}\ \text{mmol/L}$$

Accordingly, the fraction of the dissolved $Cd^{2+}$ is:

$$\frac{c(Cd^{2+})}{c(Cd_{total})} = \frac{4.7 \times 10^{-6}\ \text{mmol/L}}{1 \times 10^{-5}\ \text{mmol/L}} = 0.47$$

We can conclude from the result that an increase of the sorbent concentration at constant total Cd concentration leads to a decrease of the fraction of dissolved (i.e., nonsorbed) $Cd^{2+}$.

**Problem 12.4**

The fraction of organic carbon is:

$$f_{oc} = \frac{m_{oc}}{m_{solid}} = \frac{0.21\ \text{mg}}{1\ \text{g}} = \frac{2.1 \times 10^{-4}\ \text{g}}{1\ \text{g}} = 2.1 \times 10^{-4}$$

With $f_{oc}$, the organic carbon normalized sorption coefficient, $K_{oc}$, can be calculated:

$$K_{oc, meas} = \frac{K_{d, meas}}{f_{oc}} = \frac{2.3 \, L/kg}{2.1 \times 10^{-4}} = 1.095 \times 10^4 \, L/kg$$

$$\log K_{oc, meas} = \log (1.095 \times 10^4) = 4.04$$

By using the correlation of Baker et al., we find with $\log K_{ow} = 4.5$:

$$\log K_{oc, pred} = 0.903 \log K_{ow} + 0.094 = 4.16$$

The predicted $\log K_{oc}$ (4.16) is very close to the measured value (4.04). If we compare the nonlogarithmic values, we find:

$$K_{oc, pred} = 1.445 \times 10^4 \, L/kg$$

and

$$K_{oc, meas} = 1.095 \times 10^4 \, L/kg$$

The predicted value is about 32% higher than the measured value, but in the same order of magnitude.

**Problem 12.5**

The correlation proposed by Baker et al. reads:

$$\log K_{oc} = 0.903 \log K_{ow} + 0.094$$

This equation is valid only for neutral species. However, from a comparison of $pK_a$ and pH we can expect that naproxen occurs mostly as anion. Thus, we have to replace $\log K_{ow}$ by $\log D$ in the correlation:

$$\log K_{oc} = 0.903 \log D + 0.094$$

For an acid, $D$ can be calculated by:

$$D = \frac{K_{ow}}{1 + 10^{pH - pK_a}} = \frac{10^3}{1 + 10^{6 - 4.8}} = \frac{10^3}{1 + 10^{1.2}} = \frac{1\,000}{16.85} = 59.35$$

The logarithm of $D$ is:

$$\log D = 1.77$$

Alternatively, we can use the logarithmic equation:

$$\log D = \log K_{ow} - \log(1 + 10^{pH - pK_a}) = 3 - \log(1 + 10^{1.2}) = 3 - \log(16.85) = 3 - 1.23 = 1.77$$

With log $D$, we can predict log $K_{oc}$ according to:

$$\log K_{oc} = 0.903 \log D + 0.094 = (0.903)\,(1.77) + 0.094 = 1.69$$

## Problem 12.6

If the ionized species shows no significant sorption, the pH-dependent sorption coefficient can be calculated by:

$$K_{oc}(\text{pH}) = K_{oc,n}\,(1 - \alpha)$$

with

$$1 - \alpha = \frac{1}{1 + 10^{\text{pH} - \text{p}K_a}}$$

For pH = 6 and p$K_a$ = 4.31, we get:

$$1 - \alpha = \frac{1}{1 + 10^{6 - 4.31}} = \frac{1}{49.98} = 0.02$$

and the $K_{oc}$ at pH = 6 is:

$$K_{oc}(\text{pH} = 6) = (2\ 020\ \text{L/kg})\,(0.02) = 40.4\ \text{L/kg}$$

# A Appendix

## A.1 Some Important Constants

| Constant | Symbol | Value |
|---|---|---|
| Avogadro constant (number of species per mole) | $N_A$ | $6.022 \times 10^{23}$ 1/mol |
| Elementary charge (charge of a proton, negation of the charge of an electron) | $e$ | $1.602 \times 10^{-19}$ C |
| Ebullioscopic constant of water | $K_B$ | 0.513 K·kg/mol |
| Faraday constant | $F$ | 96 485.34 C/mol |
| Cryoscopic constant of water | $K_F$ | −1.86 K·kg/mol |
| Universal gas constant | $R$ | 8.3145 J/(mol·K) |
| | | 0.083145 bar· L/(mol·K) |
| Vacuum permittivity | $\varepsilon_0$ | $8.854 \times 10^{-12}$ A·s/(V·m) |

## A.2 Some Important Logarithm Rules

Logarithmic quantities (e.g., log $a$, log $c$, log $K$) are frequently used in hydrochemistry and, consequently, also in this textbook. Therefore, it seemed useful to compile the main definitions and rules that are needed to work with logarithms.

**Definitions:**
The logarithm of a number $x$ is the exponent $n$ of a basis $b$ if $x$ is expressed as power of $b$:

$$x = b^n \;\longrightarrow\; \log_b x = n$$

If the basis $b$ is 10, the logarithm is referred to as decimal (decadic, common) logarithm:

$$x = 10^n \;\longrightarrow\; \log_{10} x = n$$

Note that the decimal logarithm is often written simply as log $x$ instead of $\log_{10} x$. This simplified notation is also used in this book.

If the basis $b$ is the number $e$ (Euler number, $e = 2.71828$), the logarithm is referred to as natural logarithm, written as ln $x$:

$$x = e^n \;\longrightarrow\; \log_e x = \ln x = n$$

The term $e^n$ is sometimes also written as exp($n$). The relationship between the decimal and the natural logarithm is given by:

$$\log x = \frac{\ln x}{\ln 10} = \frac{\ln x}{2.303}$$

https://doi.org/10.1515/9783110758788-014

**p notation:**

The p notation, often used in hydrochemistry, means the negative decimal logarithm of a quantity and is used for constants and activities (or concentrations):

$$pK = -\log K$$

$$pX = -\log a(X) \approx -\log c(X)$$

**Logarithm of a product:**

$$\log(xy) = \log x + \log y$$

$$\ln(xy) = \ln x + \ln y$$

**Logarithm of a quotient:**

$$\log\left(\frac{x}{y}\right) = \log x - \log y$$

$$\ln\left(\frac{x}{y}\right) = \ln x - \ln y$$

**Logarithm of the *p*-th power of a number:**

$$\log x^p = p \log x$$

$$\ln x^p = p \ln x$$

**Inverse properties:**

$$\log 10^n = n \qquad 10^{\log x} = x$$

$$\ln e^n = n \qquad e^{\ln x} = x$$

## A.3 List of Important Equations

The following list is a compilation of the most important equations from the main text. The equations are arranged according to the chapters where they were introduced and where the reader can find a detailed interpretation.

### Chapter 2: Structure and Properties of Water

Density, $\rho$:

$$\rho = \frac{m}{V}$$

Dynamic viscosity ($\eta$) and kinematic viscosity ($v$):

$$\eta = v\,\rho$$

Molar heat capacity, $c_p$:

$$c_p = \left(\frac{\partial H}{\partial T}\right)_p \qquad \bar{c}_p = \frac{\Delta H}{T_2 - T_1}$$

Capillary rise, $h$, in a capillary with the inner radius $r$:

$$h = \frac{2\,\sigma\cos\theta}{\rho\,g\,r}$$

Vapor pressure of small droplets with the radius $r$, $p$(droplets):

$$\frac{p(\text{droplets})}{p_0} = \exp\left(\frac{2\sigma\,V_m}{r\,R\,T}\right)$$

Coulomb's law:

$$F_C = \frac{1}{4\,\pi\,\varepsilon}\frac{q_1\,q_2}{r^2} = \frac{1}{4\,\pi\,\varepsilon_0\,\varepsilon_r}\frac{q_1\,q_2}{r^2}$$

where $c_p$ is the molar heat capacity, $F_C$ is the Coulomb force (electrostatic interaction force between point charges), $g$ is the gravity of Earth, $H$ is the enthalpy, $h$ is the capillary rise, $m$ is the mass, $p$(droplets) is the vapor pressure of small droplets, $p_0$ is the vapor pressure (bulk liquid), $q_1$, $q_2$ are charges, $R$ is the gas constant, $r$ is the radius or the distance, $T$ is the absolute temperature, $V$ is the volume, $V_m$ is the molar volume, $\varepsilon$ is the permittivity of the medium, $\varepsilon_0$ is the vacuum permittivity ($8.854 \times 10^{-12}$ A $\cdot$ s/(V $\cdot$ m)), $\varepsilon_r$ is relative permittivity (dielectric constant), $\eta$ is the dynamic viscosity, $v$ is the kinematic viscosity, $\theta$ is the contact angle, $\rho$ is the density, and $\sigma$ is the surface tension.

## Chapter 3: Concentrations and Activities

Mass concentration, $\rho^*$:

$$\rho^*(X) = \frac{m(X)}{V}$$

Molar concentration (molarity), $c$:

$$c(X) = \frac{n(X)}{V}$$

Substance amount (number of moles), $n$:

$$n(X) = \frac{m(X)}{M(X)}$$

Equivalent concentration, $c(1/z\,X)$:

$$c\left(\frac{1}{z}X\right) = \frac{n\left(\frac{1}{z}X\right)}{V} = z\,c(X)$$

Molal concentration (molality), $b$:

$$b(X) = \frac{n(X)}{m_{\text{solvent}}}$$

Mass fraction, $w$:

$$w(X) = \frac{m(X)}{m_{\text{solution}}} = \frac{m(X)}{\sum m}$$

Mole fraction, $x$:

$$x(X) = \frac{n(X)}{n_{\text{solution}}} = \frac{n(X)}{\sum n}$$

Note: For gaseous mixtures, $y$ is used as the symbol for the mole fraction.

Partial pressure, $p_i$:

$$p_i = y_i\,p_{\text{total}}$$

Relationship between partial pressure, $p_i$, and molar concentration, $c_i$, in the gas phase:

$$p_i = \frac{n_i}{V}\,R\,T = c_i\,R\,T$$

Ionic strength, $I$:

$$I = 0.5 \sum_i c_i\,z_i^2 \qquad \text{(exact definition)}$$

$$I\,(\text{mol/L}) = \frac{\kappa_{25}\,(\text{mS/m})}{6\,200} \qquad \text{(empirical correlation)}$$

Ion balance, condition of electrical neutrality:

$$\sum_{\text{cations}} c_i\,z_i = \sum_{\text{anions}} c_i\,z_i$$

Activity, $a$:

$$a = \gamma c$$

Activity coefficient, $\gamma_z$:

$$\log \gamma_z = -A\,z^2 \sqrt{\frac{I}{\mathrm{mol/L}}} \qquad \text{(Debye–Hückel equation)}$$

$$\log \gamma_z = -A\,z^2 \frac{\sqrt{\dfrac{I}{\mathrm{mol/L}}}}{1 + 1.4\sqrt{\dfrac{I}{\mathrm{mol/L}}}} \qquad \text{(Güntelberg equation)}$$

$$\log \gamma_z = -A\,z^2 \left( \frac{\sqrt{\dfrac{I}{\mathrm{mol/L}}}}{1 + \sqrt{\dfrac{I}{\mathrm{mol/L}}}} - 0.3\,\frac{I}{\mathrm{mol/L}} \right) \qquad \text{(Davies equation)}$$

where $A$ is the parameter in the activity coefficient equations (0.5 in the practically relevant temperature range), $a$ is the activity, $b$ is the molality, $c$ is the molar concentration, $I$ is the ionic strength, $M$ is the molecular weight (molar mass), $m$ is the mass, $n$ is the substance amount (number of moles), $p_i$ is the partial pressure, $p_{total}$ is the total pressure, $R$ is the gas constant, $T$ is the absolute temperature, $V$ is the volume, $w$ is the mass fraction, $x$ is the mole fraction (liquid phase), $y$ is the mole fraction (gas phase), $z$ is the ion charge (absolute value), $\gamma$ is the activity coefficient, $\kappa_{25}$ is the electrical conductivity measured at 25 °C, and $\rho^*$ is the mass concentration.

## Chapter 4: Colligative Properties

Relative vapor pressure lowering, $\Delta p / p_{01}$:

$$\frac{p_{01} - p_1}{p_{01}} = \frac{\Delta p}{p_{01}} = x_2$$

Boiling point elevation, $\Delta T^{LV}$:

$$\Delta T^{LV} = b\,K_B$$

Freezing point depression, $\Delta T^{SL}$:

$$\Delta T^{SL} = b\,K_F$$

Osmotic pressure, $\pi$:

$$\pi = cRT$$

where $b$ is the molality, $c$ is the molar concentration, $K_B$ is the ebullioscopic constant, $K_F$ is the cryoscopic constant, $p_{01}$ is the vapor pressure of the pure solvent, $p_1$ is the vapor pressure of the solvent in the solution, $R$ is the gas constant, $T$ is the absolute temperature, $T^{LV}$ is the boiling point, $T^{SL}$ is the freezing point, $x_2$ is the mole fraction of the dissolved substance, and $\pi$ is the osmotic pressure.

Note: The given equations are valid for single solutes. In the case of multicomponent solutions, the sum of all solute concentrations, molalities, or mole fractions has to be used.

## Chapter 5: Chemical Equilibrium

Law of mass action – thermodynamic equilibrium constant, $K^*$, and conditional equilibrium constant, $K$:

$$\nu_A\, A + \nu_B\, B \rightleftharpoons \nu_C\, C + \nu_D\, D$$

$$K^* = \frac{(a_C)_{eq}^{\nu_C} (a_D)_{eq}^{\nu_D}}{(a_A)_{eq}^{\nu_A} (a_B)_{eq}^{\nu_B}} = \frac{(\gamma_C\, c_C)_{eq}^{\nu_C} (\gamma_D\, c_D)_{eq}^{\nu_D}}{(\gamma_A\, c_A)_{eq}^{\nu_A} (\gamma_B\, c_B)_{eq}^{\nu_B}}$$

$$K = \frac{(c_C)_{eq}^{\nu_C} (c_D)_{eq}^{\nu_D}}{(c_A)_{eq}^{\nu_A} (c_B)_{eq}^{\nu_B}} = K^* \frac{(\gamma_A)^{\nu_A} (\gamma_B)^{\nu_B}}{(\gamma_C)^{\nu_C} (\gamma_D)^{\nu_D}}$$

Gibbs energy of reaction, $\Delta_R G$:

$$\Delta_R G = \Delta_R H - T \Delta_R S$$

$$\Delta_R G = \Delta_R G^0 + RT \ln Q$$

$$\Delta_R G^0 = -RT \ln K^*$$

$$\Delta_R G = -RT \ln K^* + RT \ln Q = RT \ln \frac{Q}{K^*}$$

Standard Gibbs energy of reaction, $\Delta_R G^0$:

$$\Delta_R G^0 = \sum_i \nu_i\, G^0_{f,i}(\text{products}) - \sum_i \nu_i\, G^0_{f,i}(\text{reactants})$$

Equilibrium constant of a reverse reaction, $K^*_{reverse}$:

$$K^*_{reverse} = \frac{1}{K^*_{forward}}$$

Equilibrium constant of an overall reaction, $K^*_{\text{overall}}$:

$$K^*_{\text{overall}} = K^*_1 \, K^*_2 \, \cdots \, K^*_n$$

Temperature dependence of the equilibrium constant ($\Delta_R H^0 =$ constant):

$$\ln \frac{K^*_{T_2}}{K^*_{T_1}} = \frac{\Delta_R H^0}{R} \left( \frac{1}{T_1} - \frac{1}{T_2} \right)$$

where $a$ is the activity, $c$ is the molar concentration, $G^0_{f,i}$ is the standard Gibbs energy of formation, $\Delta_R G$ is the Gibbs energy of reaction, $\Delta_R G^0$ is the standard Gibbs energy of reaction, $\Delta_R H$ is the enthalpy of reaction, $\Delta_R H^0$ is the standard enthalpy of reaction, $K^*$ is the thermodynamic equilibrium constant, $K$ is the conditional equilibrium constant, $K^*_{T_1}$ is the equilibrium constant at temperature $T_1$, $K^*_{T_2}$ is the equilibrium constant at temperature $T_2$, $Q$ is the reaction quotient, $R$ is the gas constant, $\Delta_R S$ is the entropy of reaction, $T$ is the absolute temperature, and $v_i$ is the stoichiometric factor.

## Chapter 6: Gas–Water Partitioning

Henry's law for a compound A, expressed with the Henry constant, $H$:

$$c(A) = H(A) \, p(A)$$

Temperature dependence of the Henry constant:

$$\ln \frac{H}{H_0} = \frac{\Delta H^0_{\text{sol}}}{R} \left( \frac{1}{T_0} - \frac{1}{T} \right)$$

Henry's law for a compound A, expressed with a dimensionless distribution constant, $K_c$:

$$c_{\text{aq}}(A) = K_c(A) \, c_g(A)$$

Relationship between the Henry constant, $H$, and the distribution constant, $K_c$:

$$K_c(A) = H(A) \, R \, T$$

Approximate estimation of the Henry constant, $H$:

$$H = \frac{c_s}{p_v}$$

Gas-phase mass concentration in a closed water/gas system, $\rho^*_g$:

$$\rho^*_g = \frac{m_{\text{total}}}{(V_g + K_c \, V_{\text{aq}})}$$

Aqueous-phase mass concentration in a closed water/gas system, $\rho_{aq}^*$:

$$\rho_{aq}^* = \frac{m_{total}}{\left(\dfrac{V_g}{K_c} + V_{aq}\right)}$$

where $c_{aq}$ is the aqueous-phase molar concentration, $c_g$ is the gas-phase molar concentration, $c_s$ is the aqueous solubility, $H$ is the Henry constant, $H_0$ is the Henry constant at standard temperature (25 °C), $\Delta H_{sol}^0$ is the standard enthalpy of solution, $K_c$ is the distribution constant, $m_{total}$ is the total mass in the closed system, $p$ is the partial pressure, $p_v$ is the vapor pressure, $R$ is gas constant, $T$ is the absolute temperature, $T_0$ is the standard temperature (25 °C = 298.15 K), $V_{aq}$ is the volume of the aqueous phase, $V_g$ is the volume of the gas phase, $\rho_{aq}^*$ is the aqueous-phase mass concentration, and $\rho_g^*$ is the gas-phase mass concentration.

## Chapter 7: Acid/Base Equilibria

Ion product (dissociation constant) of water, $K_w^*$, and definitions of pH and pOH:

$$K_w^* = a(H^+)\, a(OH^-) \qquad\qquad pK_w^* = -\log K_w^*$$

$$pH = -\log a(H^+) \approx -\log c(H^+)$$

$$pOH = -\log a(OH^-) \approx -\log c(OH^-)$$

$$pK_w^* = pH + pOH$$

Acidity constant, $K_a^*$, of the acid HA:

$$HA \rightleftharpoons H^+ + A^-$$

$$K_a^* = \frac{a(H^+)\, a(A^-)}{a(HA)} \qquad\qquad pK_a^* = -\log K_a^*$$

Basicity constant, $K_b^*$, of the base B:

$$B + H_2O \rightleftharpoons BH^+ + OH^-$$

$$K_b^* = \frac{a(BH^+)\, a(OH^-)}{a(B)} \qquad\qquad pK_b^* = -\log K_b^*$$

Relationship between $K_a^*$ and $K_b^*$ for a conjugate acid/base pair:

$$K_a^*\, K_b^* = K_w^*$$

$$pK_a^* + pK_b^* = pK_w^*$$

pH or $c(H^+)$ calculation for an acid solution:

$$c^3(H^+) + K_a\, c^2(H^+) - (K_a\, c_0(\text{acid}) + K_w)\, c(H^+) - K_w\, K_a = 0 \quad \text{(general)}$$

$$c^2(H^+) + K_a\, c(H^+) - K_a\, c_0(\text{acid}) = 0 \quad \text{(for } c(A^-) \gg c(OH^-))$$

$$c(H^+) = \sqrt{K_a\, c_0(\text{acid})} \quad \text{(for weak acids)}$$

$$pH = -\log c_0(\text{acid}) \quad \text{(for very strong acids)}$$

pOH or $c(OH^-)$ calculation for a base solution:

$$c^3(OH^-) + K_b\, c^2(OH^-) - (K_b\, c_0(\text{base}) + K_w)\, c(OH^-) - K_w\, K_b = 0 \quad \text{(general)}$$

$$c^2(OH^-) + K_b\, c(OH^-) - K_b\, c_0(\text{base}) = 0 \quad \text{(for } c(HB^+) \gg c(H^+))$$

$$c(OH^-) = \sqrt{K_b\, c_0(\text{base})} \quad \text{(for weak bases)}$$

$$pOH = -\log c_0(\text{base}) \quad \text{(for very strong bases)}$$

pH of a buffered solution after introducing a strong acid with $c(H^+)$ or a strong base with $c(OH^-)$:

$$pH = pK_a + \log \frac{c(\text{buffer base}) - c(H^+)}{c(\text{buffer acid}) + c(H^+)}$$

$$pH = pK_a + \log \frac{c(\text{buffer base}) + c(OH^-)}{c(\text{buffer acid}) - c(OH^-)}$$

Speciation of a conjugate acid/base pair:

$$f(\text{acid}) = 1 - \alpha$$

$$f(\text{base}) = \alpha$$

with

$$\alpha = \frac{1}{10^{pK_a - pH} + 1}$$

Acid and base neutralizing capacities to pH = 4.3 and pH = 8.2 (most general definitions):

$$ANC_{4.3} = Alk_T = m = c(HCO_3^-) + 2\,c(CO_3^{2-}) + c(OH^-) - c(H^+)$$

$$ANC_{8.2} = Alk_P = p = c(CO_3^{2-}) + c(OH^-) - c(CO_2) - c(H^+)$$

$$BNC_{4.3} = Aci_M = -m = -c(HCO_3^-) - 2\,c(CO_3^{2-}) - c(OH^-) + c(H^+)$$

$$BNC_{8.2} = Aci_P = -p = -c(CO_3^{2-}) - c(OH^-) + c(CO_2) + c(H^+)$$

Relationship between $c(\text{DIC})$, $m$, and $p$:

$$m - p = c(\text{CO}_2) + c(\text{HCO}_3^-) + c(\text{CO}_3^{2-}) = c(\text{DIC})$$

where $Alk_P$ is the phenolphthalein (or carbonate) alkalinity, $Alk_T$ is the total alkalinity, $ANC_{4.3}$ and $ANC_{8.3}$ are the acid neutralizing capacities to pH = 4.3 and pH = 8.2, respectively, $Aci_M$ is the mineral acidity, $Aci_P$ is the phenolphthalein (or CO$_2$) acidity, $a$ is the activity, $BNC_{4.3}$ and $BNC_{8.2}$ are the base neutralizing capacities to pH = 4.3 and pH = 8.2, respectively, $c$ is the molar concentration, $c(\text{DIC})$ is the concentration of dissolved inorganic carbon (carbonic acid species), $c_0$ is the initial concentration, $f(\text{acid})$ is the fraction of the acid in a conjugate acid/base pair, $f(\text{base})$ is the fraction of the base in a conjugate acid/base pair, $K_a$ is the conditional acidity constant, $K_a^*$ is the thermodynamic acidity constant, $K_b$ is the conditional basicity constant, $K_b^*$ is the thermodynamic basicity constant, $K_w$ is the conditional dissociation constant of water, $K_w^*$ is the thermodynamic dissociation constant of water, $m$ is the $m$ value, $p$ is the $p$ value, and $\alpha$ is the degree of protolysis.

## Chapter 8: Precipitation/Dissolution Equilibria

Solubility product, $K_{sp}^*$, of a compound $C_m A_{n(s)}$:

$$C_m A_{n(s)} \rightleftharpoons m\,C^{i \cdot n+} + n\,A^{i \cdot m-}$$

$$K_{sp}^* = a^m(C^{i \cdot n+})\,a^n(A^{i \cdot m-}) \qquad\qquad pK_{sp}^* = -\log K_{sp}^*$$

Relationship between solubility and solubility product for a compound $C_m A_{n(s)}$:

$$c_s = \sqrt[m+n]{\frac{K_{sp}}{m^m\,n^n}}$$

$$c_s = \frac{c(C^{i \cdot n+})}{m} = \frac{c(A^{i \cdot m-})}{n}$$

where $a$ is the activity, $c$ is the molar concentration, $c_s$ is the solubility, $K_{sp}$ is the conditional solubility product, $K_{sp}^*$ is the thermodynamic solubility product, $i$ is an integer multiplier, and $m$ and $n$ are stoichiometric factors.

## Chapter 9: Calco-Carbonic Equilibrium

Tillmans equation:

$$c(\text{CO}_2) = \frac{K_T}{f_T}\,c^2(\text{HCO}_3^-)\,c(\text{Ca}^{2+})$$

with

$$K_T = \frac{K_{a2}^*}{K_{a1}^* K_{sp}^*}$$

and

$$f_T = \frac{1}{\gamma_1^6}$$

Langelier equation:

$$pH_{eq} = -\log K_{La} - \log f_{La} - \log c(HCO_3^-) - \log c(Ca^{2+})$$

with

$$\log K_{La} = \log K_{a2}^* - \log K_{sp}^*$$

and

$$\log f_{La} = 5 \log \gamma_1$$

Saturation index:

$$SI = pH_{meas} - pH_{eq}$$

where $c$ is the molar concentration, $f_{La}$ is the overall activity coefficient in the Langelier equation, $f_T$ is the overall activity coefficient in the Tillmans equation, $K_{a1}^*$ is the acidity constant of $CO_{2(aq)}$, $K_{a2}^*$ is the acidity constant of $HCO_3^-$, $K_{La}$ is the Langelier constant, $K_{sp}^*$ is the solubility product of $CaCO_{3(s)}$, $K_T$ is the Tillmans constant, $pH_{meas}$ is the measured pH, $pH_{eq}$ is the equilibrium pH (calculated from the Langelier equation), $SI$ is the saturation index, and $\gamma_1$ is the activity coefficient of a univalent ion.

## Chapter 10: Redox Equilibria

Redox half-reaction – equilibrium constant, $K^*$:

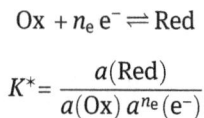

$$Ox + n_e\, e^- \rightleftharpoons Red$$

$$K^* = \frac{a(Red)}{a(Ox)\, a^{n_e}(e^-)}$$

Redox half-reaction – redox intensity, pe, and standard redox intensity, $pe^0$:

$$pe = -\log a(e^-)$$

$$pe^0 = \frac{1}{n_e} \log K^*$$

$$pe = pe^0 + \frac{1}{n_e} \log \frac{a(Ox)}{a(Red)}$$

$$pe = pe^0 + \frac{1}{n_e} \log \frac{\Pi\, a^v(Ox)}{\Pi\, a^v(Red)} \qquad \text{(generalized form)}$$

pH-dependent redox half-reaction – standard redox intensity for a given pH, $pe^0(pH)$:

$$Ox + n_e\, e^- + n_p\, H^+ \rightleftharpoons Red$$

$$pe^0(pH) = pe^0 - \frac{n_p}{n_e} pH$$

Redox intensity, pe, and redox potential, $E_H$:

$$E_H = E_H^0 + \frac{2.303\, R\, T}{n_e\, F} \log \frac{\Pi\, a^v(Ox)}{\Pi\, a^v(Red)} \qquad \text{(Nernst equation)}$$

$$pe = \frac{F}{2.303\, R\, T} E_H$$

$$pe^0 = \frac{F}{2.303\, R\, T} E_H^0$$

Speciation of a redox couple:

$$f(Ox) = \frac{1}{1 + 10^{n_e(pe^0 - pe)}}$$

$$f(Red) = 1 - f(Ox)$$

Speciation of a redox couple that is influenced by the pH:

$$f(Ox) = \frac{1}{1 + 10^{n_e[pe^0(pH) - pe]}}$$

$$f(Red) = 1 - f(Ox)$$

Complete redox reaction – equilibrium constant, $K^*$:

$$Ox_1 + Red_2 \rightleftharpoons Red_1 + Ox_2$$

$$K^* = \frac{a(Ox_2)\, a(Red_1)}{a(Red_2)\, a(Ox_1)}$$

$$\log K^* = n_e(pe_1^0 - pe_2^0) = \log \frac{a(Ox_2)\, a(Red_1)}{a(Red_2)\, a(Ox_1)}$$

Gibbs energy, $\Delta_R G$, and standard Gibbs energy, $\Delta_R G^0$, of a redox reaction:

$$\Delta_R G = -2.303\, n_e\, R\, T(\mathrm{pe}_1 - \mathrm{pe}_2)$$

$$\Delta_R G^0 = -R\, T \ln K^* = -2.303\, n_e\, R\, T(\mathrm{pe}_1^0 - \mathrm{pe}_2^0)$$

where $a$ is the activity, $E_H$ is the redox potential, $E_H^0$ is the standard redox potential, $F$ is the Faraday constant, $f(\mathrm{Ox})$ is the fraction of the oxidant, $f(\mathrm{Red})$ is the fraction of the reductant, $\Delta_R G$ is the Gibbs energy of reaction, $\Delta_R G^0$ is the standard Gibbs energy of reaction, $K^*$ is the thermodynamic equilibrium constant, $n_e$ is the number of electrons, $n_p$ is the number of protons, pe is the redox intensity, $\mathrm{pe}^0$ is the standard redox intensity, $\mathrm{pe}^0(\mathrm{pH})$ is the standard redox intensity at a given pH, $R$ is the gas constant, $T$ is the absolute temperature, and $v$ is the stoichiometric factor.

## Chapter 11: Complex Formation

Individual complex formation constant, $K^*$, and overall complex formation constant, $\beta^*$:

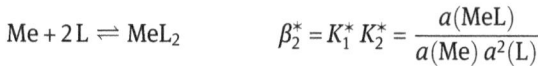

$$\mathrm{Me} + \mathrm{L} \rightleftharpoons \mathrm{MeL} \qquad K_1^* = \frac{a(\mathrm{MeL})}{a(\mathrm{Me})\,a(\mathrm{L})}$$

$$\mathrm{MeL} + \mathrm{L} \rightleftharpoons \mathrm{MeL}_2 \qquad K_2^* = \frac{a(\mathrm{MeL}_2)}{a(\mathrm{MeL})\,a(\mathrm{L})}$$

$$\mathrm{Me} + 2\mathrm{L} \rightleftharpoons \mathrm{MeL}_2 \qquad \beta_2^* = K_1^*\, K_2^* = \frac{a(\mathrm{MeL})}{a(\mathrm{Me})\,a^2(\mathrm{L})}$$

Note: The equations above are written for two ligands (L) as an example. For more than two ligands, the equations have to be extended in analogous manner.

Relationship between dissociation and formation constants:

$$K_{\mathrm{diss}}^* = \frac{1}{K_{\mathrm{form}}^*}$$

where $a$ is the activity, $K^*$ and $K_{\mathrm{form}}^*$ are individual complex formation constants, $K_{\mathrm{diss}}^*$ is the individual complex dissociation constant, and $\beta^*$ is the overall complex formation constant.

## Chapter 12: Sorption

Sorbed amount of sorbate, $q$, or sorbed mass of sorbate, $q^*$:

$$q = \frac{n_{\mathrm{sorbate}}}{m_{\mathrm{sorbent}}} \qquad \text{or} \qquad q^* = \frac{m_{\mathrm{sorbate}}}{m_{\mathrm{sorbent}}}$$

Material balance (batch process, sorbent initially free of sorbate):

$$q_{eq} = \frac{V}{m_{sorbent}} (c_0 - c_{eq})$$

$$q_{eq}^* = \frac{V}{m_{sorbent}} (\rho_0^* - \rho_{eq}^*)$$

Linear sorption isotherm:

$$q = K_d\, c$$

Nonlinear Langmuir isotherm:

$$q = \frac{q_m\, K_L\, c}{1 + K_L\, c}$$

Nonlinear Freundlich isotherm:

$$q = K_F\, c^n$$

Material balance for the distribution of a compound C between the liquid phase and the solid phase S:

$$C + S \rightleftharpoons S\text{–}C$$

$$c(C_{total}) = c(C) + q(C)\, \rho^*(S)$$

Protonation and deprotonation of surface OH groups ($\equiv$SOH) and the related conditional acidity constants $K_{a1}$ and $K_{a2}$:

$$\equiv SOH + H^+ \rightleftharpoons \equiv SOH_2^+$$

$$\equiv SOH_2^+ \rightleftharpoons \equiv SOH + H^+$$

$$K_{a1} = \frac{c_{ss}(\equiv SOH)\, c(H^+)}{c_{ss}(\equiv SOH_2^+)}$$

$$\equiv SOH \rightleftharpoons \equiv SO^- + H^+$$

$$K_{a2} = \frac{c_{ss}(\equiv SO^-)\, c(H^+)}{c_{ss}(\equiv SOH)}$$

Conditional and intrinsic acidity constants, $K_a$ and $K_a^{int}$:

$$K_a = K_a^{int}\, \exp\left(\frac{-\Delta z_s\, F\, \Psi_s}{R\, T}\right)$$

$$\log K_a = \log K_a^{int} - \frac{\Delta z_s\, F\, \Psi_s}{2.303\, R\, T}$$

Point of zero charge and acidity constants of the surface OH groups:

$$pH_{pzc} = 0.5(pK_{a1} + pK_{a2}) = 0.5(pK_{a1}^{int} + pK_{a2}^{int})$$

Organic carbon normalized sorption coefficient, $K_{oc}$:

$$K_{oc} = \frac{K_d}{f_{oc}}$$

$$f_{oc} = \frac{m_{oc}}{m_{solid}}$$

General $\log K_{oc}$–$\log K_{ow}$ correlation for neutral organic sorbates:

$$\log K_{oc} = a \log K_{ow} + b$$

pH-dependent $n$-octanol–water partition coefficient, $\log D$, for acids and bases:

$$\log D(\text{acid}) = \log K_{ow} - \log(1 + 10^{pH - pK_a})$$

$$\log D(\text{base}) = \log K_{ow} - \log(1 + 10^{pK_a - pH})$$

pH-dependent sorption coefficient of dissociating solutes:

$$K_d(pH) = (K_{d,n} - K_{d,i})(1 - \alpha) + K_{d,i}$$

Retardation factor:

$$R_d = \frac{v_w}{v_c}$$

Relationship between the retardation factor and the (linear) sorption coefficient:

$$R_d = 1 + \frac{\rho_B}{\varepsilon_B} K_d$$

where $a$, $b$ are empirical parameters in the $\log K_{oc}$–$\log K_{ow}$ correlation, $c$ is the molar concentration, $c_{ss}$ is the concentration of the surface sites, $D$ is $n$-octanol/water partition coefficient of an ionized solute at a specific pH, $F$ is the Faraday constant, $f_{oc}$ is the fraction of organic carbon in the sorbent, $K_a$ is the acidity constant, $K_{a1}$ is the acidity constant for the deprotonation of the protonated surface OH groups ($\equiv SOH_2^+$), $K_{a2}$ is the acidity constant for the deprotonation of the surface OH groups ($\equiv SOH$), $K_a^{int}$ is the intrinsic acidity constant of the surface OH groups, $K_d$ is the sorption (distribution) coefficient of the linear isotherm, $K_{d,n}$ is the sorption coefficient of neutral species, $K_{d,i}$ is the sorption coefficient of ionized species, $K_F$ is the sorption coefficient of the Freundlich isotherm, $K_L$ is the sorption coefficient of the Langmuir isotherm, $K_{oc}$ is the organic carbon normalized sorption coefficient, $m$ is the mass, $n$ is the substance amount, $n$ is the exponent in the Freundlich isotherm, $pH_{pzc}$ is the pH at the point of zero charge, $pK_a$ is the acidity exponent ($= -\log K_a$), $q$ is the sorbed amount (sorbent loading), $q^*$ is the sorbed mass, $q_m$ is the maximum sorbed amount in the Langmuir isotherm, $R$ is the gas constant, $R_d$ is the retardation factor, $T$ is the absolute temperature, $V$ is the volume, $v_w$ is the mean pore water velocity, $v_c$ is the velocity of the concentration front, $z_s$ is the charge of the surface sites, $\alpha$ is the degree of dissociation, $\varepsilon_B$ is the effective porosity, $\rho_B$ is the bulk density, $\rho^*$ is the mass concentration, and $\Psi_s$ is the surface potential.

# Nomenclature

## Preliminary notes

**Note 1:** In the parameter list, general dimensions are given instead of special units. The dimensions for the basic physical quantities are indicated as follows:

| | |
|---|---|
| I | electric current |
| L | length |
| M | mass |
| N | amount of substance (mol) |
| T | time |
| $\Theta$ | temperature |

Additionally, the following symbols for derived types of measures are used:

| | |
|---|---|
| E | energy ($L^2\,M\,T^{-2}$) |
| F | force ($L\,M\,T^{-2}$) |
| P | pressure ($M\,L^{-1}\,T^{-2}$) |
| U | voltage ($L^2\,M\,I^{-1}\,T^{-3}$) |

**Note 2:** Empirical parameters like *a, b, c, A, B, C* or similar are not listed here. They are explained in context with the respective equations in the text.

## English alphabet

| | |
|---|---|
| *A* | surface area ($L^2$) |
| *A* | parameter in the Debye-Hückel equation (0.496 at 10 °C, 0.509 at 25 °C) |
| $A_m$ | specific surface area ($L^2\,M^{-1}$) |
| $Aci_M$ | mineral acidity ($N\,L^{-3}$) |
| $Aci_P$ | $CO_2$ acidity ($N\,L^{-3}$) |
| $Alk_P$ | phenolphthalein (or carbonate) alkalinity ($N\,L^{-3}$) |
| $Alk_T$ | total alkalinity ($N\,L^{-3}$) |
| $ANC_{4.3}$ | acid neutralizing capacity to pH 4.3 ($N\,L^{-3}$) |
| $ANC_{8.2}$ | acid neutralizing capacity to pH 8.2 ($N\,L^{-3}$) |
| *a* | activity ($N\,L^{-3}$) |
| $BNC_{4.3}$ | base neutralizing capacity to pH 4.3 ($N\,L^{-3}$) |
| $BNC_{8.2}$ | base neutralizing capacity to pH 8.2 ($N\,L^{-3}$) |
| *b* | molality ($N\,M^{-1}$) |
| *C* | capacitance per unit surface area ($I\,T\,L^{-2}\,U^{-1}$) |
| *CH* | carbonate hardness ($N\,L^{-3}$) |
| *c* | molar concentration ($N\,L^{-3}$) |
| | *subscripts:* |
| | 0  initial |
| | aq  in the aqueous phase |
| | eq  equilibrium |

https://doi.org/10.1515/9783110758788-015

| | |
|---|---|
| g | in the gas phase |
| s | saturation (solubility) |
| std | standard |
| total | total concentration |
| $c_m$ | alternative symbol for molality (see also $b$) |
| $c_p$ | molar heat capacity ($E\ N^{-1}\ \Theta^{-1}$) |
| $c_{ss}$ | concentration of surface sites ($N\ M^{-1}$ or $N\ L^{-2}$) |
| $D$ | $n$-octanol–water partition coefficient of ionized species (dimensionless) |
| $E_H$ | redox potential (U) |

*superscript:*

| | |
|---|---|
| 0 | standard redox potential |
| $F$ | Faraday constant (96 485 C/mol) |
| $F_C$ | electrostatic force, Coulomb force (F) |
| $f$ | fraction (dimensionless) |

*subscript:*

| | |
|---|---|
| oc | of organic carbon |
| $f$ | summarized activity coefficients in the equations of the calco-carbonic equilibrium (dimensionless) |

*subscripts:*

| | |
|---|---|
| a1 | first dissociation step of the carbonic acid |
| a2 | second dissociation step of the carbonic acid |
| La | Langelier equation |
| sp | solubility product of calcite |
| T | Tillmans equation |
| $G$ | Gibbs energy (E) or molar Gibbs energy ($E\ N^{-1}$) |

*subscripts:*

| | |
|---|---|
| f | formation |
| sol | dissolution |

*superscript:*

| | |
|---|---|
| 0 | standard |
| $g$ | gravity of Earth ($9.81\ m/s^2$) |
| $H$ | enthalpy (E) or molar enthalpy ($E\ N^{-1}$) |

*subscripts:*

| | |
|---|---|
| cond | condensation |
| freez | freezing |
| fus | fusion |
| hyd | hydration |
| sol | dissolution |
| vap | vaporization |

*superscript:*

| | |
|---|---|
| 0 | standard |
| $H$ | Henry constant ($N\ L^{-3}\ P^{-1}$) |
| $H_{inv}$ | Henry constant in the inverse form of Henry's law ($P\ L^3\ N^{-1}$) |
| $H_x$ | Henry constant in Henry's law written with the liquid-phase mole fraction ($P^{-1}$) |
| $h$ | height (L) |
| $i$ | van't Hoff factor (dimensionless) |
| $K$ | conditional equilibrium constant (unit depends on the specific law of mass action) |

*subscripts:*

a acidity
b basicity
sp solubility product
w water dissociation

$K^*$ thermodynamic equilibrium constant (unit depends on the specific law of mass action)

*subscripts:*

a acidity
b basicity
diss complex dissociation
form complex formation
sp solubility product
w water dissociation

*superscript:*

int intrinsic

$K_B$ ebullioscopic constant ($\Theta$ M N$^{-1}$)
$K_c$ distribution constant (concentration ratio) in Henry's law (dimensionless)
$K_d$ distribution coefficient in the linear sorption isotherm (L$^3$ M$^{-1}$)

*subscripts:*

i ionized species
n neutral species

$K_F$ cryoscopic constant ($\Theta$ M N$^{-1}$)
$K_F$ parameter of the Freundlich isotherm ((N M$^{-1}$)/(N L$^{-3}$)$^n$ or (M M$^{-1}$)/(M L$^{-3}$)$^n$)
$K_L$ parameter of the Langmuir isotherm (L$^3$ N$^{-1}$ or L$^3$ M$^{-1}$)
$K_{La}$ Langelier constant (L$^3$ N$^{-1}$)
$K_{oc}$ organic carbon normalized distribution coefficient (L$^3$ M$^{-1}$)

*subscripts:*

i ionized species
n neutral species

$K_{ow}$ n-octanol-water partition coefficient (dimensionless)
$K_T$ Tillmans constant (L$^6$ N$^{-2}$)
$K_{xy}$ distribution constant (mole fraction ratio) in Henry's law (dimensionless)
$M$ molecular weight (M N$^{-1}$)
$m$ alternative symbol for molality (see also *b*)
$m$ *m* alkalinity (N L$^{-3}$)
$m$ mass (M)

*subscripts:*

aq in the aqueous phase
g in the gas phase
oc of organic carbon
solid solid material
total total mass

$NCH$ non-carbonate hardness (N L$^{-3}$)
$n$ amount of substance, number of moles (N)
$n$ parameter (exponent) of the Freundlich isotherm (dimensionless)
$n_e$ number of electrons in a redox equation (dimensionless)
$n_p$ number of protons in a redox equation (dimensionless)
$ON$ oxidation number (dimensionless)

| | |
|---|---|
| $p$ | pressure or partial pressure (P) |
| | *subscripts:* |
| | 0      vapor pressure of a pure bulk liquid |
| | crit    critical pressure |
| | total   total pressure |
| | tp      pressure at the triple point |
| | v       vapor pressure |
| $p$ | $p$ alkalinity (N L$^{-3}$) |
| pH | negative decimal logarithm of the proton activity (dimensionless) |
| | *subscripts:* |
| | c       at the crossover point |
| | pzc    at the point of zero charge |
| pe | redox intensity, negative decimal logarithm of the electron activity (dimensionless) |
| | *subscript:* |
| | c       at the crossover point |
| | *superscript:* |
| | 0      standard redox intensity |
| p$K$ | negative decimal logarithm of the conditional equilibrium constant (dimensionless) |
| | *subscripts:* |
| | a      acidity |
| | b      basicity |
| | sp     solubility product |
| | w     water dissociation |
| p$K^*$ | negative decimal logarithm of the thermodynamic equilibrium constant (dimensionless) |
| | *subscripts:* |
| | a      acidity |
| | b      basicity |
| | sp     solubility product |
| | w     water dissociation |
| pOH | negative decimal logarithm of the hydroxide ion activity (dimensionless) |
| $Q$ | reaction quotient (same unit as the related equilibrium constant) |
| $Q_s$ | surface charge (N M$^{-1}$) |
| $q$ | charge (I T) |
| $q$ | sorbed substance amount (sorbent loading) (N M$^{-1}$) |
| | *subscript:* |
| | eq     equilibrium |
| $q^*$ | sorbed mass (sorbent loading) (M M$^{-1}$) |
| | *subscript:* |
| | eq     equilibrium |
| $q_m$ | maximum sorbent loading, parameter of the Langmuir isotherm (M M$^{-1}$ or N M$^{-1}$) |
| $R$ | universal gas constant (8.3145 J/(mol·K), 8.3145 Pa·m$^3$/(mol·K), 0.083145 bar·L/(mol·K)) |
| $R_d$ | retardation factor (dimensionless) |
| $r$ | radius (L) |
| $r$ | distance (L) |
| $S$ | entropy (E $\Theta^{-1}$) or molar entropy (E $\Theta^{-1}$ N$^{-1}$) |

subscript:
sol      dissolution
$SI$      saturation index (dimensionless)
$T$      absolute temperature ($\Theta$)
superscripts:
LV      boiling point (liquid/vapor phase transition)
SL      melting point (solid/liquid phase transition)
$TH$      total hardness ($N\ L^{-3}$)
$U$      molar lattice energy ($E\ N^{-1}$)
$V$      volume ($L^3$)
subscripts:
aq      of the aqueous phase
g      of the gas phase
$V_m$      molar volume ($L^3\ N^{-1}$)
$v_c$      velocity of the concentration front ($L\ T^{-1}$)
$v_w$      mean pore water velocity ($L\ T^{-1}$)
$w$      weight fraction (dimensionless)
$x$      mole fraction in the liquid phase (dimensionless)
$y$      mole fraction in the gas phase (dimensionless)
$z$      charge number (dimensionless)
subscripts:
s      surface layer
$\beta$      beta layer

## Greek alphabet

$\alpha$      degree of protolysis or dissociation (dimensionless)
$\beta$      alternative symbol for mass concentration (see also $\rho^*$)
$\beta_n$      conditional overall complex formation constant (unit depends on the number of ligands, $n$)
$\beta_n^*$      thermodynamic overall complex formation constant (unit depends on the number of ligands, $n$)
$\gamma$      activity coefficient (dimensionless)
subscripts:
1      univalent ion
z      ion with the charge $z$
$\varepsilon$      permittivity ($I^2\ T^4\ L^{-3}\ M^{-1}$ or $I\ T\ L^{-1}\ U^{-1}$)
$\varepsilon_B$      effective porosity of the aquifer material (dimensionless)
$\varepsilon_r$      relative permittivity (dimensionless)
$\varepsilon_0$      vacuum permittivity ($8.854 \times 10^{-12}$ A·s/(V·m))
$\eta$      dynamic viscosity ($M\ L^{-1}\ T^{-1}$)
$\vartheta$      Celsius temperature (degree)
subscripts:
crit      critical temperature
tp      temperature at the triple point
$\varphi$      osmotic coefficient (dimensionless)
$\kappa_{25}$      electrical conductivity measured at 25 °C ($I\ U^{-1}\ L^{-1}$ or $I^2\ T^3\ L^{-3}\ M^{-1}$)

| | |
|---|---|
| $v$ | kinematic viscosity ($L^2\,T^{-1}$) |
| $v$ | number of ions within a formula unit of a dissociating substance (dimensionless) |
| $v_i$ | stoichiometric coefficient in a reaction (dimensionless) |
| $\pi$ | osmotic pressure (P) |

*subscripts:*
ideal  ideal solution
real   real solution

| | |
|---|---|
| $\theta$ | contact angle (degree) |
| $\rho$ | density ($M\,L^{-3}$) |

*subscript:*
B     bulk (aquifer material)

| | |
|---|---|
| $\rho^*$ | mass concentration ($M\,L^{-3}$) |

*subscripts:*
aq    aqueous phase
g     gas phase

| | |
|---|---|
| $\sigma$ | surface free energy (surface tension) ($F\,L^{-1}$ or $M\,T^{-2}$) |
| $\sigma_s$ | surface charge density ($I\,T\,L^{-2}$) |
| $\psi$ | electrical potential (U) |

*subscripts:*
s     surface layer
β     beta layer

# Bibliography

Amiri F, Rahman MM, Börnick H, Worch E: Sorption behaviour of phenols on sandy aquifer material during flow through experiments: the effect of pH. Acta Hydrochim Hydrobiol 32 (2004) 214–224.

Baker JR, Mihelcic JR, Luehrs DC, Hickey JP: Evaluation of estimation methods for organic carbon normalized sorption coefficients. Water Environ Res 69 (1997) 136–145.

Benjamin MM: Water Chemistry. McGraw–Hill, New York (2002).

DIN 38404-10: German standard methods for the examination of water, wastewater and sludge – Physical and physico-chemical parameters (group C). Part 10: Calculation of the calcite saturation of water (C 10). DIN Deutsches Institut für Normung e.V., Berlin (2012).

Gerstl Z: Estimation of organic chemical sorption by soils. J Contam Hydrol 6 (1990) 357–375.

Girvin DC, Scott AJ: Polychlorinated biphenyl sorption by soils: measurement of soil–water partition coefficients at equilibrium. Chemosphere 35 (1997) 2007–2025.

Hassett JJ, Banwart WL, Griffen RA: Correlation of compound properties with sorption characteristics of nonpolar compounds by soil and sediments: concepts and limitations. In: Francis, CW, Auerbach SI (eds): Environmental and Solid Wastes: Characterization, Treatment and Disposal. Butterworth Publishers, Newton, MA (1983), 161–178.

Karickhoff SW, Brown DS, Scott TA: Sorption of hydrophobic pollutants on natural sediments. Water Res 13 (1979) 241–248.

Kenaga EE, Goring CAI: Relationship between water solubility, soil sorption, octanol–water partitioning, and concentration of chemicals in biota. In: Eaton JG, Parrish PR, Hendricks AC (eds): Aquatic Toxicology ASTM STP 707. American Society for Testing and Materials, Philadelphia, PA (1980), 79–115.

Millero FJ: Physical Chemistry of Natural Waters. Wiley-Interscience, New York (2001).

NOAA National Oceanic and Atmospheric Administration. Resource collections: Oceans and Coasts, (2022a), retrieved on March 10, 2022 from https://www.noaa.gov/education/resource-collections/ocean-coasts/ocean-acidification.

NOAA National Centers for Environmental Information. Climate at a Glance: Global Time Series, (2022b), retrieved on March 10, 2022 from https://www.ncdc.noaa.gov/cag/.

Sabljić A, Güsten H, Verhaar H, Hermans J: QSAR modelling of soil sorption, improvements and systematics of log $K_{oc}$ vs. log $K_{ow}$ correlations. Chemosphere 31 (1995) 4489–4514.

Schwarzenbach RP, Westall J: Transport of nonpolar organic compounds from surface water to groundwater. Laboratory sorption studies. Environ Sci Technol 15 (1981) 1360–1367.

Sigg L, Stumm W: Aquatische Chemie (Aquatic Chemistry). vdf Hochschulverlag, Zürich (2011).

Stumm W: Chemistry of the Solid–Water Interface. Wiley–Interscience, New York (1992).

Trenberth KE, Smith L, Qian TT, Dai AG, Fasullo J: Estimates of the global water budget and its annual cycle using observational and model data. J Hydrometeorol 8 (2007) 758–769.

van Gestel CAM, Ma W, Smit CE: Development of QSARs in terrestrial ecotoxicology: earth-worm toxicity and soil sorption of chlorophenols, chlorobenzenes and dichloroaniline. Sci Tot Environ 109/110 (1991) 589–604.

Worch E: Drinking Water Treatment: An Introduction. De Gruyter, Berlin/Boston (2019).

Worch E, Grischek T, Börnick H, Eppinger P: Laboratory tests for simulating attenuation processes of aromatic amines in riverbank filtration. J Hydrol 266 (2002) 259–268.

https://doi.org/10.1515/9783110758788-016

# Index

https://doi.org/10.1515/9783110758788-017

www.ingramcontent.com/pod-product-compliance
Lightning Source LLC
Chambersburg PA
CBHW080911220326
41598CB00034B/5544

9 783110 758764